The

'Indispensable Handbook' to the Optical Lantern

Elibron Classics
www.elibron.com

Elibron Classics series.

ISBN 1-4021-4119-X (paperback)
ISBN 1-4212-7529-5 (hardcover)

This Elibron Classics Replica Edition is an unabridged facsimile of the edition published in 1888 by Iliffe and Son, London.

THE

"INDISPENSABLE HANDBOOK"

TO THE

OPTICAL LANTERN.

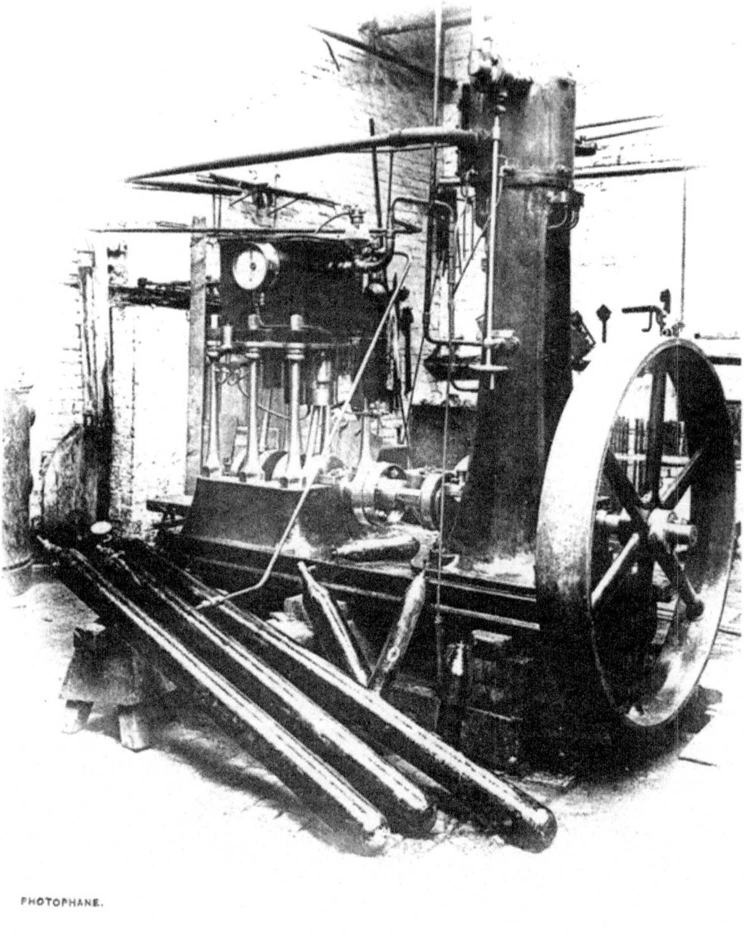

PHOTOPHANE.

Howard Lane & Co.'s Gas Compressing Engine.

THE "INDISPENSABLE HANDBOOK" TO THE OPTICAL LANTERN:

A COMPLETE CYCLOPÆDIA ON THE SUBJECT OF OPTICAL LANTERNS, SLIDES, AND ACCESSORY APPARATUS.

COMPILED AND EDITED BY

WALTER D. WELFORD,

SUB-EDITOR OF *Photography*

AND

HENRY STURMEY

(*Camera Club*),

EDITOR OF *Photography* AND *The Cyclist*.

PROFUSELY ILLUSTRATED.

LONDON:

ILIFFE AND SON, 98, FLEET STREET, E.C.

1888.

Preface.

THE growing popularity of the Optical Lantern as a means of instruction and amusement, and the very large variety of apparatus now placed on the market in connection with its use, have induced us to make it the subject of the present volume, which is the fifth of the series of Indispensable Technical Handbooks which have now for the past ten years been successively issued by our publishers. Treating on the subject of the construction, varieties and use of the Bicycle, the Tricycle, the Safety Bicycle, and Photographic Apparatus respectively, these Handbooks have proved wonderfully popular, and have gone through several editions. With the hope, therefore, that we may be able to produce on the Optical Lantern a work which shall prove at least as useful and acceptable as its predecessors, we have taken the matter up, and here present to—we hope—an appreciative public the results of our labours.

W. D. WELFORD, COMPILER.
HENRY STURMEY, EDITOR.

Contents.

SECTION III.—SLIDES.

SECTION IV.—GENERAL.

Compiler's Notes.

IN issuing this work, the outcome of several months' laborious compilation, I wish first to thank personally those makers of articles mentioned in the work for their kind assistance. There are but few unrepresented, and these have only themselves to blame. No stone has been left unturned to make the various Parts and Sections as complete as possible. With the full consciousness of having done my utmost, by visit, letter, and postcard, to have included, the specialities of each and every manufacturer, I rest content with that assertion.

The Frontispiece.—The compressed gas cylinder is rapidly coming to the front—in fact, it only requires an extension of the facilities of getting it refilled to send the gas-bag into obscurity. The frontispiece, produced by the Photophane process, shows us the gas compressor as in daily use at Messrs. Howard, Lane & Co.'s works. A few notes will be interesting. On entering the works, one is struck at the number of cylinders, or, as they term them, vessels, lying about in heaps and piles. Some are handy little things, almost small enough to put in the pocket, whilst huge 150 feet monsters are leaning against the walls. These are principally for balloon work. A balloon requires about 10,000 feet of gas, so that the cylinders have to hold a fair number of cubic feet or it would take so many. The firm make their oxygen from potass but the hydrogen in different ways—by the action of a special kind of sulphuric acid on steel, by ordinary acid on zinc, and the decomposition of water by electr . y. The gas, as made, passes into the holder which lies above the co pressor. It is pumped out by a three-stage compressing engine into the cylinders. There are in reality three pumps, each increasing the pressure. As regards speed, 150 feet of hydrogen takes ten minutes. The engine will compress up to two tons per inch, but the cylinders are usually filled to 150 atmospheres, or about one ton to the inch The majority of gas vessels are made with a seam, but they have also some made out of flat plate, by a process of stamping and drawing, and, of course, seamless. One made this way was shown us, nine inch diameter and four feet long.

Addresses of Firms.—To avoid too frequent repetition, I have thought it best to omit the addresses of the firms mentioned in the body of the work, but to give them collectively on a special page at the end.

The Slides.—In this is attempted what, I believe, has not been done before, to give some idea of the sets published, and what they are about. Any shortcomings in that part I must ask to be excused, owing to the exceeding difficulty of the task, at the same time sincerely hoping the dea may prove of use and interest.

The work has been largely a labour of love, though, perhaps, just a trifle in excess of the former. Any suggestions as to additions, alterations, errors, &c , will be acceptable for use in the case of a second edition With these few remarks I lay down a pen that has been running over paper since May last with an earnest sigh of relief, and can only hope my efforts will prove of utility to the ever-increasing army of lanternists.

WALTER D. WELFORD.

47, HAGLEY ROAD, EDGBASTON,
November 1888.

Section I.

LANTERNS.

PART I.

Oil Lanterns.

THE Optical Lantern, known perhaps more popularly as the Magic Lantern, is familiar to everyone in its effects, and few of us have not at some period of our lives witnessed a lantern entertainment. The Optical Lantern, however, of the present day has quite grown out of its position as a scientific toy only, and the many improvements which modern science has introduced in its construction, together with the adaptation to its purpose of photography, microscopy, and other sciences have made it to-day an instrument of the highest use, both for educational and scientific purposes, and the great attention which has been given to its construction of late, together with the increase in its use as an educational medium, have brought so many commercial competitors into the field as to render the compilation and production of the present work not only feasible but necessary. It is not our purpose here to enter into a dissertation upon the history of the lantern, but to consider its construction, its use, and its management. As will be gleaned by the briefest of glances through the pages of the present work, varieties of the Optical Lantern may be had from a mere nothing up to almost any amount which the enthusiast cares to spend upon it; but whether the lantern be a toy at a shilling or an elegant limelight triunial at a hundred guineas the principle is the same. The rays of an artificial light are collected and passed through a transparent picture, which by the divergence of the rays through a lens is thrown, brilliantly illuminated and magnified, upon a screen. The necessities, therefore, of lantern construction are—firstly, a box or case in which the light is contained and prevented from escaping except in the one direction required; a lamp or other artificial light; a condenser or lens, the purpose of which is to concentrate the light, and throw its rays evenly and uniformly through the slide; and a lens or objective for magnifying the picture thus secured. In the oil lanterns, a reflector will also be required.

Of the first, we may say that a great variety exists, from the simple tin body or box holding the light of the cheap toy concern, through various stages of japanned and otherwise ornamented and neatly-finished constructions of the same material in the cheaper lanterns, the Russian iron bodies of the medium class —which Russian iron, by the way, neither rusts nor tarnishes—to the elegant construction of wood, iron, and brass which makes up the handsomer and more expensive instrument. Brass, of course, both looks better and lasts longer than tin, and is more taking in appearance than iron, and when wood is used in the construction of a lantern, it is as an outer casing only, the internal part or body proper being usually constructed of Russian iron with a channel or air passage left between the iron and wood for the free passage of a current of air which keeps the body of the lantern cool, and prevents the cracking of the wood which otherwise would take place. In shape, lanterns are now-a-days chiefly made oblong, though cheaper varieties of what are termed "the old pattern" are still made higher than either their length or breadth, the reason of this mainly being that they are constructed to contain a lamp which requires a somewhat tall glass chimney, and therefore necessitates height in the lantern. The shape and get up of the lantern is, however, a secondary consideration to the other constructional points. The quality of the light, the condenser, and the objective are the chief points for consideration, and a lantern, the body of which is of the cheapest and commonest material, if fitted with good lenses, will give far better results than one of the most handsome external appearance with faulty ones.

Of the light or lamp we shall speak at length in another section; suffice it to say here, that in the cheaper form of lantern a colza or sperm oil lamp is used, whilst in the better class oil lantern paraffin is burnt with a combination of from two to as many as five wicks. The highest form of light is that known as the limelight, the various systems of which are treated upon in the chapter devoted to illuminants. Few lamps burning colza or sperm oil are fit to rank much higher than toys, the power of the light emitted by such lamps being comparatively small, and consequently they are not suited to show pictures of more than 5 or 6 feet in diameter. A very good light is, however, obtained by a skilful combination of wicks for the burning of paraffin, and very fair results with a picture of from 8 to 10 feet in diameter can be obtained with a good lantern, the power of the best form measured by the light of a standard candle being from 60 to 80. As, however, the power of the limelight ranges from 200 to 400 candles, and even higher still, it will be seen at once how vastly superior this light is to either of the others, although of course, being more expensive and entailing a little more trouble in working, it is not so well suited as paraffin for family and semi-private entertainments.

The construction of the condenser will also be treated upon in a later chapter, and we will only explain here that its purpose is to collect as many of the rays of light as possible, and pass them through the picture in an even manner. Perhaps the commonest form of condenser for cheap lanterns is the plano-convex lens, which is identical with that used in the well-known "bull's-eye" lantern; but as much greater advantages are obtained by the use of a double condenser, that is, by the use of two lenses of different construction placed together, such is to be preferred, the best being what is termed the double plano-convex, in which two plano-convex lenses are mounted with their plain sides outside, and the convex sides facing and nearly touching one another. Other forms will be spoken of in the chapter devoted to this purpose.

Of the lens or objective we shall also speak at length in a separate chapter. Here we will merely mention that its object is to receive the rays of light passing from the condenser through the picture, and to magnify and project them on to the screen. The objective is really the most important point to be considered in the whole lantern, as upon it depends in a great measure the quality of the picture. It must possess a flat field—that is to say, the definition or sharpness must be as good at the sides of the picture as in the centre. It must be free from distortion; in other words, straight lines must appear upon the screen as such, and not curved in the centre, and it must also be what is termed achromatic—that is to say, there must be no prismatic colouring about the disc of light thrown through it on the screen. In many of the toy lanterns and cheaper ones of higher pretensions, single lenses, which are little better than "bits of glass," are used; but to get the best results as much care must be taken in the construction of a lantern objective as in that of a high-class photographic lens, perhaps one of the best kinds of objective being precisely similar in construction to what is termed the "portrait" lens of the photographer. The lens or objective is fitted to the lantern in a tube some inches in front of the condenser, and if the lantern and the screen could always be placed at exactly the same distance from each other, it could easily be made a fixture, but as such an exact adjustment is not possible, the lens requires to be adjusted, or, as it is termed, "focussed" instead. The position of the lantern and screen having been fixed, the lens is moved until the picture appears sharp upon the screen. This movement is made in the cheaper lanterns by means of a slide, but by far the most satisfactory plan is the use of a rack and pinion, by which the focus may be obtained with the greatest nicety.

A reflector is required in all oil lanterns, and its use is to reflect or throw as many rays as possible on to the condenser. It is usually fitted inside the door of the lantern, which in oil lanterns usually forms the back, and it is important that it should be of the right curve and at the proper distance from the wick, as well, of course, as being thoroughly polished and kept scrupulously clean from dirt and dulness. In the centre is usually a small hole glazed with a piece of blue glass, through which the operator can inspect the condition of the wicks, and see what adjustment may be necessary. With lanterns constructed for the limelight this door is usually at the side, and some have doors both at the back and side. The lantern slide—of the making and various designs of which we shall speak later on—is carried in a channel immediately in front of the condenser, and, as slides vary in width, it is important that this channel should be provided with a spring which will hold the slide firmly in position, as without this it is next to impossible to obtain proper focus and clear definition of the whole picture. Some lanterns are specially constructed to facilitate the changing of the slides, and these are worth consideration.

For the production of what are termed "dissolving views," in which one picture gradually fades away whilst another as gradually appears, the one picture as it were fading into the other, a double lantern or "biunial" is a necessity. The biunial differs in no way in its construction from the single lantern; it is simply two twin lanterns sometimes placed one above the other as in the majority of limelight biunials. With most oil biunials the lanterns are placed side by side, adjustment being made so that the circles of light thrown by each lantern on the screen exactly coincide. The dissolving of views with the oil lantern is effected by means of what is termed a "fan dissolver," which consists of a doubly-serrated plate of tin or other thin sheet metal. This is so adjusted that whilst one lens is open the other is covered by the dissolver. Upon moving it the serrated portion of one end gradually passes over the face of the open lens, whilst the serrated portion of the other end as gradually opens up the lens hitherto closed, thus opening one lantern and closing the other at the same time, and mingling the two pictures with each other, the dissolver being moved until the lantern which is first opened is entirely covered, and that which was at first closed is entirely open and the picture thrown by it fully visible upon the screen. With limelight lanterns a different system of dissolving is adopted, which will be treated upon elsewhere.

Although we have spoken of the slide and implied that it is of necessity a transparent object, by the use of a prism or reflector placed at a proper angle opaque objects can also be shown in the lantern. Thus the reflection of a picture or photograph, a living organism, or the works of a watch in motion may be thrown upon the screen, enlarged, of course, and beautifully illuminated, and a description of several lanterns specially constructed to this end will be found in the following pages. The various devices in slides and the methods of showing them will be fully dealt with in the section devoted to that purpose.

The Cheaper Oil Lanterns.

The difficulty that meets us at the very outset is simply at what price to commence the details of cheap lanterns. Magic lanterns can be bought at toy shops, we believe, for 1/- each, with "hand painted" slides complete. But it must be obvious that such things are mere toys for children; we have, therefore, limited the price to £1 as the lowest of any use to the reader, and as certainly the first to come within the scope of this work. The difficulty might have been obviated by dealing only with mineral oil lanterns, leaving out the old fashioned Phantasmagoria, burning colza, sperm and olive oils. But these are still included in makers' catalogues, which proves some sort of a demand still to exist; and this fact, coupled with another, viz , that the prices run up as high as £5, compels their inclusion. Under the heading of "The Cheaper Oil Lanterns' will be found those costing from £1 to £3 3/- but not above the last-named sum

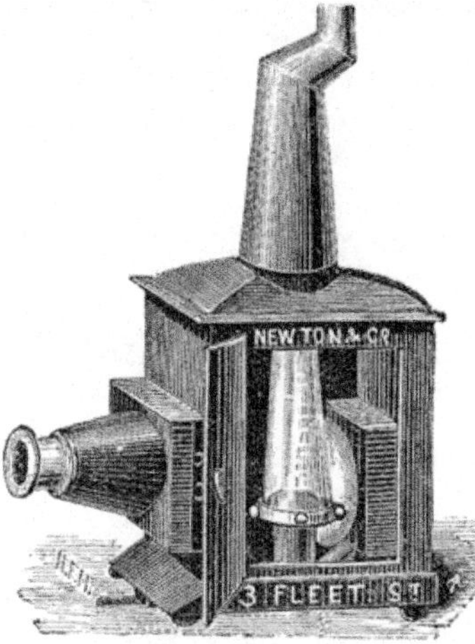

No. 1.

1.—The Phantasmagoria.

Details.—Before the advent of the Sciopticon, these were the oil lanterns made, sold and used. Generally classed and priced by numbers from No 1 upwards, they were sold by everyone. A few prices will be sufficient.

2.—Newton's No. 5.

Details.—3in. condenser, brass focussing tube, solarised argand lamp for burning colza oil, silvered reflector, and japanned tinned body, giving a picture from four to five feet diameter. Complete in case .. £1 15 0

3.—Newton's Phantasmagoria.

Details.—Lenses 3¼in. diameter, double combination achromatic front lenses with rack and pinion adjustment to the focus tube, fountain solarised argand lamp and reflector. Packed in case £3 3 0

4.—Medland's No. 6 (J. B. Medland).

Details.—English-made lantern, brass sliding front, 3in. condenser, solar lamp. In black box £1 5 0

5.—Medland's No. 7.

Details.—Same as above, but with 3in. compound achromatic condenser, compound front lens, and solar argand lamp £2 0 0

6.—Medland's No. 8.

Details.—The same, but with 3½in. condensers and best fountain lamp £3 3 0

7.—Ottway's No. 5 (J. Ottway & Son).

Details.—3in. condenser, brass focussing tube, solarised argand lamp for burning colza oil, silvered reflector, for giving a picture five feet in diameter. Complete in case £1 10 0

8. Ottway's Phantasmagoria.

Details.—Double condensers 3½in. diameter, double combination front lenses, rack and pinion adjustment, solar fountain lamp. Packed in case £3 3 0

9.—Theobald's No. 6 (J. Theobald & Co.)

Details.—This is the ordinary shape, but burns mineral oil, and is moreover a cheap lantern £1 8 0

10.—Theobald's No. 7.

Details.—The largest size the firm now make in the old shape £2 5 0

11.—Steward's No. 6 (J. H. Steward).

Details.—Stout japanned body, brass focussing tube, spring stage for holding slides, improved paraffin lamp, brass mounts, chimney, silvered reflector, wicks and stick; 3in. condenser, giving a disc up to eight feet in diameter £1 10 0

12.—Steward's No. 7.

Details.—3½in. condenser, japanned tin stage for firmly holding slides; brass focussing tube, carrying front lenses; improved solar argand lamp, chimney, silvered reflector, etc., giving a disc eight feet in diameter £1 17 6

13.—Noakes's No. 6a (D. Noakes & Son).

Details.—3½in. condenser, paraffin lamp, sliding front, strong japanned body £1 7

Leaving the old shape, we proceed to the various types of lanterns that have of recent years been introduced and improved upon. They are constructed generally much on the same lines, but vary in detail.

14.—The Guinea (Archer & Sons).

Details.—Condensers 3¼in. diameter, object lenses in brass screw cell, spring slide holder large enough to take mechanical slides 4½in. wide, powerful circular wick paraffin lamp, &c., will show a picture from six to nine feet in diameter on the screen £1 1 0

Extras.—3½in condensers, 5/-

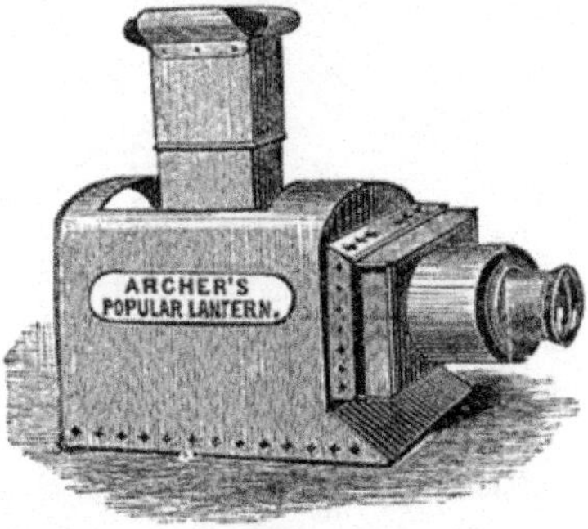

No. 15.

15.—The Popular (Archer & Sons).

Details.—Body of japanned tin, strongly made, and ventilated to keep perfectly cool, best 4in. double condensers, plano-convex, mounted in brass screw cells, achromatic object lenses, mounted in brass, patent three-wick lamp with toughened glass £2 2 0

Remarks.—An exceedingly good little instrument for the money. Messrs. Archer say it will show a 12-feet picture.

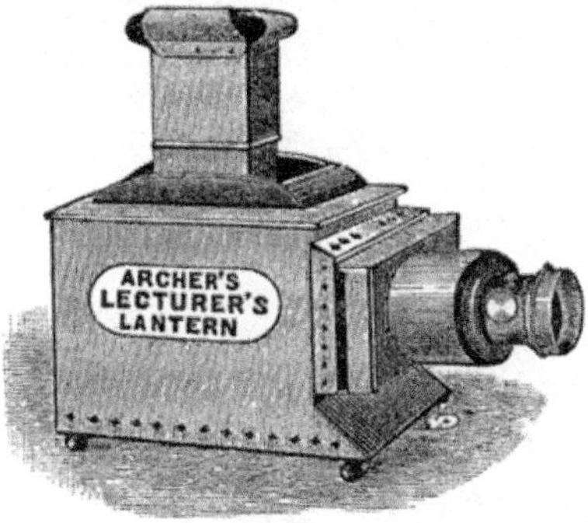

No. 16.

16.—The Lecturer's (Archer & Sons).

Details.—With best 4in. plano-convex condensers, mounted in brass ventilated screw cells, best three-wick lamp, with toughened glass, double achromatic portrait combination object lenses, giving a flat field, and sharp to the edge, rack and pinion focussing.

Extras.—Case, 2/6

Remarks.—This is an extremely portable lantern, and very cheap indeed.

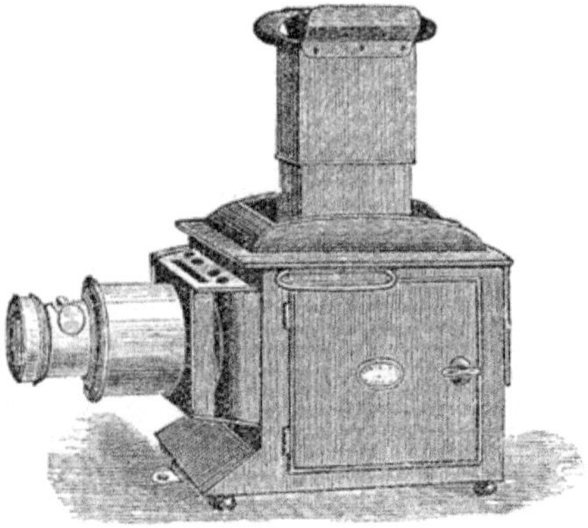

No. 17.

17.—The Excelsior (ARCHER & SONS).

Details.—Three wick lamp, improved slide holder, lamp is put in or taken out of the body of the lantern without removing the iron chimney, brass front tube draws off for the use of accessory apparatus, lime-light can be used when required without any alteration, doors both sides as well as at the back for convenience. Complete in case, with lock and key, leather handle, screws, &c., forming stand.. .. £3 0 0

Extras.—Four-wick lamp (wicks 2in. wide), 5/- Dallmeyer's pattern new system object lens, 5/- Mahogany body, 10/- Extra top, 1/6 Lime-light tray, 1/6

Remarks—In every respect a first-class oil lantern.

18.—The Economic (THOMAS ARMSTRONG & BROTHER).

Details.—Russian iron body, two-wick lamp, 3½in. condensers, single achromatic objective. Complete in box £2 15 0

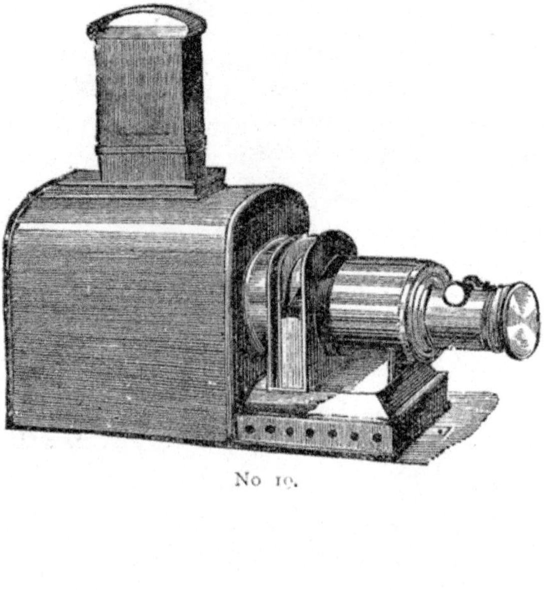

No 19.

19.—The Amateur (W. BANKS).

Details.—Russian iron body, 4in. condensers, three-wick refulgent lamp, spring stage and registering carrier, portrait objective with rack and pinion. In box complete £2 5 0

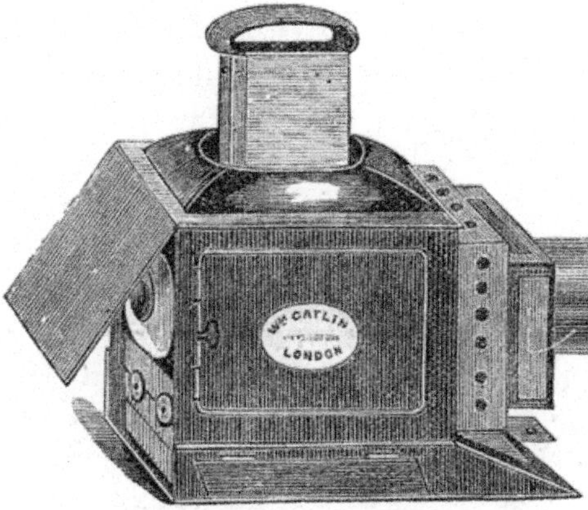

No. 20.

20.—The Triple Wick (W. CATLIN).

Details.—Russian iron, three-wick lamp, 3½in. condensers, portrait combination front lens with rack and pinion £3 3 0

Extras.—Same lantern, with four-wick lamp and 4in. condensers, 7/-

21.—Cubley & Preston's Mineral Oil (CUBLEY & PRESTON).

Details.—Ventilated Russian iron body, 3½in. condensers, achromatic front lens, nickel-plated, three-wick lamp. Russian iron case with leather handle £3 3 0

Extras.—4in. condensers and rack adjustment to front lens, 17/- Five-wick lamp, 20 -

Remarks.—Well made, though of ordinary pattern. The appearance is enhanced by the nickel-plated front lens.

22.—The Co-operative Educational Duplexicon (HUGHES, Kingsland).

Details.—3½in. standard sized condenser, front lens in brass sliding tube for focussing; the body made of best Russian iron, japanned, spring slide holders, having large two-wick lamp with pinions, and the patent hood, with conoidal plates that resist any heat, silver reflector .. £1 10 0

Remarks —A useful and very cheap lantern.

23.—The Co-operative Educational Triplexicon (HUGHES, Kingsland).

Details —3½in. double condensers, achromatic front lens with rack and pinion, triplex lamp (see "Lamps"), opens at the back.

Extras.—4in. condensers, 8/- ½-plate combination front lens, 10/6

24.—The Co-operative Three and Four Wick (HUGHES, Kingsland).

Details.—Body and foundation base shaped feet of Russian iron, 4in. double condensers, front lens has rack and pinion, three or four-wick lamps Case included £3 3 0

Remarks.—Mr. Hughes, in his catalogue, says, "These are the usual trade lanterns, and known under various extraordinary titles."

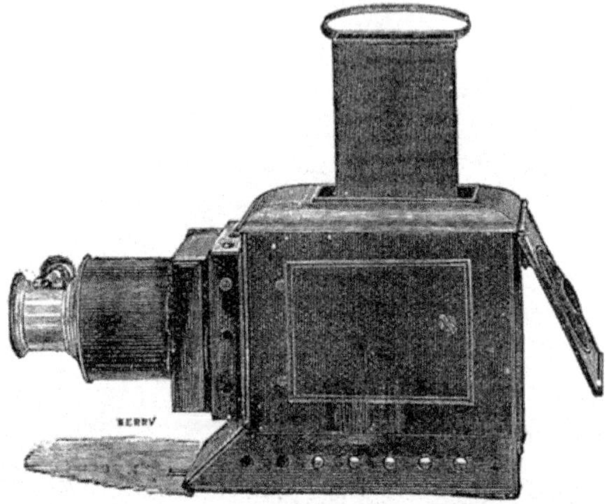

No. 25.

25.—The Marvel (MASON & CO)

Details.—Japanned body, 4in. compound condensers, achromatic double combination front lens with rack adjustment, three-wick lamp with 1½in. wicks. In box.

Best quality £2 10 0
Cheap ,, ... £1 15 0

Remarks.—A plain, straightforward little lantern.

26.—The Pioneer (NOAKES & SON).

Details.—Best quality japanned tin, sliding front tube, 4in. condenser, double combination achromatic objective with rack and pinion, lamp comes out without removing chimney, three-wick lamp. Portable case £2 10 0

Extras.—Four-wick lamp, 5/-

Remarks.—The usual style, but none the less a good article.

27.—The Phonopticon (OTTWAY & SON).

Details.—Double combination achromatic front lenses, rack and pinion, 3½in. condensers, improved patent three-wick lamp. Portable case £3 3 0

Extras.—4in. condensers, 9/6

Remarks.—A lantern that has many admirers, the name being well known.

28.—Pexton's No. 1 (Chatham Pexton).

Details.—Tin lantern with back and side doors, plano-convex condensing lenses, 4in diameter, double combination achromatic front lenses with rack and pinion adjustment, powerful refulgent three-wick lamp for burning paraffin oil £2 15 0

Extras.—Polished brass tube and spring slide holder, 10/-

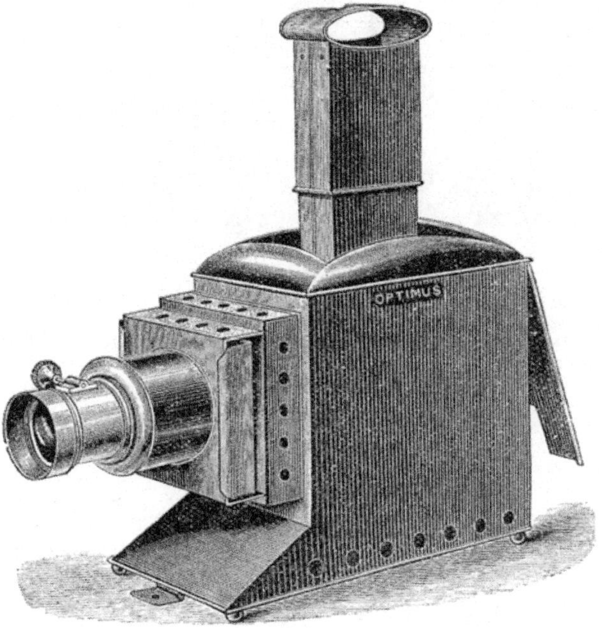

No. 29.

29.—The Optimus (Perken, Son & Rayment).

Details—Japanned tin, telescopic sliding fronts fitted with achromatic lenses, of photographic combination, with rack-work adjustment, two-wick refulgent lamp, and 3½in. condenser. Complete in box £1 13 6

Extras.—4in. condenser and three-wick lamp, 2/- Mahogany body, lined tin, 13/6 Finished brass stage and telescopic draw tube, 36/-

Remarks—An extremely cheap instrument. Readers will note the very small difference in price for an extra ½in. condenser and three-wick lamp, viz., 2/- only. Very good front lenses are used, the definition being really good. In the mahogany body form the mahogany is kept well away from the tin, thus giving excellent ventilation all round.

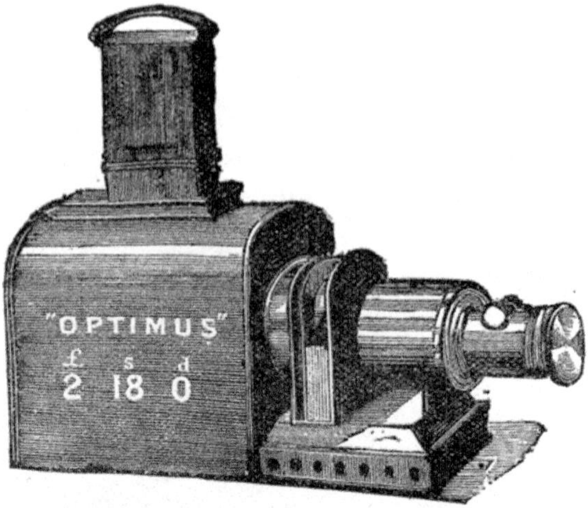

No. 30.

30.—The Optimus (PERKEN, SON, & RAYMENT).

Details.—Russian iron body, 4in. compound condensers, portrait combination front lens £2 12 0

Remarks.—The price has recently been reduced, so that the illustration is not correct as regards that particular. Another cheap lantern.

31.—Place's No. 1 (J. PLACE).

Details.—Condensers 3½in., mounted in brass cell, 2in. diameter front lens, paraffin lamp £1 5 0

Remarks.—Suitable for parlour or small schoolroom use. Well worth the price.

32.—Place's No. 2 (J. PLACE).

Details.—Condensers 4in. diameter, mounted in brass cell, compound achromatic focus lens with rack and pinion; paraffin lamp, opening at back for limelight £2 10 0

Remarks.—This lantern is good enough for a small exhibition.

33.—Pumphrey's No. M (A. PUMPHREY).

Details.—Japanned tinned lantern, 3½in. condensers, achromatic focus lens 1¾in diameter, patent triplex lamp. In box £2 10 0

Extras.—4in condenser 10/-

34.—The Sunlight (W. TYLAR).

Details.—Japanned tin, portrait combination front lens, three-wick lamp, sliding front tube.. .. .. £2 10 0

Extras.—Four-wick lamp, 5/-

35.—Watson's Cheap Phantasmagoria (W. WATSON & SONS).

Details.—Japanned body, 4in. condensers, portrait combination objective with rack and pinion, three-wick lamp £2 10 0

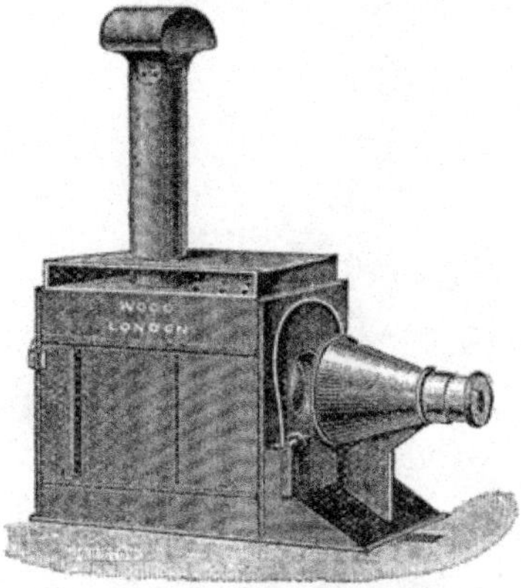

No. 36.

36.—The Exhibition (E. G. Wood).

Details.—Body of japanned tin, fitted with a single 3in. condensing lens and brass-mounted objective. The lamp is fitted with two wicks, and burns mineral oil £1 12 6

Extras.—Case, 3 6

Remarks.—Suitable for school or home use only.

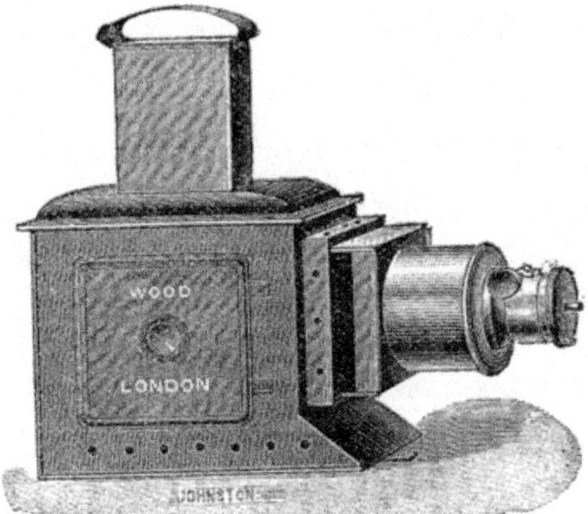

No. 37.

37.—The Victoria (E. G. Wood).

Details.—Body of japanned tin, double condenser 4in diameter, and brass-mounted objective. The lamp is constructed for burning mineral oil, fitted with three wicks, and does not need a glass chimney. In case complete £2 5 0

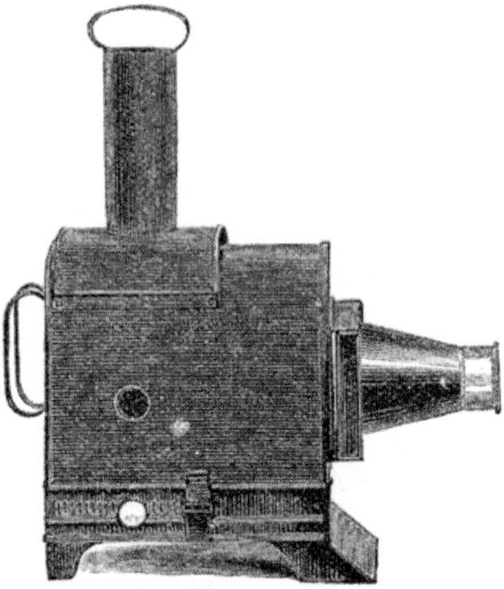

No. 38.

38.—The Exhibition (Wholesale Firm).

Details.—Tin japanned body, 3in. condenser, mineral oil lamp. In black box £1 14 0

Extras.—3½in. condenser, 3/6.

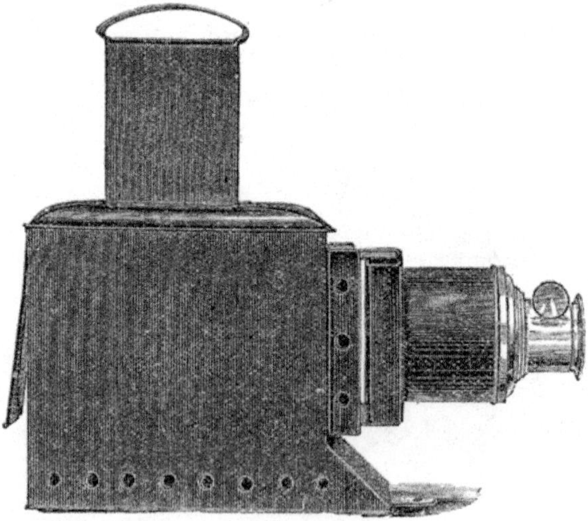

No. 39.

39.—Cheap Tin (Wholesale Firm).

Details.—Japanned tin body, dome-shaped top, japanned tin stage, front tube and sliding tube, with brass O.G.; 4in. plano-convex compound condenser in brass cell; double combination achromatic front lens in brass mount, with rack and pinion, three-wick Russian iron paraffin lamp with jointed Russian iron chimney. In black box, leather handle .. £1 16 0

Extras.—Four-wick lamp, 3/-

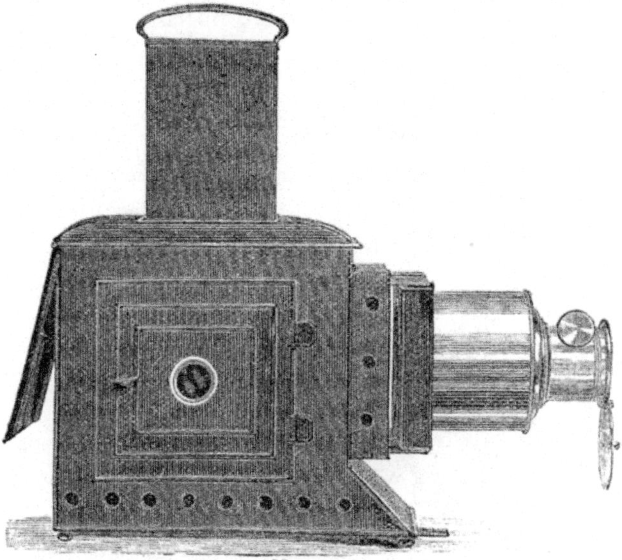

No. 40.

40.—Superior Tin (Wholesale Firm).

Details.—Japanned tin body with dome-shaped top, side door with blue glass sight hole, japanned tin stage, and with front tube and sliding O.G. tubes, all brass; improved three-wick lamp; 4in. plano-convex compound condenser in brass cell; double combination achromatic front lens in brass mount, with large size lens to the back combination, rack and pinion focussing, sliding shutter or flasher. Stained and polished wood box, leather handle £2 5 0

Extras.—Four-wick lamp, 3/-

Remarks.—This lantern, having all brass front tubes and side door, is worth the extra money.

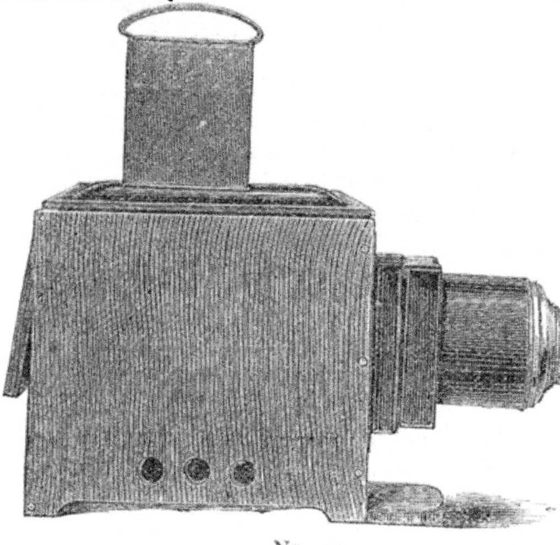

No. 41.

41.—Cheap Mahogany Body (Wholesale Firm).

Details.—Polished mahogany body, lined with japanned tin, dome-shaped-top, japanned tin stage, front tube and sliding tube with brass O.G.; 4in. plano-convex compound condenser in brass cell; double combination achromatic front lens in brass mount, rack and pinion; three-wick lamp. Black box with leather handle.. .. .. £2 12 6

Extras.—Three-wick lamp, 3/-

Remarks.—Handsome looking and cheap.

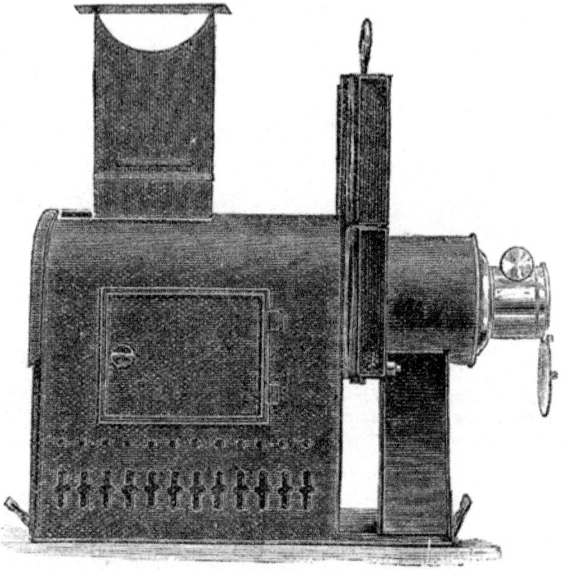

No. 42.

42.—The Cheap Metamorphoser (Wholesale Firm)

Details —This is a cheaper form of this well-known lantern, brought out this year. In most particulars it is the same as No. 82 (which see), but of course not so well finished £3 3 0

Extras.—Four-wick lamp, 3/-

Better Class Oil Lanterns.

We are quite aware that the above title may be cavilled at when the division is seen to be one of pounds, shillings, and pence. Possibly the title would be more truthful in some cases were it to read "More expensive oil lanterns." But, as Captain Corcoran says, "One must draw the line somewhere." And certainly in our work the exigencies of systematic arrangement are alone responsible for this arbitrary £3 3/- limit. We shall commence with a antern that in reality ought to have preceded all those burning mineral oil. But the price being over the limit, it must be described at the commencement of the above division. This work being more of a commercial than a historical nature, dates, legends and romance must play second fiddle to hard c

43.—The Sciopticon (Sciopticon Company).

Details.—We enter into a very detailed description of this instrument, as it was the first mineral oil lantern introduced, the forerunner of all those of recent date, and which partake more or less of its features. On that account, therefore, a description will be interesting and useful. It was the invention of Mr L. Marcy, of Philadelphia, introduced into England by the late Mr. Walter Woodbury, and is sometimes known as the "Woodbury Sciopticon." To commence with the different parts forming the instrument.

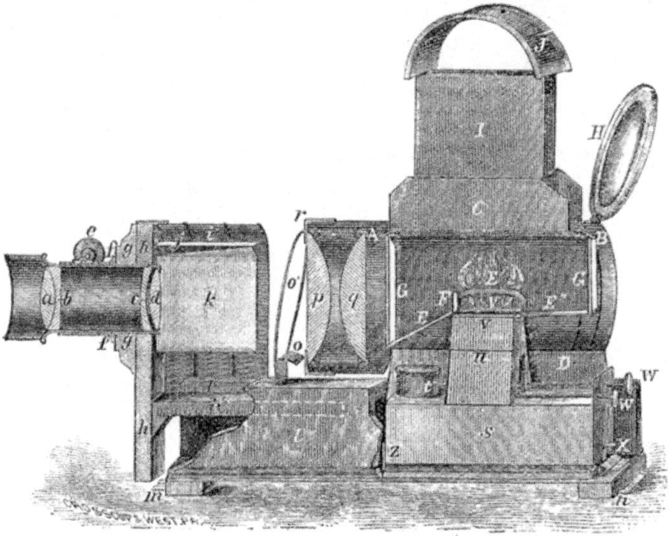

No. 43.

Details.—The lenses, mountings, &c., are shown in section. What is left of the frame and cylinder, the lamp, chimney, reflector, &c., are shown in perspective. The parts are as follow:—

a b—Front combination of the objective cemented together.

c d—Back combination separated by a ring. If the cells holding these combinations are unscrewed, and the lenses removed, they must be returned in the same order and position as seen in the diagram.

e—Milled head for adjusting the focus, when necessary.

f f—Flange attached to the projecting wooden ring, *g g*.

*h h h*¹—Wooden frame of the extension front, *h*¹ sliding in a groove within the body frame, *l l*.

l l—Portion of the wooden frame, the rest being mostly cut away to show the lamp.

m—Claw attached to the front foot

n—Flange under the back foot. On the top of the Sciopticon case, or box in which it is carried, and which can be placed upon a stand or table to elevate the instrument to proper height while in use, are two round-headed screws, slightly raised, and at the distance apart of *m n*; *m* clings to one, and *n* slips under the other, thus holding the instrument firmly in place. When a pair are used for dissolving views the fronts are thus held in a fixed position, while the rear ends may be spread apart till the discs on the screen coincide.

*o o*¹—Stage and spring carrying the slides, &c , to be shown. This is fixed to a base-board, which together with the hood is moved in front of the condenser to its proper focus. See Fig. 1.

p q—Condensing lenses.

r—Brass ring, holding the condenser cells suspended in the cylinder, so as not to be anywhere in contact with it. To remove the condenser, the extension front and the stage are drawn off

s—Lamp for paraffin oil. It holds a pint, or enough to last about four hours.

t—Nozzle to admit the oil. It is large, so that if a wick is carelessly turned down into the lamp it can be fished out with a bent wire.

u—Side of one of the two tubes, showing how the conduction of heat downwards is counteracted by breaking the connection in the metal.

v v—Tops of the two tubes. They carry wicks an inch and a half wide. The wicks are pushed down the tubes till they are caught by the ratchet-wheels and drawn down. Should a loose thread of the wick get clogged in the wheels it must be drawn and cut off.

w w—Buttons for adjusting the wicks; both are turned *inward* to raise the wicks, and *outward* to draw them down.

A B—Portions of the cylinder not cut away, seen beyond the condenser and flame-chamber.

C—Portion of the cylinder turned up, to give free ventilation all about the flame-chamber.

D—Portion of the cylinder turned down and supported by the wooden frame.

*E E*¹—Bottom of the flame-chamber. It is not supported by contact with the lamp, thus avoiding the conduction of heat downwards. *E*¹ answers to the deflecting cap of a common lamp. *E* slopes so as not to shade the light from the condenser.

F—Narrow strip of glass, held in a socket before the flame, to give upward direction to heated air.

*G G*¹—Front and back glasses of flame-chamber.

H—Reflector used also to close the rear of the cylinder The centre of concavity is at *E*¹, so that reflected rays are thus made to coincide with incident rays from *E*¹ to the condenser.

I—Chimney (telescopic) to regulate the draught.

J—Chimney cap, for darkening the outlet

The focussing arrangement of the Sciopticon is somewhat different to the ordinary run of oil lanterns. It may be briefly described as focussing with the slide instead of the lens. Mr. George Smith thus refers to it in the *Sciopticon Manual:* "The principle is well known in optics, but, curiously enough, it has hitherto always been overlooked in the construction of optical lanterns Our system will be better understood by considering the functions of the different lenses. The object of the condensing lens is to collect the rays of light and throw them into the front lens, by which they are transmitted to the screen. As these rays collected by the condenser issue from it in the form of a cone, there is a point where the front lens might be fixed. If a slide is now introduced between the condenser and front lens an image of it will be thrown on a screen, the size of which will be determined by the position of the slide between the lenses. If the different lenses are of such a combination of focal length that a picture of the required size is produced when the *whole* of the light collected by the condensers is concentrated on the slide, it is evident that the maximum of illumination is produced on the screen. This is accomplished in the improved Sciopticon by our new focussing arrangement."

Altogether, the Sciopticon, though first in the field, has not by any means been distanced in the race, and is at the present time a sound optical, as well as convenient lantern. Special care appears to be taken that the lenses and condensers are thoroughly good. Complete, in stained box.. .. .. £4 10 0

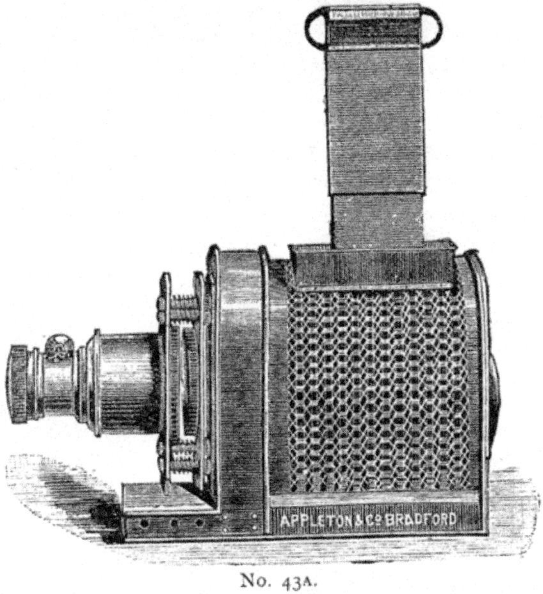

No. 43A.

43a.—Appleton's Russian Iron (Appleton & Co)

Details.—Ornamented or fancy Russian iron body; 4in. condensers; achromatic front lens with rack and pinion adjustment; three-wick lamp; best blow through jet for limelight £3 15 0

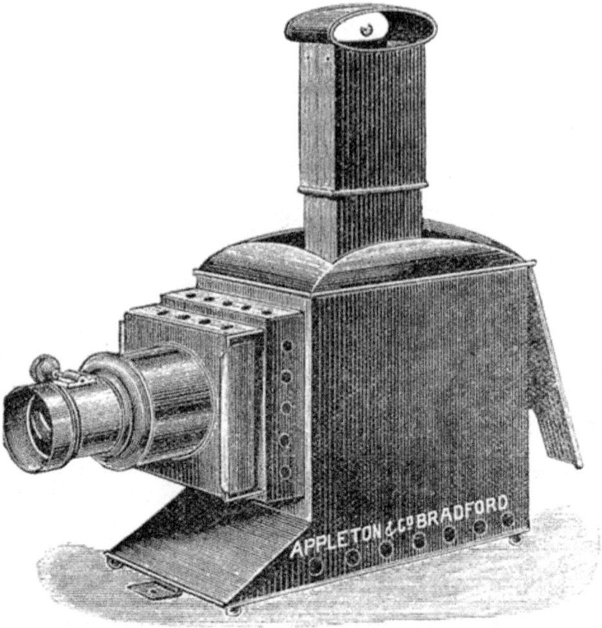

No. 43B

43b.—Appleton's Cheap Mahogany (APPLETON & Co.

Details.—Same as 43a, but with mahogany body, instead of the Russian iron £3 15

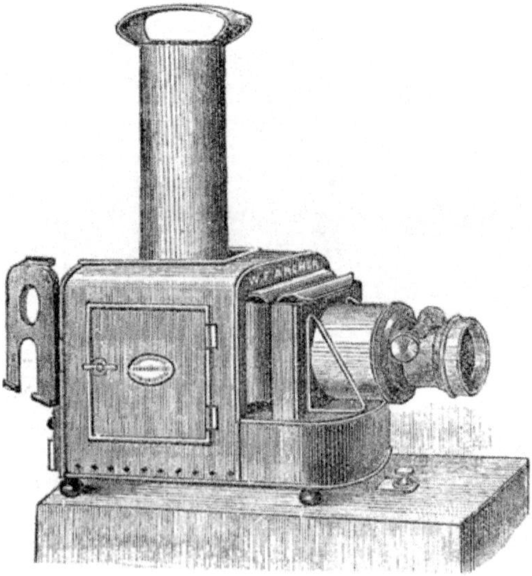

No. 44.

44.—The Photinus (Archer & Sons).

Details.—Japanned tinned body, lined to ensure ventilation, doors both sides and at back; lamp takes right out without removing chimney; slides may be put in at top of holder as well as at sides; 4in condenser, double combination portrait objective, Dallmeyer's new pattern, with rack and double pinion, four-wick lamp of special design. Complete in case forming stand £5 0 0

Extras.—Mahogany body, 20/- Russian iron top and sliding tray for limelight, 3/6

Remarks.—A lantern which has met with most pronounced success since its introduction by Mr. Archer; in every respect a really excellent apparatus.

45.—The Manchester (Armstrong & Brother).

Details.—Russian iron body; 4in. condensers, double achromatic portrait combination front lens with rack and pinion, three-wick lamp. In box £3 15 0

Extras.—Mahogany body, 25/-

Remarks.—A good plain instrument, the ordinary mineral oil lantern style The mahogany body form varies slightly in design from the other.

46.—The Technical (Cubley & Preston).

Details.—Russian iron body, supported by four nickel-plated columns on a mahogany base; two fronts, detachable. One has two openings for the lens, so that either transparent or opaque objects may be exhibited The other front is nickel-plated, and has a separate removable stage for showing chemical or other experiments. This stage can be extended to a length of 10in., and has an attachment for raising and lowering the exhibit. The lamp has five wicks, 2¼in. wide, with extra large oil cistern and silvered reflector, the front lens is double achromatic, with rack and pinion in nickel-plated mounts; 4in. condenser.. .. .. £7 7 0

Extras.—An electrical decomposing trough (see "Scientific Accessories"), which is specially adapted to this lantern, 21/-

Remarks.—A very useful all-round lantern, perhaps most suitable for school or society demonstrations It has been adopted by the French Government for use in their technical schools.

47.—The Pamphengos (Hughes, Kingsland).

Details.—Doors both sides, having entire brass fronts with brass spring slide holders, which admit of the slides being inserted at the top as well as the sides, hinged door at the back, with blue glass sight hole fixed in brass screw flange to enable the operator to observe and regulate the flame while exhibiting. To the fronts are attached strong brass draw tubes, which can be extended from 4 to 7 inches, with solid brass screw flange for adapting the front lenses, which are double combination, 2in. in diameter, specially curved, &c., giving a very flat field. The lenses are mounted in brass screw cells, and are of medium focus, to enable them to be used at suitable distances in various sized rooms. To the outer jacket are fitted rackwork, with double pinions for working either side, which is a convenience when limited to area, also 4in improved double achromatic condensers, fixed in brass screw cells, with openings at top to allow air to pass between to prevent condensation on the lenses Black japanned case, with leather handle, having the appearance of a small travelling portmanteau £6 6 0

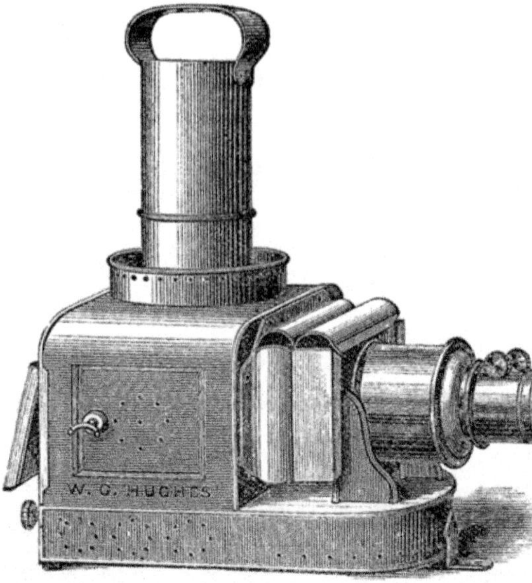

No. 47.

Remarks.—No instrument has attracted greater attention than Mr. Hughes's Pamphengos, and it well deserves the encomiums so freely passed upon it. It is very handsome in appearance, convenient to use, and when worked properly, gives a power of light not excelled except by the limelight itself. A description of the special lamp used will be found further on. We cannot, perhaps, do better than quote the opinion of Mr. Lewis Wright, as given by him in the *English Mechanic*:—" I satisfied myself perfectly that the new lamp fully illuminated a coloured circular photograph slide of the standard size, on a disc 12ft. in diameter, uniformly all over to the edges, and as well as I have several times seen done by an oxy-calcium jet But a 12ft. disc, well lighted, certainly surprised me. This, however, was not the main object of my visit. I took with me a 'lantern polariscope,' with reflecting bundle of glass, as described some time ago in these columns. The large end was too small to fit any lantern Mr. Hughes had, being fitted for 3¼in. condensers, while his were 4in. All we could do was to prop it up on packing, and insert as centrally as possible into the larger flange-nozzle: and the disadvantage in want of exact centring must have been perceptible. But such as it was, I proved by experiment with (1) selenite and (2) crystallisation slides, 1⅝in. diameter, and (3) with crystals of calcite, nitre, and topaz, that the lamp had power enough to project these well on a disc of 5ft. in diameter. I never expected to see this attained with any oil light, and that it can be done will place many experiments, before impracticable without gas apparatus, within the reach of country teachers and others. That is my reason for recording as briefly as possible the result of my trial of the new lamp.'

48.—Second Quality Pamphengos (HUGHES, Kingsland).

Details—Strong japanned tin body, 4in. double condensers, spring slide holders, the patent Pamphengos lamp, with the improved annealed blown coniodal plates, double combination portrait lenses in brass mounts, with rack and pinion. Japanned portmanteau case £4 4 0

Remarks.—Secondary rather in outward appearance than in working power.

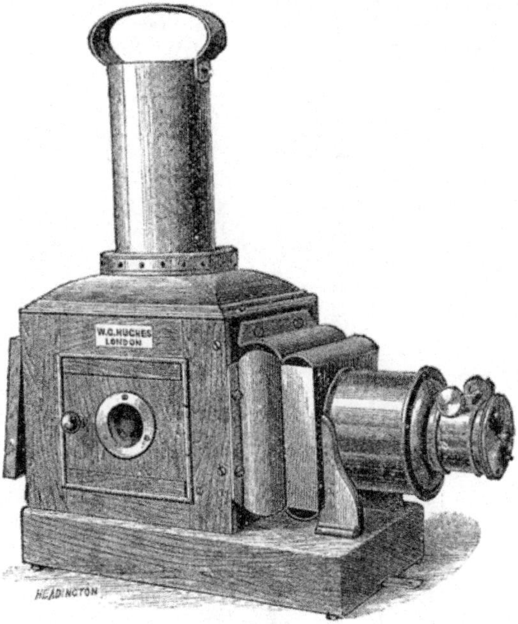

No 49.

49.—New Pattern Pamphengos (HUGHES, Kingsland).

Details.—Same optical arrangements, but with a new form of mahogany body, with solid plinth foot, two doors, blue glass sight holes, and circular brass flanges £7 11 0

Remarks.—The elegant appearance and the brass fronts are retained in this form, the mahogany not interfering

50.—The Circulatory Pamphengos (HUGHES, Kingsland)

Details—An ordinary pattern lantern, fitted with a new method of changing slides, very simple in action, working automatically. The frame itself revolves on an axis, and at each turn moves the same to and fro into such a position that when it reaches home it is fixed rigidly and securely on either side by the help of springs and catches suitably arranged for its reception. The shutter or screen in front of the nozzle of lantern opens from a centre so that the rays of light transmitted may not be shadowed, and is dependent upon a lever arm which, from the commencement of the frame rotating, allows the shutter to remain closed, and when it has passed round to the other side releases it again, and so places each picture on the screen almost simultaneously, without those in front having witnessed the manœuvre. A special point respecting this invention is this, the frame, revolving as it does on an axis and not in a grooved sliding frame, is not liable to fix, as the friction and resistance is nothing, the merest touch will set it going, and so prevent any failure during an exhibition. All slides can be shown with it, either framed or unframed. For the latter two special carriers are provided, while for slipping slides there is a spring guard to prevent the glasses shooting out, and in the case of chromatropes the handles merely allowed to drop parallel with the frame. All slides are inserted perpendicularly, and not upside down as in the ordinary way .. £9 9 0

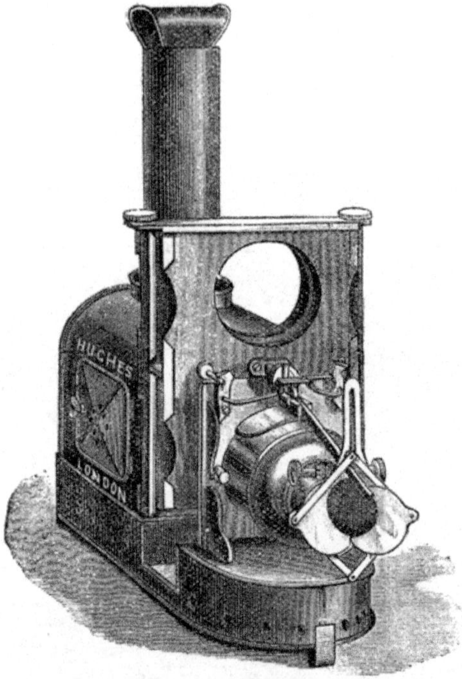

No. 50.

Remarks.—The arrangement being entirely of polished brass adds to outward appearances, and as regards ease in working the lantern, the advantage is great.

51.—Hughes's Triplexicon (Hughes, Kingsland).

Details.—Japanned body and lamp, with 4in. double achromatic condensers in brass screw cells, and double combination front lenses (specially curved) 2in diameter, which allow more light to pass through, giving a very flat field, consisting of four distinct lenses arranged together, mounted in double brass screw cells, the brass front draws out to 6in., with

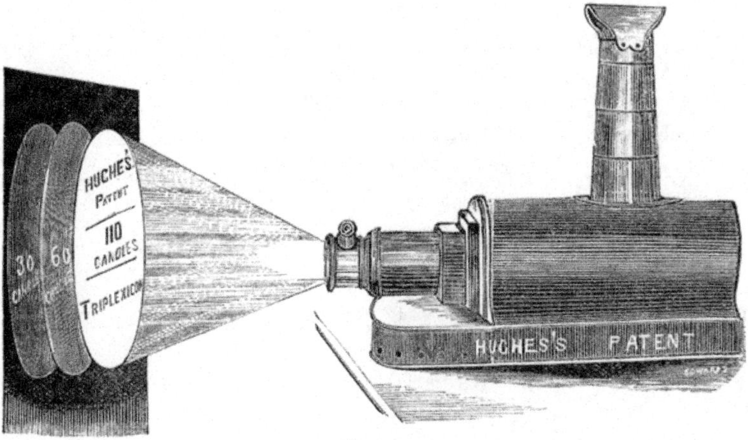

No. 51.

brass screw flange, to which are screwed the brass rack sliding tubes, with double pinions for working either side, improved spring slide holders which admit of the slides being inserted at the top as well as sides; groove to receive a jet. Packed in japanned case with leather handle .. £5 5 0

Extras.—Mahogany b.dy, 15/-

52.—Second Quality Triplexicon (Hughes, Kingsland)

Details.—Japanned body with the ordinary achromatic 4in. double condensers, spring slide holders, japanned sliding tubes to draw out to 6in., double combination portrait lenses, brass mounts, with rack and pinion. In japanned case £3 10 0

Extras.—Mahogany body

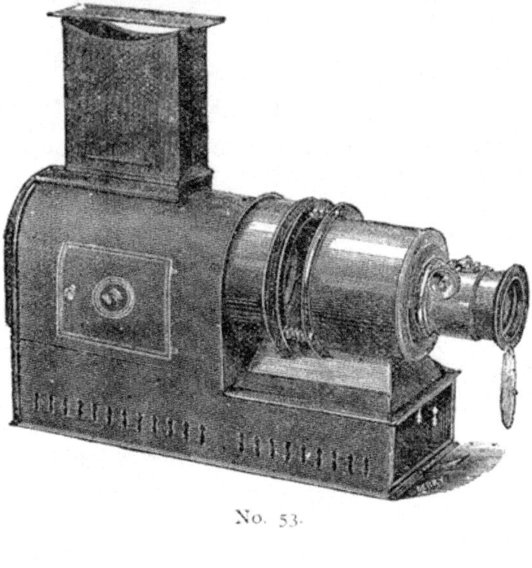

No. 53.

53.—The Imperial (MASON & Co.)

Details.—Russian iron body, 4in. compound condensers; achromatic double combination front lens, with rack adjustment; four-wick lamp with 2in. wicks. Packed in case 14in. x 9in. x 5in. £3 10 0

Remarks.—Well made and giving good light.

54.—Medland's Mahogany Body (J. B. MEDLAND).

Details.—Mahogany body, lined with tin, and japanned iron stage, relieved with brass fittings; sliding front with brass O.G. and portrait lens objective, with three-wick lamp £4 10 0

Extras.—Four-wick lamp, 6/-

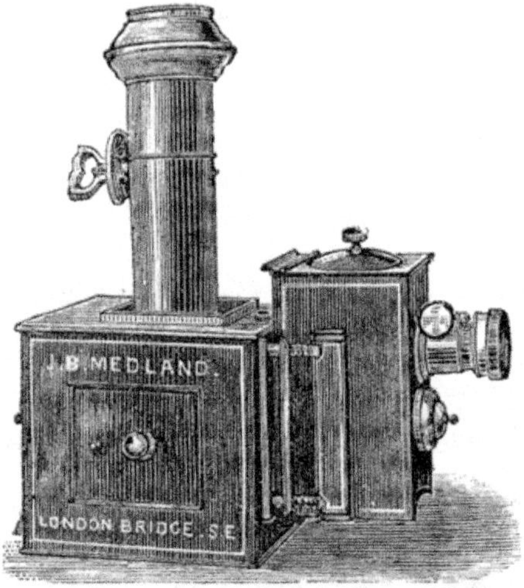

No. 55.

55.—The Pentaphane (J. B. MEDLAND).

Details.—As will be seen, the shape is somewhat different—the square front carrying the objective being an arrangement for exhibiting opaque objects, which is generally accomplished by a separate piece of apparatus, the aphengescope, the use of which necessitated altering the position of the lantern, whereas in this case it is affected almost momentarily. For exhibiting transparencies, the objective is screwed into the bottom flange, shown in the engraving, with a cap screwed in to exclude the light when not in use. The light from the lamp passes through the condenser in the ordinary way to the objective. When required for opaque objects, such as C.D.V., medal, works of a watch, &c, all that is required to be done is simply to screw the objective into the top flange as shown in the engraving. The flat mirror shown at the bottom is then brought into use, and fixed at a proper angle to throw the light from the condenser back on to the object by a piece of metal drilled with a number of holes, and a pin fixed beneath. The object is then reflected through the objective on to the screen. The difficulty of the accurate adjustment of the wicks, so as to avoid the smoke, &c., is obviated by a simple but effective arrangement of a fan or shutter placed in the iron chimney of the lamp, as shown in the engraving, which regulates the current of air, so that when the flames are left at a moderate height all level, by turning the fan, the current of air is increased, and the combined flames drawn upwards and towards the centre, where the greatest amount of illumination is of course required. 4in condensers and portrait combination objective. Light iron case, 15in. x 7½in x 13in. The price includes two carriers for opaque objects .. .. £5 0 0.

Extras.—Nickelled cone front to take the place of the square opaque arrangement, 12/-

Remarks.—A very useful lantern, and one most suitable for mixed entertainments. The principal advantages are the chimney, fitted with an entirely novel arrangement, whereby the light of the lamp can be adjusted to a nicety, by simply turning a shutter fixed therein. The rapid and easy way that it may be converted into an apparatus for the exhibition of opaque objects, without interfering with the ordinary arrangement for transparencies. The shape of front which gives so much space in front of the condenser for manipulating in experimental lectures, such as the trough experiments, electrical and other scientific purposes.

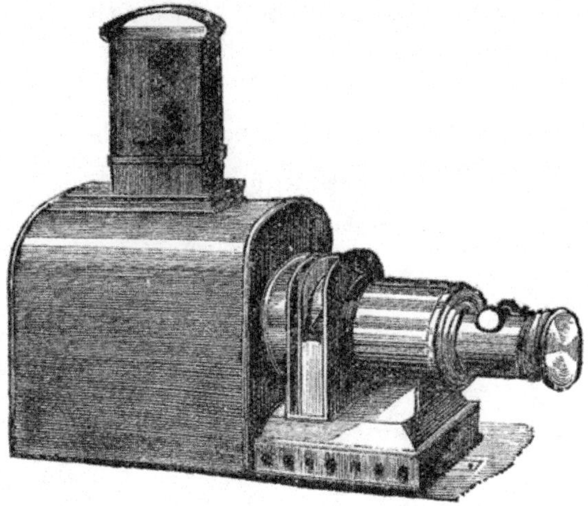

No. 56.

56.—The Perfection (J B Medland).

Details.—Body of Russsian iron, 4in. compound condenser, portrait combination front, rack adjustment, patent triple wick mineral oil lamp in improved box, with flask, funnel, etc., with nickelled cone front.. £3 12 6

Extras.—Brass stage and sliding front, 7/6. Four-wick lamp, 6/-

57.—Newton's Improved Phantasmagoria (Newton & Co)

Details.—Best quality plano-convex condensing lenses, 3½in. diameter; double combination achromatic front lenses of large diameter and best construction, with rack adjustment; fitted with new patent refulgent three-wick lamp (see "Lamps") for burning paraffin oil. Complete in stained wood polished case, 6½in x 19½in x 11½in, with lock and key £3 13 6

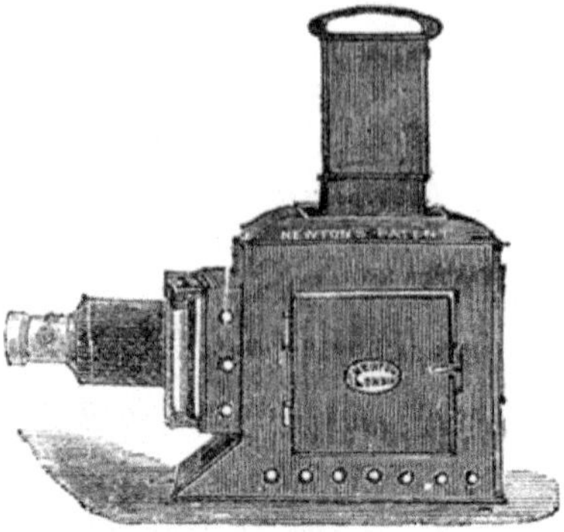

No. 57.

Extras.—4in. condensers, 10/6. Four-wick lamp, 6/-

Remarks.—The name of the firm is guarantee for good workmanship, and they obtained a silver medal at the Inventions Exhibition for this lantern. An advantage is the door at side as well as back, most useful for limelight work. This lantern has been adopted by the London School Board.

58.—Newton's Mahogany Phantasmagoria (Newton & Co.)

Details.—Mahogany body, lined with tin to prevent heating, with best quality plano convex condensing lenses, 4in. diameter; double combination achromatic front lenses, of large diameter, with rack adjustment; fitted with patent refulgent three-wick lamp for burning paraffin oil. Complete in stained wood polished case, with lock and key £5 5 0

Extras —Four-wick lamp, 5/-

59.—The Pioneer (Noakes & Son).

Details.—Polished mahogany body, lined with tin; 4in. condensers; compound achromatic front lens, with rack and pinion; three-wick lamp £3 10 0

Extras.—Four-wick lamp, 5/-

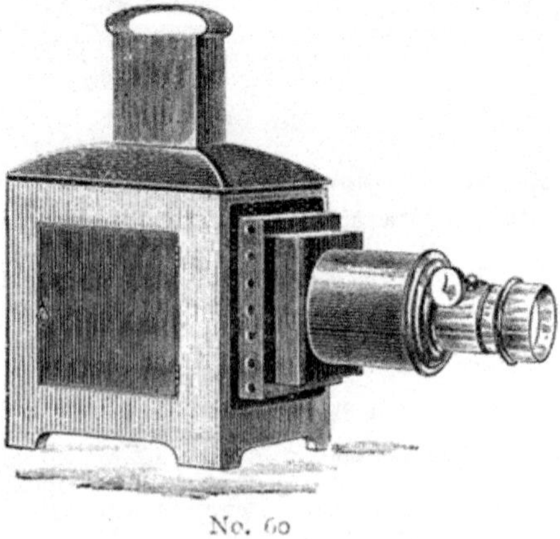

No. 60

60.—The Invicta No. 1 (Noakes & Son).

Details—Polished mahogany body, lined with tin; 4in compound condensers, double combination achromatic objective, japanned tin stages, brass O G, with three-wick lamp. In case complete £3 10 0

Extras.—Four-wick lamp, 5/-

Remarks—Soundly constructed, though without any particular specialities. A good, plain, and useful instrument.

61.—The Invicta No. 2 (Noakes & Son).

Details—Same as the No 1, but with side door, brass flange, window with sight hole and foot-board, three-wick lamp £4 5 0

Extras.—Four-wick lamp, 5/-

Remarks.—A very well finished article, the additions adding considerably to the general appearance.

62.—The Mahogany Phonopticon (Ottway & Son).

Details.—Same as No 27, but with mahogany body lined with tin, 4in. condensers Stained wood case with lock and key £4 10 0

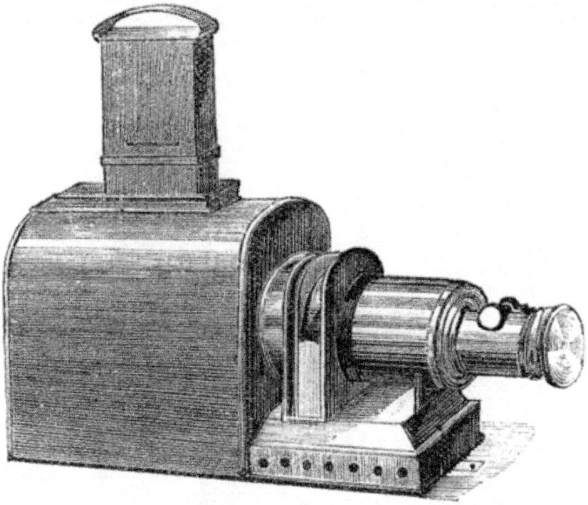

No. 63

63.—The Standard Educational Phonopticon (Ottway & Son).

Details.—Body of Russian iron, stage and sliding front of brass, 4in. compound condensers, achromatic portrait combination objective with rack and pinion, patent three-wick lamp Portable tin case £4 4 0

Extras—Four-wick lamp, 6/-

Remarks—The front and stages being all of polished brass gives this lantern a most attractive outward appearance.

64.—Pexton's No. 2 (Chatham Pexton).

Details—Same as No 28, but having polished mahogany body £3 5 0

Extras—Polished brass tube and spring slide holder, in place of the japanned tin, 10/- With cheap form complete brass front, 20/- Four-wick lamp, 5/

65.—Pexton's No. 6 (Chatham Pexton).

Details.—Polished mahogany body, doors each side, brass flanges with coloured glasses let into centre of same, lined with tin, three-wick lamp black japanned top, 4in. double condensers, massive brass front open all round, rack and pinion adjustment, fitted with best lens and flashing shutter £5 10 0

Extras.—Four-wick lamp, 5/-

Remarks.—This is a first-class lantern of extra good appearance, and most carefully constructed for strength and finish.

66.—Place's No. 3 (J Place).

Details.—Polished mahogany body, lined with sheet iron, 4in. condensers, brass telescopic sliding tube, compound achromatic focus lens with rack and pinion, paraffin lamp, opening at back for limelight £3 15 0

Remarks.—One of the best made lanterns we have seen at the price

67.—Pumphrey's No. Na (Alf. Pumphrey).

Details.—Polished mahogany body, lined with metal, brass front and fittings, 4in. condensers, achromatic front lens with rack and pinion, patent triplex lamp £4 0 0

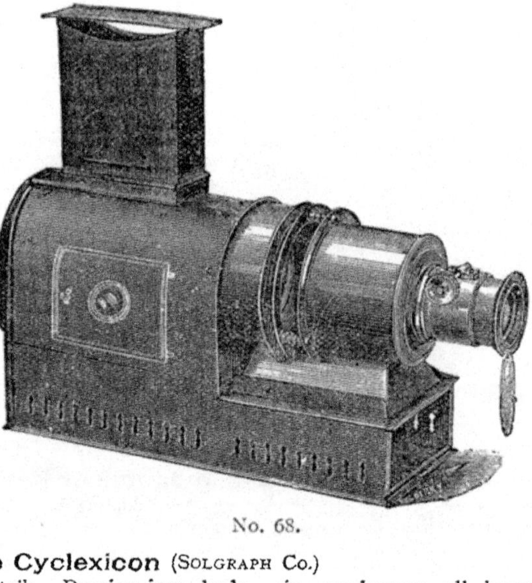

No. 68.

68.—The Cyclexicon (Solgraph Co.)

Details.—Russian iron body, 4in. condensers, all brass stage and rack adjustments, achromatic portrait front lens, three wick lamp. In case £3 17 6

69.—The Photogenic (J. H. Steward).

Details.—3½in. condensers, rack-work focussing adjustment to brass front, carrying achromatic lenses, sliding body tube, improved three-wick patent lamp, to burn paraffin or any of the best mineral oils. In strong portable case £3 10 0

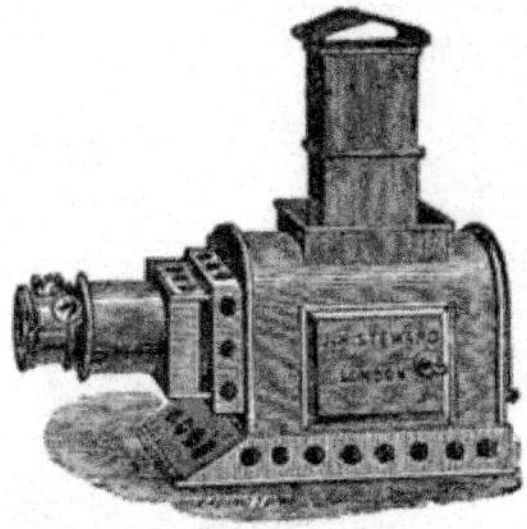

No. 69.

Extras.—4in. condenser, 10/- Mahogany body, 17/6

70.—The Mahogany Photogenic (J. H. Steward).

Details.—Same as No. 69, but having mahogany body, with brass fronts and stages, 4in. condensers £5 10 0

Remarks—These lanterns give a splendid nine or ten feet picture. A feature is the retention of the original form of stage, as in the old Phantasmagoria lanterns, thus having a firm stage or support for the proper holding of movable slides This point is not always attained in the new shapes at present on the market.

71.—The Special Photogenic (J H. Steward).

Details.—Japanned body, 4in. compound condensers, portrai combination achromatic front lens, of large diameter, with rack and pinion, spring stage, sliding tube, three-wick lamp, telescopic iron chimney. Complete in box with clamp screw £3 3 0

Remarks.—A plain, strong lantern, handy to use, and efficient.

72.—The Parisian (Theobald & Co.)

Details.—Five-wick lamp, 4in. double condensers, double focussing lenses (½-plate) and rack front, giving a 12ft. to 14ft. diameter picture, lamp fitted with draught shutter, and additional apparatus given to fit on to the front of lantern for showing opaque objects. Complete in box .. £6 0 0

Extras.—Mahogany body, 15/-

73.—The Sunlight (W. Tyler).

Details.—Polished mahogany body, lined with tin, dome-shaped top, japanned tin stage with sliding tube and brass O G., 4in. compound condenser in brass cell, plano-convex lenses, double combination achromatic front objective, rack and pinion adjustment £3 10 0

74.—Tyler's Special Russian Iron (W. Tyler).

Details.—Body of Russian iron, riveted throughout with copper rivets, stage, front tube and sliding O.G brass, mounted on solid brass foot, 4in. compound condenser in brass cell (two plano-convex lenses), double combination achromatic front objective, with rack and pinion adjustment, three-wick lamp (wick 2in wide), with deep reservoir, blue glass sight hole in door, lamp can be removed without taking off chimney. The whole packs into box 14in. × 9in. × 5½in. £3 10 0

Extras.—Four-wick lamp, 5/- Stage and front, nickelled, 5/-

Remarks.—Well made and strong; the usual pattern, however. Illustration may be seen further on at No. 120.

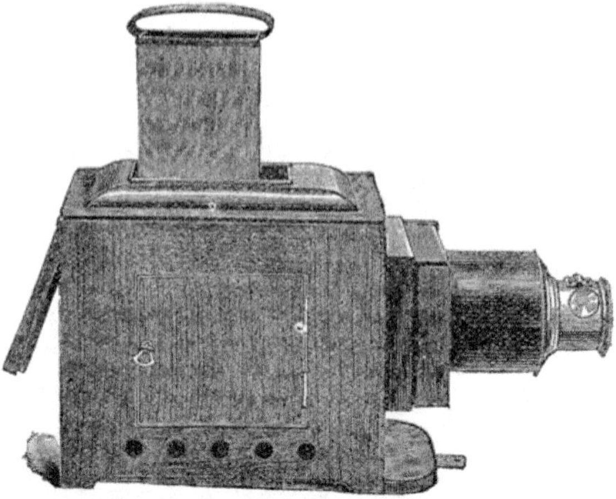

No. 75.

75.—The Perfection (W. Tyler).

Details.—Polished mahogany body, lined with tin; japanned tin stage with brass fittings, sliding front with brass O.G, 4in. compound condenser, portrait combination front lens, three-wick lamp. Portable tin case £4 5 0

Extras.—Four-wick lamp, 5/-

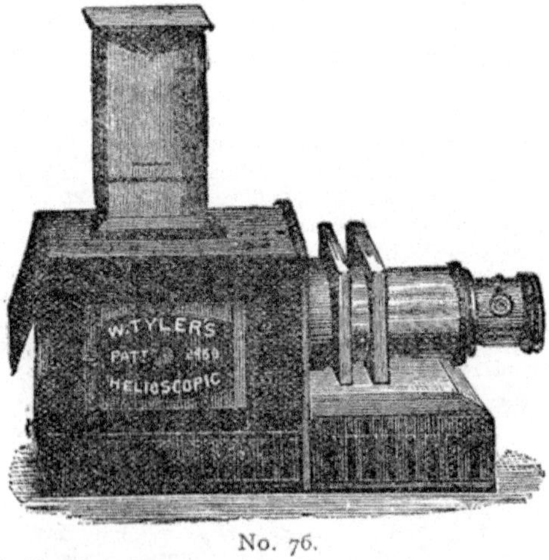

No. 76.

76.—The Helioscopic (W. Tyler).

Details.—Stoke Russian iron body, square, not round; all-brass front; telescopic chimney, packs under the front; 4in. condenser, removed from the front, not from inside; new shape stage, four-wick lamp; large diameter front lens of portrait combination with rack and pinion. In case £4 4 0

Remarks —Something a little out of the common groove. The chief advantages are - (1) The condenser removable from the front, it usually being an awkward job with the lantern all hot, to put the hand inside (2.) The stage is not only firm, but the slides work more than usually smoothly in and out.

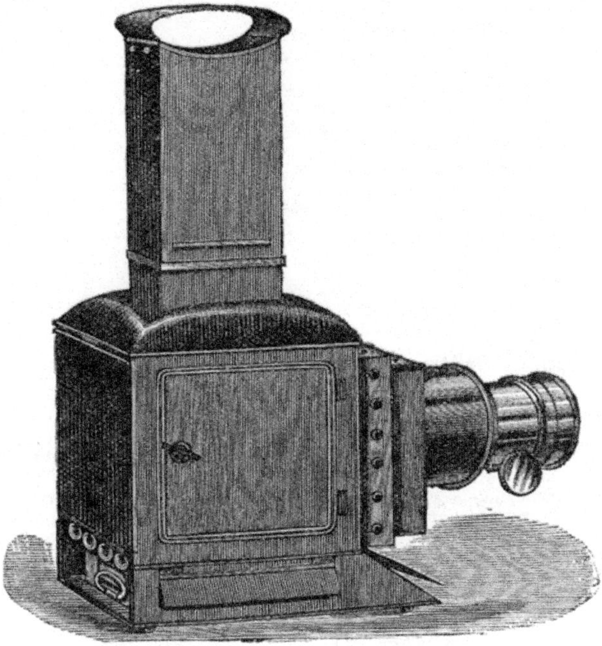

No. 77.

77.—Watson's Phantasmagoria (Watson & Sons).

Details.—Japanned tin body, four-wick lamp, 4in. condenser, brass stage, achromatic portrait front lens with rack and pinion. Case included £4 10 0

Extras.—Mahogany body, 17/6 Best finish mahogany body, brass front springs, 36/-

Remarks.—These lanterns will take a lot of beating in the way of finish and appearance.

78.—The Euphaneron (E G. Wood)

Details.—Body and pedestal of Russian plate iron, which neither rusts nor blisters; double lens condenser, 4in. diameter, of the improved form, fitted into a brass cell, and a double achromatic objective, with rack and pinion adjustment; the slide holder is mounted on a solid brass base, and is very rigid; E. G. Wood's " W " lamp, with four wicks two inches wide, the four wicks being arranged in the form of the letter W. (With this lamp no glass chimney is required, the combustion taking place in a metal chamber, and ordinary paraffin is used. The lamp is complete in itself, and is not in any way attached to the body of the instrument.) Packed in a case 14½in. by 5¾in. by 9in. and weighing about 14lbs. £4 4 0

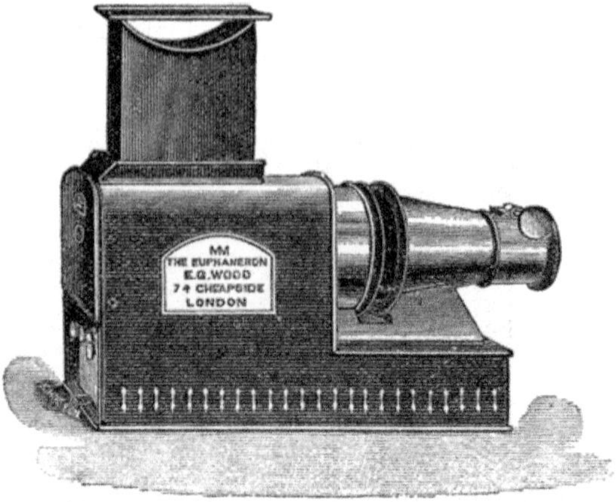

No. 78.

Remarks.—The meaning of the title is "that which shows well." But it looks well, as well as shows well. A capital lantern.

79.—Mahogany Body (Wholesale Firm).

Details.—Polished mahogany body, with panelled door, and brass-bound sight hole lined with japanned tin; dome-shaped top; japanned tin stage and sliding tube, with brass O G., and brass cell for condenser; 4in. plano-convex compound condenser in brass cell; double combination achromatic front lenses in brass mount (extra large size back lens), with rack and pinion adjustment, and fitted with a sliding shutter or flasher; four-wick Russian iron paraffin lamp, with jointed Russian iron chimney. Stained and French polished wood case, lock and key, and leather handle £3 7 6

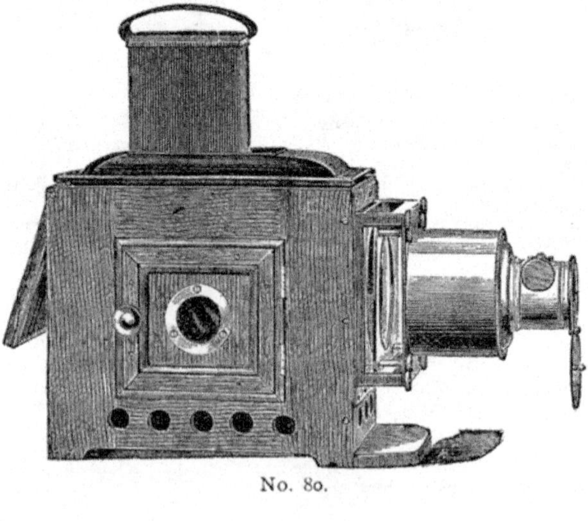

No. 80.

80.—Mahogany Body (Wholesale Firm).

Details.—Polished mahogany body, with panelled door and brass-bound sight hole, lined with japanned tin; dome-shaped top, with the entire stage and front tube and sliding tube made in brass; 4in plano-convex compound condenser in brass cell, and double combination achromatic front lenses in brass mount (with large size lenses to the back combination), rack and pinion adjustment, and fitted with a sliding flasher or shutter; four-wick Russian iron paraffin lamp, with jointed Russian iron chimney. Stained and French polished case, lock and key, leather handle £4 17 6

Extras.—Ebonised body and all the brass work nickel-plated, 7/6

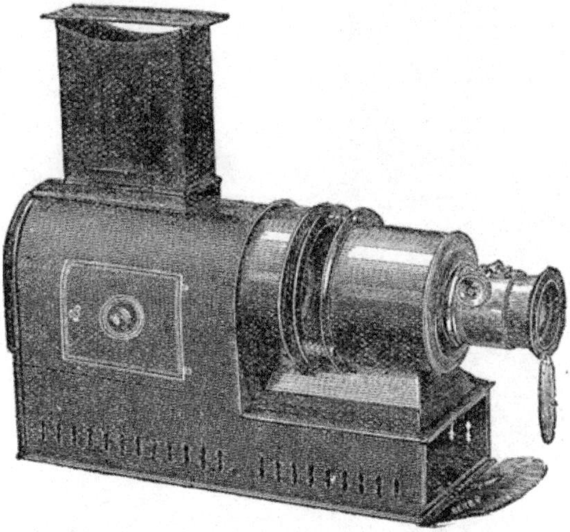

No. 81.

81.—Russian Iron (Wholesale Firm)

Details—Body of Russian iron, rivetted throughout with copper rivets, brass cell for condenser. The stage, front tube, and sliding O G. tube of best brass, and mounted on a solid brass foot, making the entire front very rigid, and the stage being open all round allows slides to be inserted from the top or the sides. Useful when chemical experiments are being tried. Side door, with blue glass sight-hole for use with the limelight. The entire brass front and stage slides on the front foot of the lantern, which allows the whole of the light from the condenser to be concentrated on the picture, so that no light may be lost even when a small picture is being exhibited. 4in. plano-convex compound condenser in brass cell, and double combination achromatic front lenses in brass mount, with rack and pinion adjustment, and fitted with a sliding shutter or flasher. Three-wick lamp, with wicks two inches wide. Very deep reservoir, which will hold sufficient oil for an exhibition of about two hours. The catch of the door of lantern body has been improved by the addition of a small support inside, so that when the door is shut the sides of the body are rigid. Back door has a hole in the centre of it, with a small circular piece of blue glass let in, so that the wicks inside the lamp can be seen and regulated from the outside when the door is shut. The body is so constructed that the lamp with the Russian iron chimney on it can be removed and replaced without taking off the chimney The chimney, when not in use, slides into the front foot of lantern body, underneath the brass front, and the front brass tce-plate forms a catch to keep the chimney in its place. Size of the box, 14 × 9 × 5½ £3 15 0

Extras.—Four-wick lamp, 4/- Stage and front tubes, nickel-plated, 7/6

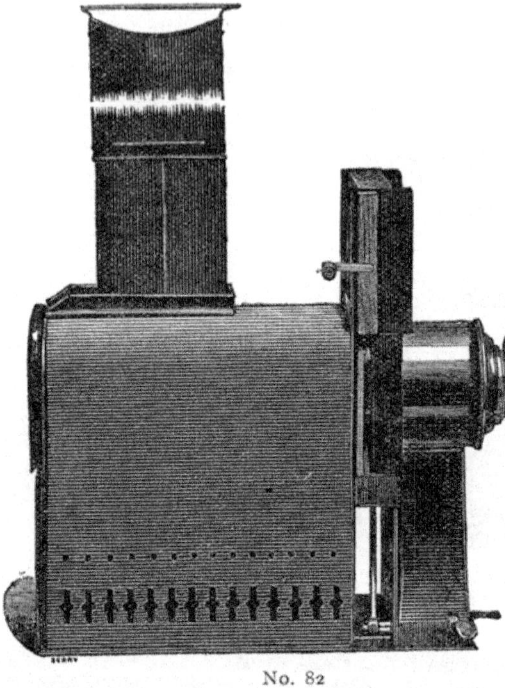

No. 82

82.—The Metamorphoser (Wholesale Firm).

Details.—Body of Russian iron, rivetted throughout with copper rivets, and the front tube and sliding O.G. tube of solid brass Side door with sight-hole for use with the limelight, 4in. plano-convex compound condenser in brass cell, and double combination achromatic front lenses in brass mount (with large size lenses to the back combination), and double pinions to the rack adjustment, and fitted with a sliding shutter or flasher, having grooves to carry shapes of coloured glass for tinting slides. New four-wick lamp, with wicks two inches wide (placed to form two wedges, thus W, this preventing the flicker of the wicks, seen on the screen, when they are placed parallel to one another), and a tall jointed Russian iron chimney, with a large oilcan and filler. The oilcan, when not in use, fits into the chimney, and the chimney then slides into the lantern body underneath the lamp, thus making the box in which the lantern is packed very small (size 15 × 11½ × 6). By means of a double changing stage, working vertically, any framed slides (such as rackwork, lever action, changing comic, &c.) can be shown in the lantern one after another without any interval, thus doing away with the appearance of the white disc of light seen on the screen while changing the slides in the stage of an ordinary single lantern. The double sliding stage is raised and lowered by a lever which enables the operator to either raise it slowly, so as to make the slides exhibited appear like a panorama, or to raise it so quickly that the change from one slide to another is almost imperceptible. Any mechanical slide can be introduced into a lecture set of photographic slides used with a carrier frame, without taking out the carrier, by simply placing the mechanical slide in the bottom stage, and raising it when required into position, and then lowering the stages again to their former position, without having, during the change, shown any white light on the screen £5 12 6

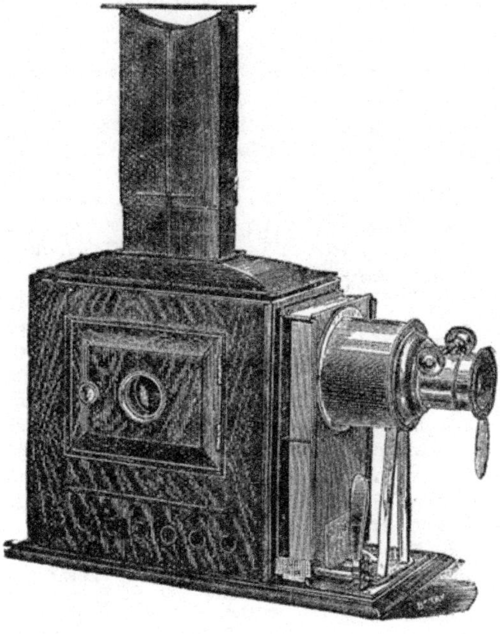

No 83.

83.—The Mahogany Metamorphoser (Wholesale Firm).

Details —Same as 82, but with polished mahogany body, with panelled door and brass-bound sight-hole, lined with tin, japanned tin dome-shape top, and with the entire stages and front tubes and supports made in brass. Stained and French-polished wood case, lock and key, leather handle £9 9 0

Remarks.—The descriptions of these two patterns speak for themselves, he lantern being a deservedly popular one, and the advantages offered great. No. 83 is especially a handsome instrument.

Pairs of Oil Lanterns for Dissolving Views.

Seeing that the lanterns we now have to deal with are, in the majority of cases at least, the same as have been described in previous pages, this division may not prove very interesting. But it will be useful, and moreover is requisite to carry out the intention of the work. At the same time it shall be made as brief as possible. Dissolving oil lanterns are not in very strong demand for two reasons—(1.) For home use and for small audiences, where an oil lantern is used at all, dissolving is not required, the pictures appearing on the screen one after the other by means of ordinary carriers, dissolving or patent carriers, or simply by plain changing of the slide. No doubt the white disc appearing between each slide is objectionable to some, but there are several ways of obviating this difficulty without recourse to a pair of lanterns. (2.) Because effects, as they are termed, are not accomplished well by oil lanterns, and this is the principal attraction of dissolving views. To put it briefly, the mere dissolving of one view into another by a mechanical method is hardly worth paying the price of two lanterns for; especially, as we have already said, special carriers and other apparatus exist for this purpose. And perhaps a further reason is that the cost of two lanterns very nearly approaches that of a limelight biunial, which, in addition to plain dissolving, gives other powers. So much against dissolving oil lanterns. But there is another side to the question—there always is in this world—and that is that dissolving can be attained without the trouble and bother of gas-making. There may also be one or two other minor advantages. Certainly, to judge from the prominence accorded the pairs of lanterns in makers' catalogues, and the efforts to make oil biunials, &c., they must meet with favour in some quarters. To keep faith, therefore, with that quarter, as well as every other quarter we can find or imagine, due particulars appear.

Commencing first with the sperm or colza oil patterns, the prices in catalogues are usually mixed up with sets of nursery tales, slides, &c., but the following may be taken as a fair sample.

84.—The following have japanned tin bodies:—

	£	s.	d.
3in. condensers, tin dissolver, in box	3	6	0
3½in. ,, ,, ,, ...	4	14	6
3in. compound condensers, tin dissolver, in box	4	17	6
3in. ,, ,, rack ,, ,, ...	6	15	0
3½in. ,, ,, ,, ,, ,, ...	8	5	0
3½in. ,, ,, ,, ,, achromatic front lenses, in box	9	9	0
4in. ,, ,, ,, ,, ,,	9	9	0
4in. ,, ,, ,, ,, ,, ,, ,,	10	10	0
4½in. ,, ,, ,, ,, ,,	11	15	0
4½in. ,, ,, ,, ,, ,, ,, ,,	13	7	6

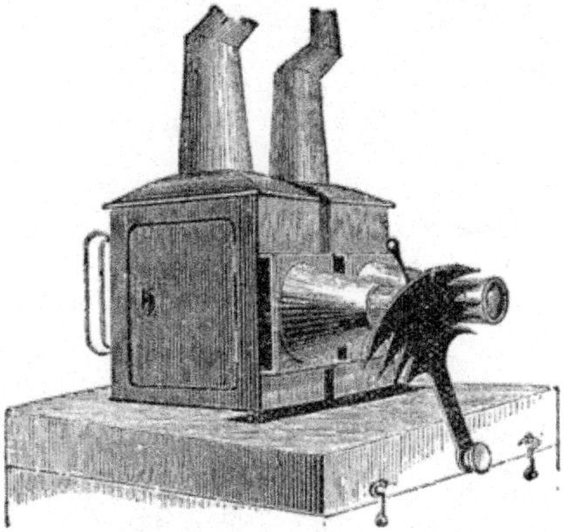

No. 84.

85.—The following have polished mahogany bodies, lined with tin, rackwork fronts, achromatic front lenses, compound condensers Complete in box :—

	£	s.	d.
3½in condensers	11	2	6
4in. ,, ...	12	7	6
4½in. ,, ...	15	0	0

A superior and ornamented Spanish mahogany body, lined with tin, brass sliding outer tubes, rack fronts, achromatic front lenses, compound condensers. In extra strong iron bound case —

	£	s.	d.
3½in. condensers	15	0	0
4in. ,, ...	16	4	0
4½in. ,, ...	22	10	0

We now come to the mineral oil lanterns. Most of those already described are obtainable in pairs. A description of the lantern itself will be found, as indicated, beside each pair.

Archer & Sons.—Complete in box with dissolver.

86.—Guinea (see 14) £2 15 0

87.—Popular (see 15) £4 15 0

88.—Lecturer's (see 16) £6 0 0

89.—Excelsior (see 17) £7 10 0

90.—Photinus (see 44) £10 10 0

91.—Cubley & Preston's Mineral Oil (CUBLEY & PRESTON).—See 21.
Details.—4in. condensers, mounted side by side on mahogany board, with black wood box, complete with dissolver £9 9 0

92.—The Co-operative Educational Duplexicon (HUGHES, Kingsland).—See 22
Details.—With dissolvers, in case £3 5 0

93.—The Co-operative Educational Triplexicon (HUGHES, Kingsland) —See 23.
Details.—Complete in case, with dissolver £4 10 0

94.—The Co-operative Three and Four Wick (Hughes, Kingsland).—See 24.

Details.—With mechanical dissolver, in case £6 10 0

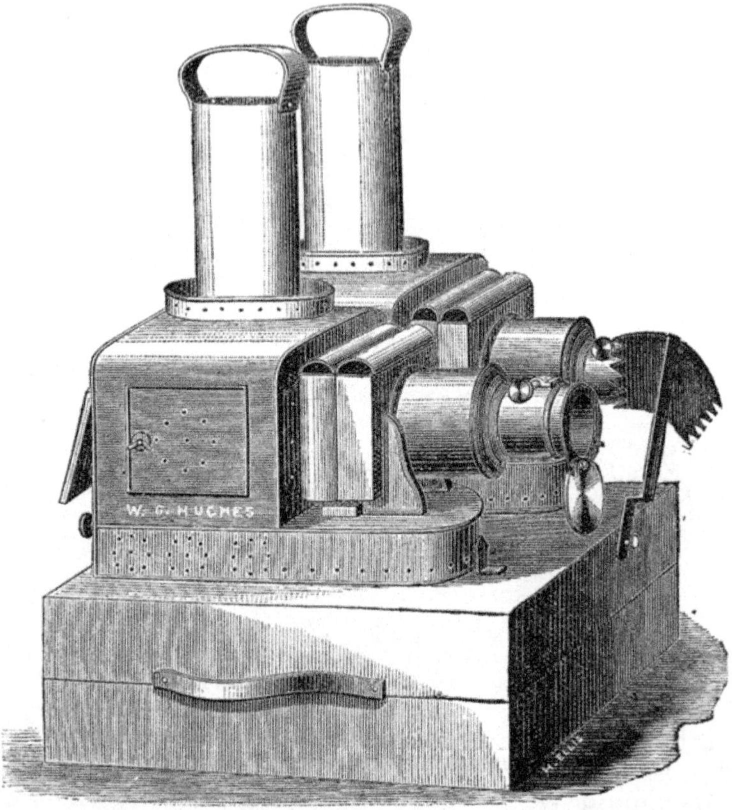

No. 95.

95.—The Pamphengos (Hughes, Kingsland).

Details.—See 47. Mechanical dissolver. Blacked japanned case, with partition and leather handle £12 12 0

Extras.—Brass dissolver with hinged fans and lever handle, also polished case and improved oilcan, 21/- Mahogany bodies, 50/-

96.—Second Quality Pamphengos (Hughes, Kingsland).

Details.—See 48. Complete in case with fan dissolver .. £8 8 0

Extras.—Mahogany bodies, 30/-

97.—Hughes's Triplexicon (Hughes, Kingsland).

Details.—See 51. Mechanical dissolver. Japanned case, with partition and leather handle £10 10 0

98.—Second Quality Triplexicon (Hughes, Kingsland).

Details.—See 52. Complete with dissolver, in case £7 10 0

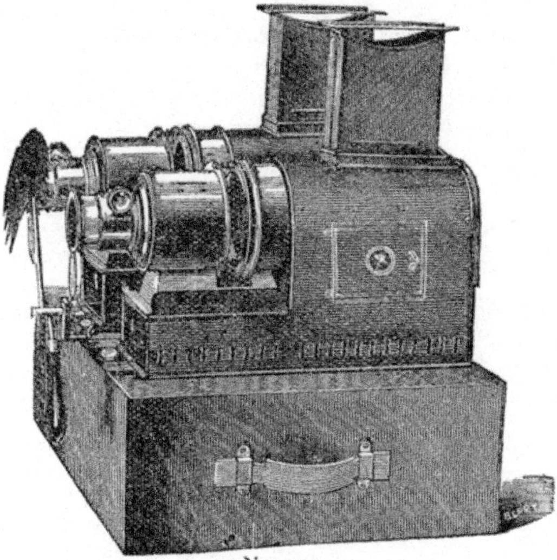

No. 99.

99.—The Imperial (Mason & Co.)
Details.—See 53. In case with fan dissolver £7 10 0

100.—Medland's Mahogany Body (J. B. Medland).
Details.—See 54. In case with dissolver £11 10 0

101.—The Perfection (J. B. Medland).
Details.—See 56. With mechanical dissolver, in travelling case £9 0 0

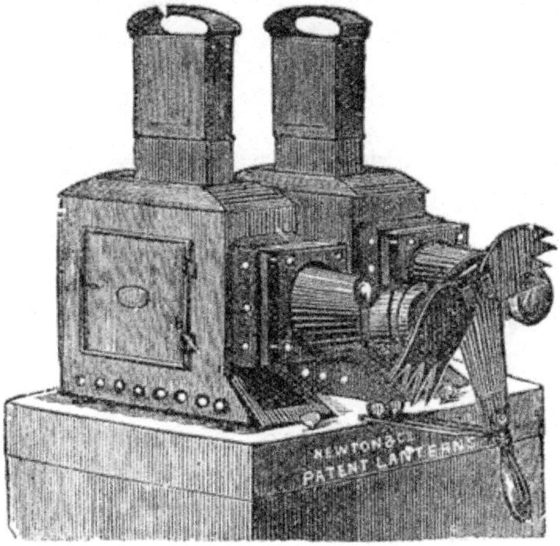

No. 102.

102.—Newton's Improved Phantasmagoria (Newton & Co.)
Details.—See 57. With dissolver, in stained polished wood case 19½ × 13 × 12 .. .. £8 8 0

103.—Newton's Mahogany Phantasmagoria (NEWTON & Co.)
Details —See 58. With dissolver, in polished wood .. £11 11 0

104.—The Invicta No. 1 (NOAKES & SON).
Details.—See 60. In case, with dissolver £7 10 0

105.—The Invicta No. 2 (NOAKES & SON).
Details.—See 61. In case, with dissolver £9 10 0

106.—The Phonopticon (OTTWAY & SON).
Details.—See 27. In box to form stand for operating, with dissolver £7 10 0

107.—The Mahogany Phonopticon (OTTWAY & SON).
See 62. In box complete, with dissolver £10 10 0

108.—The Standard Educational Phonopticon (OTTWAY & SON).
Details.—See 63. In portable case, with mechanical dissolver £11 0 0

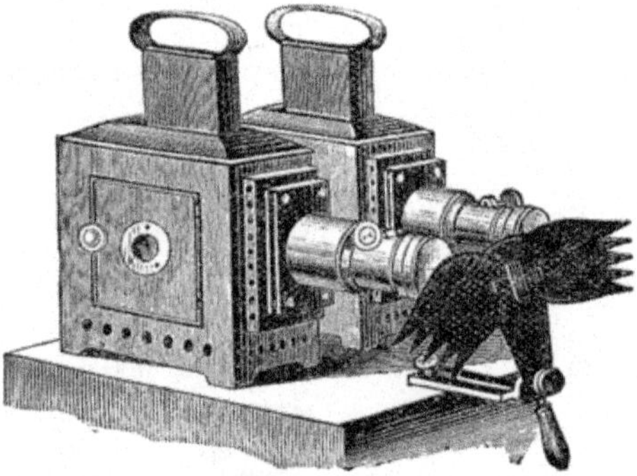

No. 110.

Chatham Pexton.
Details.— Complete, with dissolver in case.

(109)	No. 1.	See 28	.. .. ..	£6 0 0
(110)	„ 2.	„ 64	.. .. ..	7 0 0
(111)	„ 6.	„ 65	.. .. ..	11 10 0

The Optimus (PERKEN, SON & RAYMENT).
Details.—Complete in case. Dissolver with lever adjustment.

112.—Optimus (See 29) £3 15 6

113.—Optimus (See 30) £5 12 0

114.—Place's No. 3 (J. PLACE).
Details.—See 66. Complete with fan dissolver in box .. £7 10 0

115.—Pumphrey's No. M (ALF. PUMPHREY).
Details.—See 33. In box with rotating fan dissolver .. £6 0 0

116.—The Sciopticon (SCIOPTICON Co.)
Details.—See 43. In case with mechanical dissolver and slideholders. (An illustration of this appears in Section II., Part 4) £10 0 0

117.—The Cyclexicon (Solgraph Co)

Details.—See 68. With fan dissolver in case £7 15 0

118.—The Photogenic (J. H. Steward).

Details.—See 69. With dissolving fan in portable case, forming stand £8 8 0

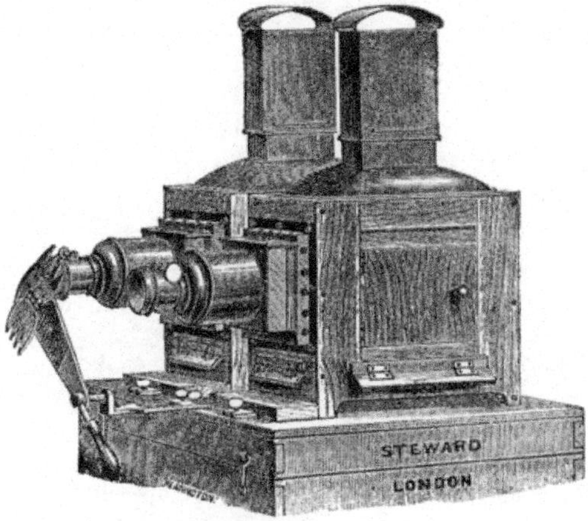

No. 119.

119.—The Mahogany Photogenic (J. H. Steward).

Details.—See 70. With dissolver in case, forming stand for exhibition £8 8 0

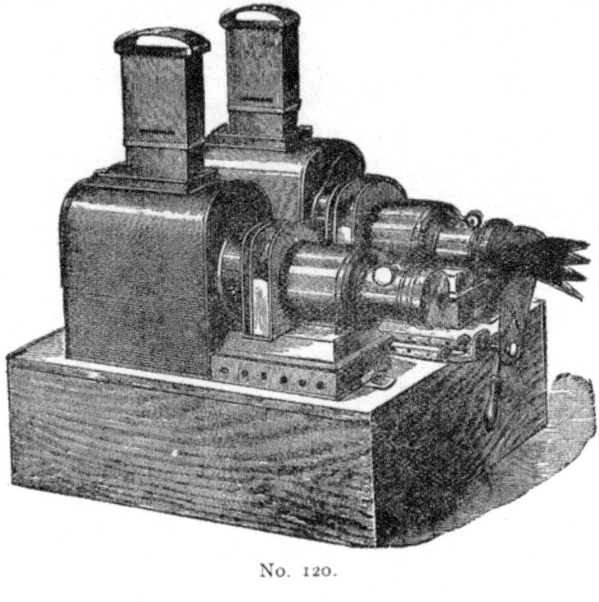

No. 120.

120.—Tyler's Special Russian Iron (W. Tyler).

Details.—See 74. New shape dissolver, with fan to fold on both sides, for flash effects; stained and varnished box, with lock and key and leather handle £7 10 0

121.—The Sunlight (W. Tyler).

Details.—See 34. Complete in box, with dissolver £5 10 0

122.—The Sunlight (W. Tyler).

Details.—See 73. Double-stained and French-polished box, with lock and key and leather handle, plates and thumbscrews for fixing lanterns, brass dissolver £7 10 0

123.—The Perfection (W. Tyler).

Details.—See 75. Complete in case, with dissolver £9 10 0

124.—Watson's Cheap Phantasmagoria (Watson & Sons).

Details.—See 35. 4in. condensers complete in case, with dissolver £9 0 0

125.—Watson's Phantasmagoria (Watson & Sons).

Details.—See 77. Polished mahogany bodies, lined with tin, dissolver and case. £10 10 0

126.—The Victoria (E G. Wood).

Details.—See 37. Complete in case, with mechanical dissolver £5 5 0

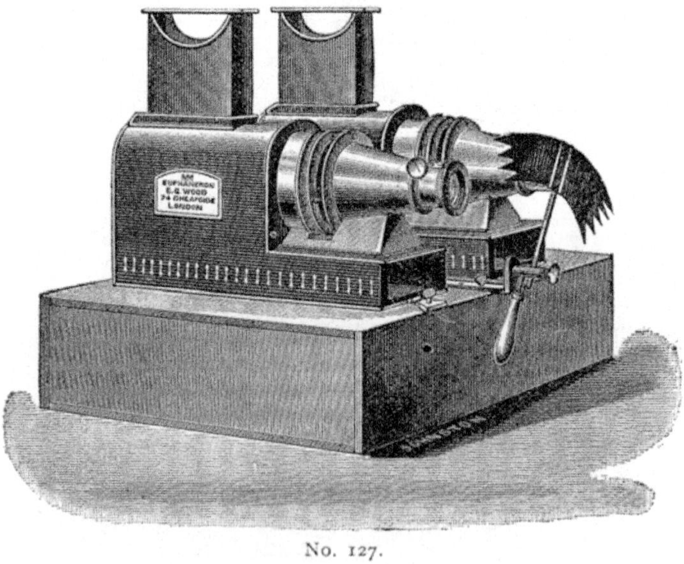

No. 127.

127.—The Euphaneron (E. G. Wood).

Details.—See 78. Complete in case with mechanical dissolver £9 9 0

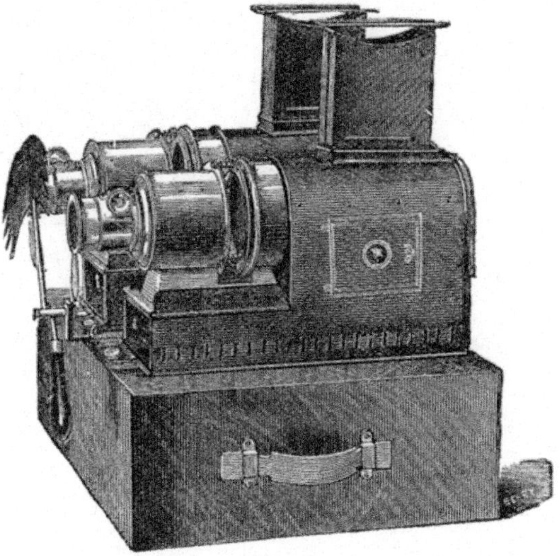

No. 128.

128.—Russian Iron (Wholesale Firm).

Details.—See 81. These lanterns are registered into pairs for dissolving views. The box is fitted with brass plates, and thumbscrews for fixing the lanterns and the dissolver in position. Each lantern has also a slot and thumbscrew at back for fixing it firmly on the box when the two are placed at the proper angle, when the two discs exactly coincide. Complete in stained and French-polished double case, with lock and key and leather handle £8 5 0

Other Patterns.

The following have brass dissolver, plates and thumbscrews, polished double case, with lock and key and handle.

	£	s.	d.
129.—Cheap Tin (See 39)	4	7	0
130.—Superior Tin (See 40)	5	5	0
131.—Cheap Mahogany (See 41)	6	0	
132.—Superior Mahogany (See 79)	7	10	
133.—Superior Mahogany (See 80)	10	10	0

134.—Helioscopic Oil Biunial (W. Tyler).

Details.—The general arrangements of each lantern are the same as 76. We have not yet seen one of these, they being hardly ready when we visited Mr. Tyler, but he described it. Of course, the great difficulty met with in oil biunials is the excessive heat; this Mr. Tyler obviates by layers of asbestos. Altogether we fancy this lantern should attract purchasers.

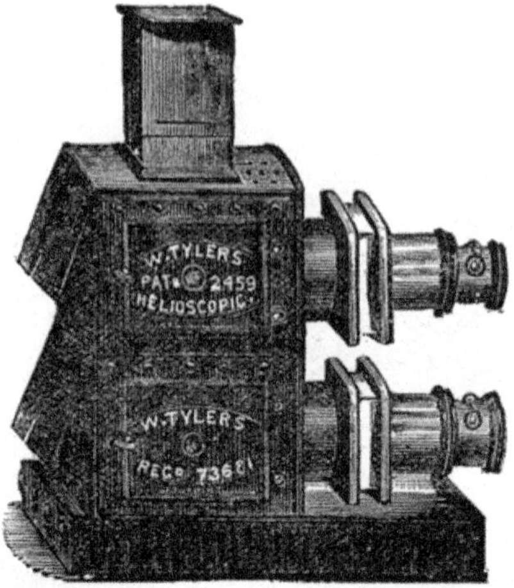

No. 134.

135.—The Combined Side by Side and Biunial (J. H. STEWARD).

Details.—4in. condenser, achromatic front lenses of large size, brass shutters, brass stages, open all round, and sliding body tube; three-wick lamps, trays and rods, for working one above another when using limelight, and side by side for oil, with dissolver and brass adjusting screws. In case £21 0 0

Remarks.—A useful lantern. Can be used with oil side by side, or as a biunial with limelight, in the latter case the bodies being short to suit the lamps. When placed one above the other, they are not inconveniently high, and the appearance is the same as an ordinary brass-fronted biunial.

Special stands or tables for converting oil lanterns into biunials will be described under the heading of tripods, supports, &c.

PART II.

Limelight Lanterns.

AS will be gleaned from the above heading, the main feature which distinguishes the lanterns described in the following pages from those already dealt with is their method of lighting, which, in place of one or other of the various kinds of oil, is the lime or oxy-hydrogen light, which is the incandescence of a small cylinder of lime, produced by a jet of oxygen or of oxygen and hydrogen mixed, playing on its surface. The various systems of producing the limelight, and the different contrivances for so doing, are fully treated of in a subsequent chapter (which should be well studied before a selection of lantern is made), so we need say no more here, save that the oxy-hydrogen jet proper or "mixed" jet is at the same time the most powerful as well as the most dangerous, whilst what is known as the "blow-through" jet, although of no more than half the power of the other, is, for all practical purposes, absolutely safe, and gives ample light for all but exhibitions on the very largest scale. As will have been noticed in the preceding chapter, many of the better-class oil lanterns are so constructed as to permit of their easy adaptation to the purposes of the limelight by the simple removal of the lantern and the substitution of the limelight jet and appliances, and, conversely, many limelight lanterns may be as easily used with the oil light for exhibitions at home, or where a large picture and volume of light is not required. The use of the limelight, it may be well here to remark, entails the additional expense and use of taps of various kinds, indiarubber supply tubes, gas-bags and pressure boards, or gas cylinders, all of which accessories are fully dealt with in their proper chapters, but which, as a rule, are not supplied, except as extras, with the lantern, so that in gauging expense these have to be considered. In some cases the jet is an extra also, and the price of the lantern covers none of the lighting appliances whatever.

Of the different varieties of the limelight lantern the most generally used is the biunial or twin lantern, but instead of being used side by side as the double oil lanterns are, they are placed one over the other, and, indeed, in the majority of cases the bodies of the two lanterns are constructed in one, though several makers make them separately, so that they can be used as separate single lantern desired, which is a distinct advantage to a person who uses a lantern for all-round purposes, and does not confine himself to regular exhibitions before public audiences. In the construction of the limelight lantern the same materials are used as in the making of oil lanterns, save that tin is at a discount, except for the lining of the cheaper sorts, and wood and brass are brought much more into requisition, the former to add finish and good looks to the instrument, and the latter to give firmness and solidity; indeed, the fronts of all the best lanterns are made wholly of brass, whilst in some the body linings, too, are of this material. Every consideration should be given to the question of ventilation in a lantern, as it is essential to the well-being of the instrument that it should not be allowed to become too hot. The leading makers are indeed fully alive to the importance of this, and construct their lanterns with an air casing or space between the wood and metal bodies, and every arrangement for the passage therein of a free current of air, whilst the heat generated in the lower lantern is taken off by similar air channels, and does not pass into the upper lantern, or, indeed, add much to its heat. An important point, too, when the heat of the most powerful limelight arrangements is considered, is the fitting of the condensers in brass cells having enough free space between the edges of the glass and metal to allow of the full expansion of the glass, without which provision they would be very liable to crack; and another little matter which should always be seen to is the provision of a small hole in the periphery of the metal cell, and between the two combinations of the condenser, for the escape of moisture. Of course, with many persons portability is quite a secondary consideration, but those who are in the habit of giving entertainments at friends' houses and in a variety of other places will find it a primary consideration, and will also find themselves well catered for by the enterprising manufacturer, who will supply them with a lantern which will go into a very small space and weigh but little.

With vertical positioned biunials of the limelight persuasion, it will be readily seen that some special provision is required for centring the discs upon the screen, and although with one or two instruments a movement is given in the body of the lantern itself, the general plan adopted is the simple method of hinging the two fronts and adjusting by screws. The plate carrying the objective, slide stage, and condenser, of the top lantern, is hinged at the bottom, close to which the front of the lower lantern is also hinged, this, as will be seen by a glance at the engravings in the following pages, at the top. At the top and bottom of these fronts respectively two screws are provided, by the adjustment of which the two discs can be made to coincide upon the screen to a nicety.

The slide stages of a biunial lantern of the best class should be open not only at the sides for the admission of the usual slides, but also both at the top and bottom, so that tanks for the illustration of chemical experiments may be inserted, and many pleasing and astonishing effects carried out. The best stages are provided with registering stops, which may be adjusted to the size of the slide used, and which will thus stop the slide when pushed in exactly at the right position for it to be centred on the screen. Another improvement, too, which has been recently introduced, and which is to be found in all the best lanterns, is a set of side windows, or, rather, doors, long and narrow, and set vertically in the sides of the lantern, close to the front. The purpose of these is to permit of the insertion behind the condenser of long slips of coloured glass, which are termed slide tinters, and by the skilful use of which many beautiful effects may be produced, such, for instance, as giving an evening and moonlight effect to a daylight picture, or providing a tinted ground for the exhibition of statuary, which is far better than throwing such pictures upon the ordinary screen. The glasses being behind the condenser are out of the focus of the objective, and thus the effect of colour attained is soft in the extreme.

Hitherto we have spoken of lanterns possessing but a single lens or objective, and consequently limited in their scope. The better-class limelight lanterns, however, are provided, not only with a single objective, but with as many as three or even more, a biunial, of course, requiring duplicate sets for each lantern. These lenses are fitted into mounts which all screw into the same flange, thus making them interchangeable, and in one or two instances the screw is replaced by a bayonet joint fitting, which enables the changing of the lenses to be effected in a very few seconds. The lenses are each of a different focus, and by their use the operator is practically independent of his situation, and can show a large or small disc as wished under almost any circumstances. With a single lens, of course, the size of the disc thrown upon the screen is dependent upon the distance of the lantern from it, and thus, in a confined situation, the operator is necessarily compelled to show only one size picture. By the use of different lenses, however, the light with each departing from the lantern at a different angle, the operator can, by the selection of one of these, obtain a larger or smaller disc at will. If he find himself in a confined space he can use a short focus lens, which will give him a larger picture than he could otherwise obtain; whilst, on the contrary, if he find himself situated at a greater distance from his screen, he can use the lens of longest focus in his possession, and obtain a disc of about the same size as he would get under the circumstances previously noted. As lenses of different foci are fitted at different distances from the slide stage, it will be seen that some arrangement must be made to allow of this being done, for the amount of adjustment obtained by the rack and pinion is comparatively small. This necessitates the use of telescopic fronts with which the best lanterns are now provided, the front being made of two or three tubes fitting inside each other like those of a telescope, and arranged in length to suit the different lenses with which the lantern is provided. In the majority of lanterns these are like those of a telescope, fitted tightly, and sliding in and out as required, the exact adjustment being obtained by means of rack and pinion at the end attached to the lens mount, although one maker constructs *each* tube with a rack and pinion, thus obtaining greater rigidity and more exact adjustment. A further improvement, too, will be found in the fitting of the objective mount with a milled head to the pinion on each side, by the use of which adjustment may be made from either side of the lantern, as it frequently happens that the operator finds himself placed in a corner where it is difficult to get at either one or the other side of the lantern, and consequently he must, if possible, adjust himself to circumstances. It is also advisable to be provided with an extra condenser with larger diameter, for use when working at long distances, as condensers which will be found quite powerful enough to thoroughly illuminate a picture ordinarily will be insufficiently powerful for very large work.

We explained in our introduction to the first chapter the method of dissolving the views into each other by the use of the twin lantern and dissolver, but the mechanical dissolver there mentioned is, after all, but a rough affair, and only suitable for use with the oil lantern, with which, indeed, it is a necessity if dissolving views are required. Dissolving with the limelight, however, is effected not by the covering of one light, and the simultaneous exposing of the other, but by the simultaneous raising and lowering of the respective lights, which necessitates the use of various kinds of combination taps or "Dissolvers" so constructed that by turning the handle of a small tap the gas is cut off from one group of jets, whilst it is increased on the other. There are several dissolvers in use, which will be found fully described in a later chapter.

All first-class multiple lanterns are now provided with what is termed the roller curtain slide, which is a long strip of metal, usually brass, fitted between the condenser and the slide, and so arranged that when pulled up or pressed down the condenser of one lantern is covered whilst the other is exposed, this really being only another method of obtaining a dissolving view effect, although with it the pictures do not merge one into the other, but appear to be behind each other. Thus, with the curtain slide down, and the bottom lantern thus shut, the plain white disc appears upon the screen. A slide representing the proscenium of a theatre, or an arrangement of curtains and blinds is placed in the slide stage of the lower lantern. The roller then draws up, and as this shuts the light off the white screen at the same time as it exposes the curtain to view, it has the effect of a curtain being dropped from the top until it covers the screen. A subject of any kind may be placed in the upper—now covered—lantern, and the movement of the curtain slide reversed, when the curtain will appear to be drawn up, disclosing to view the picture behind, an effect which is very taking. If it is desired this may be continued throughout the evening, or if dissolving views proper are preferred, it forms a very good opening and finish to the entertainment.

For flashing effects shutters are fitted which cover the fronts of the objectives, and thus shut off the light from the lanterns. They replace the caps which arc supplied with the cheaper lanterns, and are hinged so as to form a permanent attachment to the lens front. By a simultaneous and rapid movement of these the different pictures are made to appear and disappear simultaneously with a flash, which for certain subjects is preferable to either the slow dissolving or curtain effects.

A still further improvement upon the biunial system is that of the triunial, in which three lanterns are used placed one above the other. These differ in no way in their construction from the biunials, but are provided with a third lantern placed above the other two. This is usually made separate from the combined body of the other two, so that the possessor of a triunial practically has three lanterns, the triunial, biunial, and single. Although more expensive, for exhibitions where anything like effect is desired the triunial is by far the best instrument, a number of beautiful effects being obtainable by the use of a third lantern which it would be quite impossible to secure with a biunial.

In the finishing off of the larger lanterns it is usual to provide a handle of some kind, as being large and somewhat heavy affairs, they are awkward to lift otherwise, and the best form of handle for this purpose is a brass rail, fitted at the junction of the two lanterns of the biunial system, which gives ample and convenient holding, besides wonderfully setting off the appearance of the lantern. Nickel-plating is also introduced where additional finish is required, and adds much to the appearance as well as to the expense of the lantern, although, of course, as we before explained, it is quite a secondary matter so long as the working of the lantern is perfect.

We have mentioned briefly the arrangement for portability, and most lanterns now are supplied in a case which serves both as a travelling case and stand for exhibition. The elaborateness of the fitting of these is varied, but if the purchaser is going in for a complete set he will find many cases completely fitted up for the storage of the lantern as well as the slides, which, when in order for exhibition, prove extremely handy and complete. Whether, however, the case which serves to contain the lantern be used to support it for exhibition, or whether a tripod or other form of stand be adopted, it will be necessary to support the lantern in such a way that it may be tilted backwards or forwards, as the centre of the screen is, in the majority of cases, of necessity higher than the body of the lantern, and a perfectly level fitting would not obtain the desired result. With all properly constructed stands, therefore, the top to which the body of the lantern is attached is provided with a tilting screw, rachet, or other device for adjusting the level of the lantern.

Single Limelight Lanterns.

Nearly all oil lanterns, and certainly all those of recent years, such as the mineral oil series, require no alteration for the purpose of using the limelight There is usually a groove at the bottom of the body for the insertion of the jet tray, and beyond the purchase of this tray, costing from 1s. 6d. to 3s 6d , and the removal of the lamp, no other preparation is required. Indeed, many of those described in previous pages are specially fitted or arranged so that the using of the gas-jet is quite as convenient as the oil lamp , to quote one instance, viz., a door at the side, which is certainly unnecessary otherwise. It comes to this, therefore, with most makers, that their oil lantern is their single limelight one, and but few catalogue anything between the former and a biunial. But there are some patterns which, either possessing special features, or because they are strictly adapted for limelight only, must find a place in our pages.

136.—Armstrong's No. 1 (Armstrong & Brother).

Details.—Compound condensers, 3¾in ; double achromatic front lens, japanned tin body £3 10 0

137.—Armstrong's No. 2 (Armstrong & Brother).

Details.—Mahogany body, lined with tin; cavity sides, improved top; other details as No. 1 .. .. £4 10 0

138.—The Miniature (Newton & Co)

Details.—Metal body, rivetted throughout ; 4in. compound condensers; long focus objective Polished case, lock and key ; price includes the blow-through jet with ratchet lime movement £5 5 0

Remarks.—Entirely designed for portability, which will be seen by the size of lantern, which is 15½in. by 6in by 7in. Where extreme portability is required this lantern should prove satisfactory.

139.—The Institution (J. H. Steward).

Details.—Mahogany body, lined with tin ; polished bright brass stages open at top and bottom as well as sides ; 4¼in. improved condensers ; brass telescopic fronts, with three draw tubes, extending from 3in. to 12in , well made and rigid at any part ; three special achromatic front lenses, of various foci in interchangeable mounts, so that by combination the required size disc can be obtained at any distance between 10ft. and 70ft. from the screen ; rack and pinion focussing tube, and brass shutter attached to front ; oxy-hydrogen jet , mouldings and baseboard of mahogany. Travelling case £12 12 0

Remarks.—A best pattern single lantern, suitable for institutions (as its name implies), photographic or scientific societies—in fact, anywhere that the principal consideration is the plain picture on the screen, other matters being secondary.

140.—Taylor's Single Lantern (T. S. & W. Taylor).

Details.—Polished mahogany body, with glazed doors; brass stage and front; 4½in. double condensers; English portrait lens, 2in diameter, with rack adjustment. The price includes case and platinum-tipped oxy-hydrogen jet, with full adjustments £14 0 0

Remarks.—In every way a first-class lantern, the optical arrangements being in Messrs. Taylor's best style, which is saying a good deal.

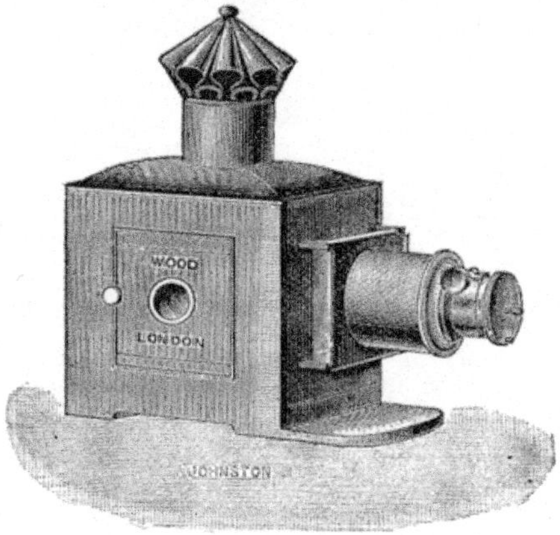

No. 141.

141.—Wood's Single Lantern (E. G. Wood).

Details.—French polished mahogany body, with japanned lining and top, 4in. condensers, and tin front; double achromatic objective. In case £4 17 6

142.—Wood's Single Lantern (E. G. Wood).

Details.—Same as 141, but of superior make, with brass front, additional and lengthening tube to objective. In box £6 6 0

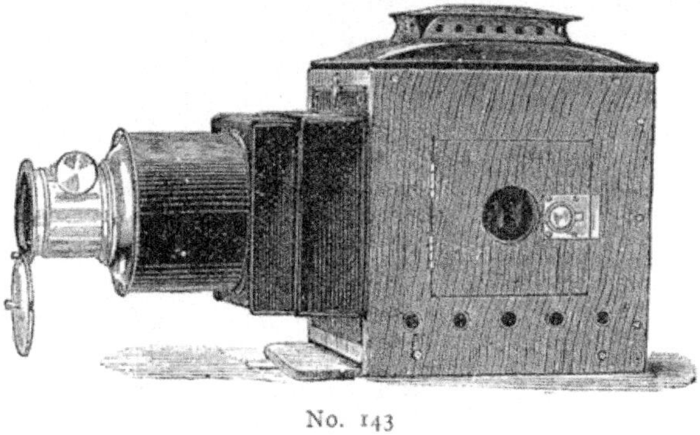

No. 143

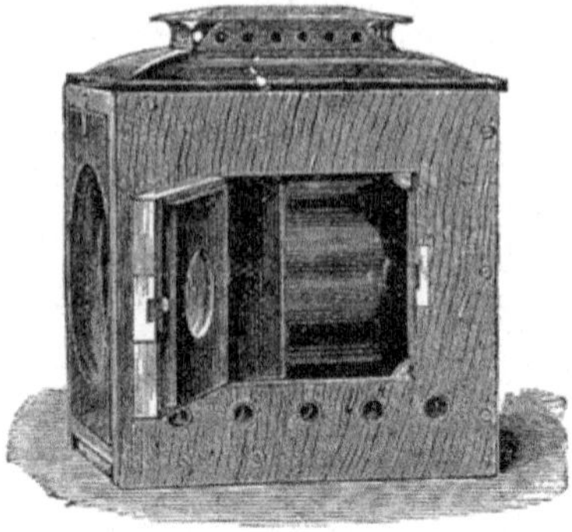

No. 143.

143.—The Pocket (Wholesale Firm).

Details.—Polished mahogany body, lined with japanned tin; Russian iron dome and top; japanned tin stage; front tube, and sliding tube with brass O.G.; fitted with 4in. plano-convex compound condenser in brass cell, double combination achromatic front lenses (with large size lenses to the back combination) in brass mount, rack and pinion adjustment, and fitted with a sliding shutter or flasher. The lantern is so constructed that when not in use the front plate carrying the stage and front tubes can be reversed and placed inside the body; size, 7½ × 7½ × 6in. Stained and polished case, with lock and key, and leather handle £3 15 0

Remarks.—A very neat and handy lantern, portability itself, as the size will testify.

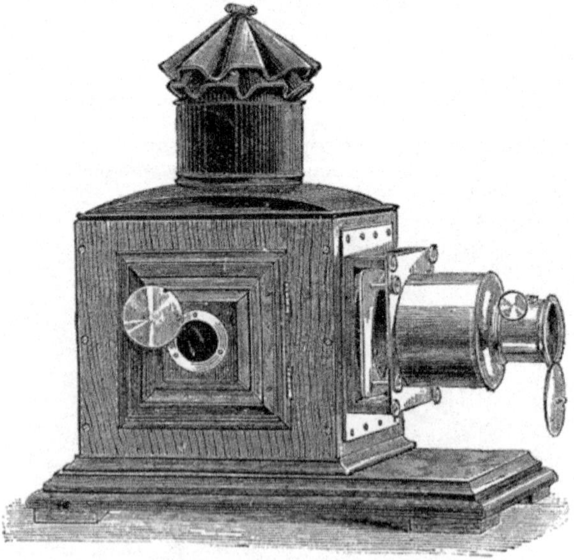

No. 144.

144.—Best Quality (Wholesale Firm).

Details.—Polished mahogany body, moulded foot and two panelled doors, brass-bound sight-holes with sliding shutters, lined with japanned tin; japanned tin dome and rose top; stage, front tube and sliding O.G. tube entirely of brass; 4in. plano-convex compound condenser in brass cell, double combination achromatic front lenses in brass mount (with large size lenses to the back combination), rack and pinion adjustment, and fitted with a sliding shutter or flasher £6 6 0

Extras.—Three telescopic drawers, and with a rack objective and four sets of double combination achromatic front lenses, £5. Stained and French-polished case, lock and key, and leather handle, 9/-

Remarks.—Strong and useful, thoroughly well made, and handsome in appearance.

Double or Biunial Limelight Lanterns.

The biunial is perhaps the best catalogued thing we have to treat of. No matter how small the price list, or how few the goods therein enumerated, the biunial is certain to be included Not that any very great differences exist either in construction, design, or method of use. But if a maker catalogues anything, it will be his biunial. There are, therefore, very many to deal with, as will be readily seen by a glance forward. The range in price is wide, and surely wide enough to suit the pockets of all. At the same time, there are a few points which must be taken into consideration when comparing one price with another, and they are.—

1st.—Japanned or mahogany body.

2nd.—Japanned or brass fronts.

3rd —Telescopic brass fronts, and the extra lenses.

4th.—Whether the price includes jets or other accessories.

There are also minor points—such as better-class objectives, and many others—but the above four are the principal causes of different prices. No 3 especially adds to the cost, including as it does extra brasswork, which has to be well finished, and extra sets of lenses. Then, as regards No. 4, the inclusion in some quotations of apparatus, deemed by others accessories, obviously must render comparison difficult. To so wide a range, too, does this open the door. Could we have given the exact prices of biunials only, matters would have been simplified; but it could not be done. The reader must, therefore, apart from the consideration of quality altogether, study what the price is stated to include. We are rather dragged into this discourse on prices by the, in some cases, extreme differences; and though hardly expecting anyone to imagine a £7 biunial equal to one at £30, making every allowance for discounts, "first hand," "direct from the manufacturers," and similar lures, yet slighter contrasts can often be judged in the way we have sketched To give an example —the "Neverbeaten" biunial is quoted at £10 10s., and the "Cantbelicked" at £11 11s, but the latter priced instrument has brass fronts. Other things being equal, the "Cantbelicked" is the cheaper instrument, for the brass fronts are worth more than the difference the price represents. So much for comparison of cost, just a word as to cheap prices. The best is oft said to be the cheapest, and by best is meant in this case the most expensive. It may be true, but it is an axiom with those who can afford the highest price, and they never forget to quote it. Other people don't stick to that idea, because they cannot. The moral of which is that the man makes his maxim to suit his pocket, not his pocket the maxim. There are many things which, whilst increasing the cost, are merely ornaments or fancy movements, not absolute necessities. The purchaser should be guided by what the compiler of this work deems the order in importance of the features of an apparatus—first and foremost, the result on the screen, next, ease in manipulation by the aid of mechanical devices or movements, and lastly, the appearance of the instrument to the outsider's gaze. Let the operator show a poor picture, and no amount of outside show will alter the adverse verdict. Let him show a first-class result, and his apparatus will be forgotten and forgiven.

The biunials mentioned in the following pages are arranged alphabetically according to the makers' names

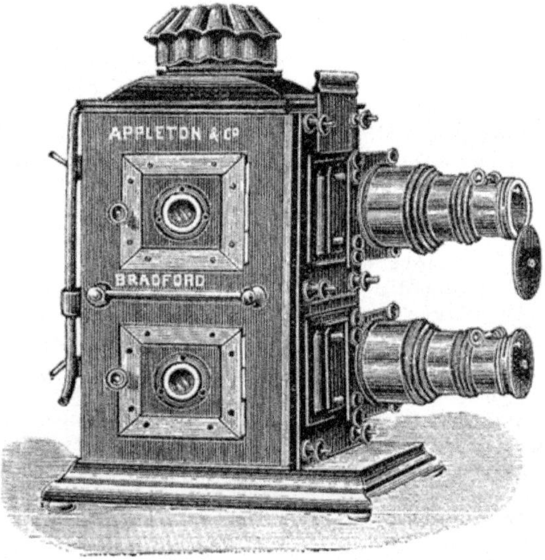

No. 145.

145.—Appleton's (Appleton & Co)

Details.—Mahogany body, with four doors; brass flanged sight holes; entire brass fronts and stages; top half takes off and fits into an extra base board—this, with an extra dome top, makes the apparatus complete either as a single or biunial; 4in. condensers; two sets of achromatic objectives, 8 and 10in. focal length; mounted in brass cells, with rack and pinion and flashing shutters, curtain diaphragm, two best blow-through jets, dissolver, supply tubes. Complete in box £18 0 0

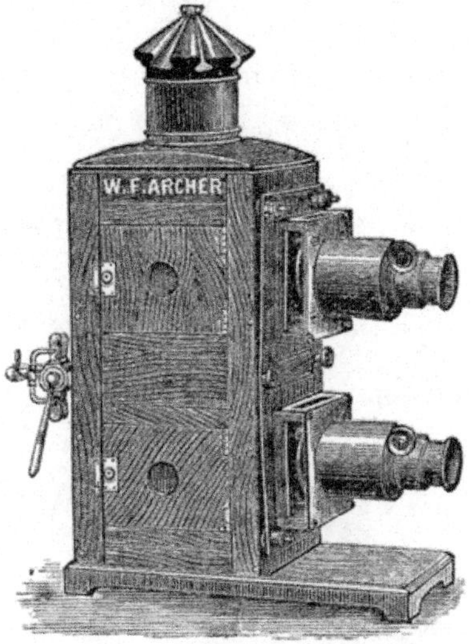

No. 146.

146.—Archer's No. 1 (Archer & Sons).

Details—Stout japanned tin body, divided into two compartments, thoroughly ventilated, and fitted with two pairs of best quality 4in. plano-convex condensers; double achromatic object lenses, with rack and pinions; screws for the proper adjusting of the fronts; removable top; sliding trays, etc £5 0 0

Remarks.—A cheap form of biunial, but good enough for anybody at the price.

147.—Archer's No. 2 (Archer & Sons).

Details.—Polished mahogany body, lined inside with iron, 4in condensers, double achromatic fronts, rack and pinions, japanned fronts, with Archer's improved slide-holders, one draw out brass tubes, removable top in two parts, &c. The whole fitting into an improved form of case.. £7 10 0

Extras—Dallmeyer's pattern new form large object lenses, with double pinions and flashing shutters, 15/- See also No. 3 " Extras."

148.—Archer's No. 3 (Archer & Sons).

Details.—Same as No 147, but with four panelled doors, brass-bound sight holes, fronts made of solid brass, fitted with diaphragm to produce rolling curtain effect, double pinions to the rackwork fronts, Dallmeyer's pattern, large object lenses, flashing shutters, one draw telescopic fronts, &c The whole fitting into a new form of case £11 10 0

Extras.—Nos. 147 and 148, made as two lanterns, for use one on top of the other as a biunial, or side by side as a pair of lanterns, or as single lanterns, and also to take three or four-wick paraffin oil lamps, £4. No. 3, with ebonised body and all the brasswork nickel-plated, £1.

Remarks.—A very good apparatus indeed, and one that is capable of the best work.

149.—Archer's No. 4 (Archer & Sons).

Details.—Same as Nos. 147 and 148, but with three-draw telescopic tubes extending to 12in.; four pairs of achromatic front lenses of different foci, for showing pictures the same size at four different distances from the screen, condensing lenses of a special form, the whole fitting into an improved form of case £17 0 0

Remarks.—The same lantern as No. 148, but the four pairs of lenses and telescopic fronts make the difference in price.

150.—Banks's Combination (W. Banks).

Details.—Mahogany or walnut body, with four doors; 4in. condensers; telescopic fronts, one draw, of brass, achromatic front lens; rack and pinion focussing of 8in. or 10in. focal length; screw adjustment centring discs, fitted with improved jets for either blow through or mixed gases as desired; body divides in centre, forming two single lanterns, which pack into separate boxes; improved dissolver. Complete in boxes with locks and keys £9 9 0

Remarks—A small and neat apparatus, very handy to carry about, and useful if one lantern only is sometimes required.

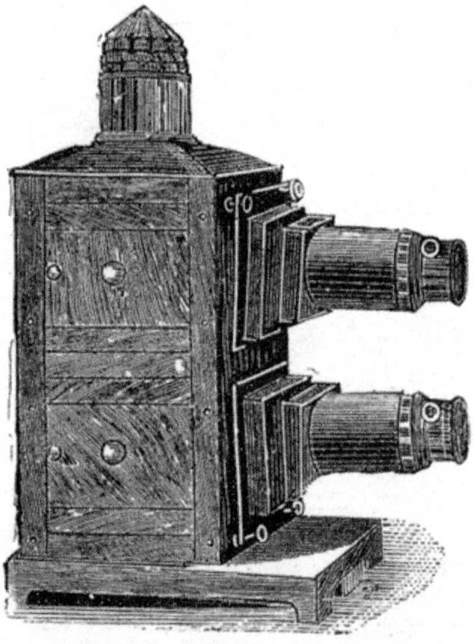

No. 151.

151.—Catlin's No. 1. (William Catlin).

Details.—Polished mahogany body, lined with stout tin; metal fronts and stages, black japanned; brass mounts and adjustments; double combination achromatic focus lenses, with rack and pinion; 3½in. condensers Complete in cases forming stand £9 9 0

Extras.—4in. condensers, 21/-

152.—Catlin's Superior (William Catlin)

Details.—Polished mahogany body, lined with tin, 4in. condensers, achromatic focus lenses, the fronts and stages entirely of brass, with large

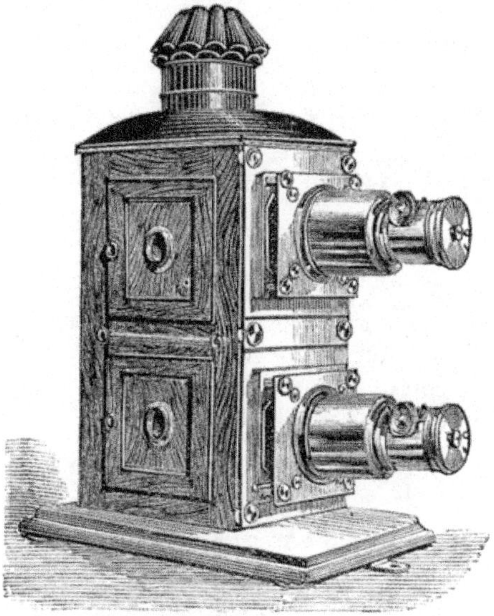

No. 152.

sliding tube for long focus; the slide opening or stage is formed with brass pillars giving access for effects on all sides; sliding diaphragm for rising curtain effect, and shutters for flashing on effects, and six way dissolving tap. Packed in case forming table for lantern £14 0 0

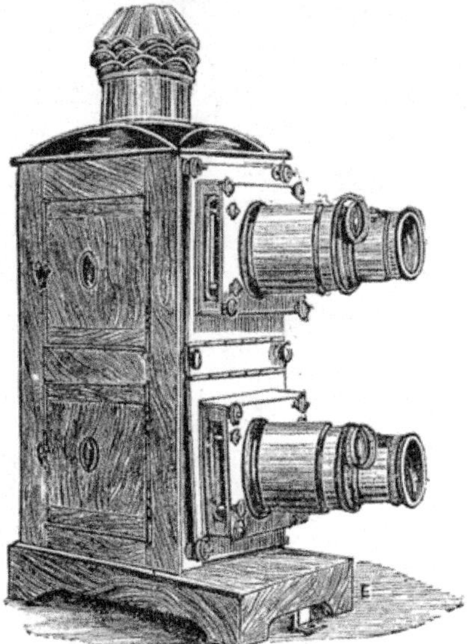

No. 153.

153.—The Standard (William Catlin).

Details.—Polished mahogany body, with brass rail handles; handsome rosewood doors, raised panels, the body lined with tin to keep perfectly cool, 4¼in. condensers, achromatic focus lenses for giving picture of equal size at five different distances; the fronts and stages entirely of brass, with three telescopic draws, to suit the various foci; sliding curtain diaphragm, shutters, six way dissolving tap. Packed in case, to form stand, having two movable sheives on which to arrange the slides £21 0 0

154.—The C.P.S. (Cubley & Preston).

Details—Russian iron body, ventilating dome top, nickel-plated doors to both sides, with sight holes; 5in. double condensers, and extra large achromatic front lenses, with racks and pinions; these are mounted on heavy brass fronts, fitted with adjusting screw and brass stage. Complete with two oxy-hydrogen gas jets, having the cogwheel adjustment for the limes. £15 15 0

Remarks—Considering the price includes two best jets, this is not at all a dear lantern. Both the condensers and objectives are extra large, and should give good light power.

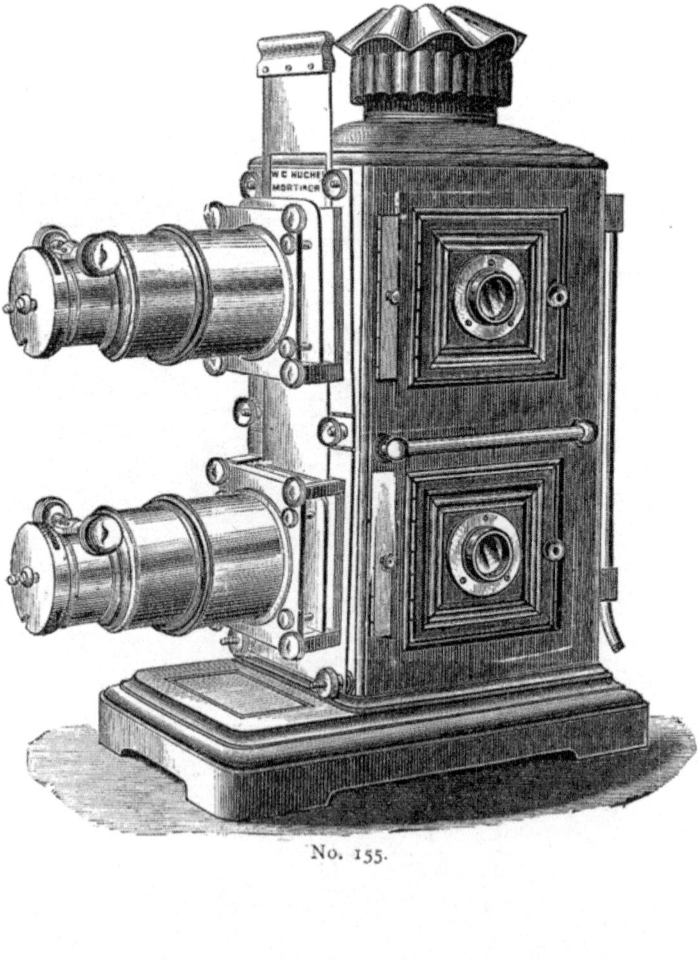

No. 155.

155.—The Grand (W. C. Hughes, Kingsland).

Details.—As this biunial possesses much the same features as the "Grand Triple," we will avoid repetition by referring the reader to the instrument, which will be found amongst the triunials The above lantern has polished mahogany body, lined with iron and japanned, but not quite so elaborate in pattern; entire solid brass fronts with improved slide holder in upper lantern, the same telescopic draws, diaphragm or flasher, for rolling curtain effects, the large rack mount with shutters. four sets of high-class objectives, as described, 4, 5, 6 and 7in. focus, an extra set for very short focus lenses, two pairs of the new form of condensers, 4in. in diameter; also an extra 4½in. ditto; two of the special oxy-hydrogen chamber jets, for the ultra effect; two Malden dissolving taps, brass supply pipes, the improved brass trays as before mentioned. The whole fitted in handsome case with compartments £65 0 0

Remarks.—None could desire a better biunial than this. Everything for comfort, simplicity in working, and handsomeness of appearance, seems to have been carefully studied.

156.—The Excelsior (W. C. Hughes, Kingsland).

Details.—Polished mahogany body, lined with iron; solid brass fronts, large sliding tubes of long focus lenses, 4in. double condensers in brass cells, two sets double combination front lenses, rack and double pinion, and flashing shutters (or three portrait combination lenses); diaphragm for rolling curtain effect; two mixed or blow through jets, and Malden double dissolver. In case £13 10 0

Extras.—Telescopic draw tubes of long focus lenses, £4 10s. Long focus lenses for various distances, 16/6 each.

Remarks—This the maker calls his "Trade Biunial." It is, however, excellent value for the money.

157.—The Co-operative (W. C. Hughes, Kingsland).

Details.—Polished mahogany bodies, plinth foot; two doors with blue glass sight holes and brass flange mountings, inner lining of iron; sliding trays and brass rods to hold jets; dome top and pretty chimney; all handsomely japanned; 4in. double condenser in brass screw cells; double combination front lenses, brass cells, jackets and rack and pinion, substantial brass fronts with brass stages and spring slide holders, brass register screws and springs for producing the coincidence of discs, brass one draw telescope; front tubes with brass flange mounting for screwing front lenses into, diaphragm for producing rolling curtain effects; pair of oxy-hydrogen safety jets with cog-wheel lime turners; improved brass dissolving tap, black twill gas bag, 38in. x 24in. x 22in.; pair of pressure boards, 38in. x 25in.; iron retort; zinc purifier; box of limes in tin canister, 10 feet of tubing, &c. Box to hold lantern, &c. £15 15 0

Remarks.—Really a wonderful lantern and apparatus at the price. It is, moreover, extremely light and portable.

158.—Hughes's Star (W. C. Hughes, Kingsland).

Details.—Body of Cornwall tin, japanned, rivetted and hard soldered, iron sliding trays; 3½in. double condensers, mounted in brass screw cells, double combination portrait lenses, with rack and pinion; spring slide-holders; brass sliding tubes, with brass flange for focussing; brass register screws and springs for exactness of registration; fitted with oxy-hydrogen safety jets.. .. .. £6 6 0

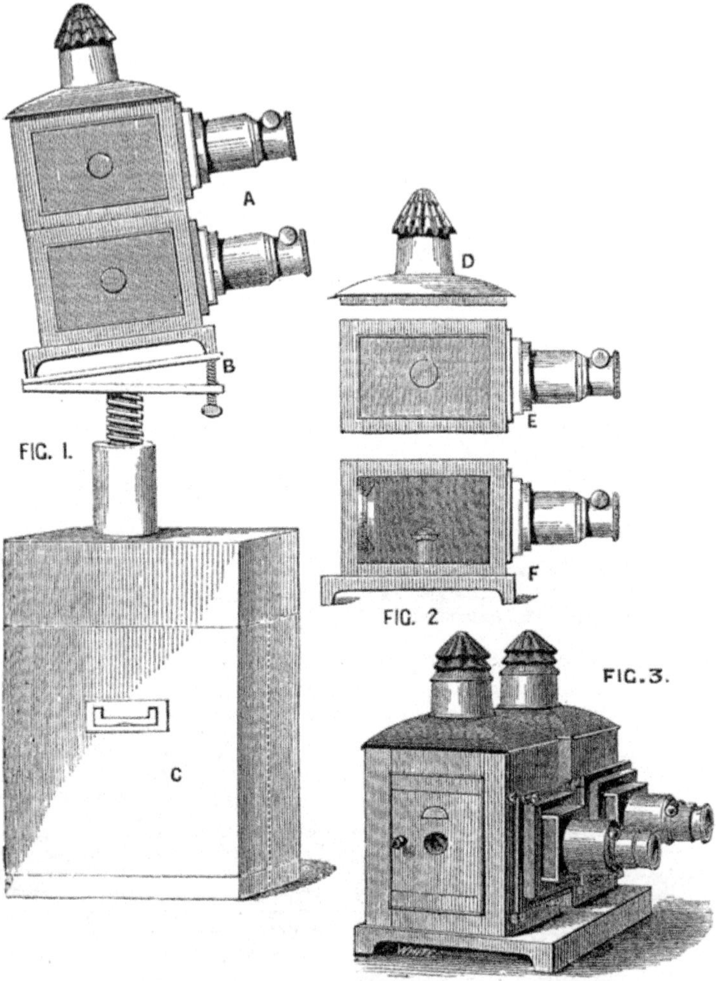

No. 158.

Extras—4in. condensers, 34/- Polished mahogany bodies, 30/- Ordinary biunial box, 14/- "Star" biunial box, as per illustration, 42/- Extra foot, chimney and top, for use with the aphengescope, or when, for other reasons, required side by side, in tin, 30/-; mahogany 40/-

Remarks.—When used side by side oil lamps may be used, thus giving the double convenience.

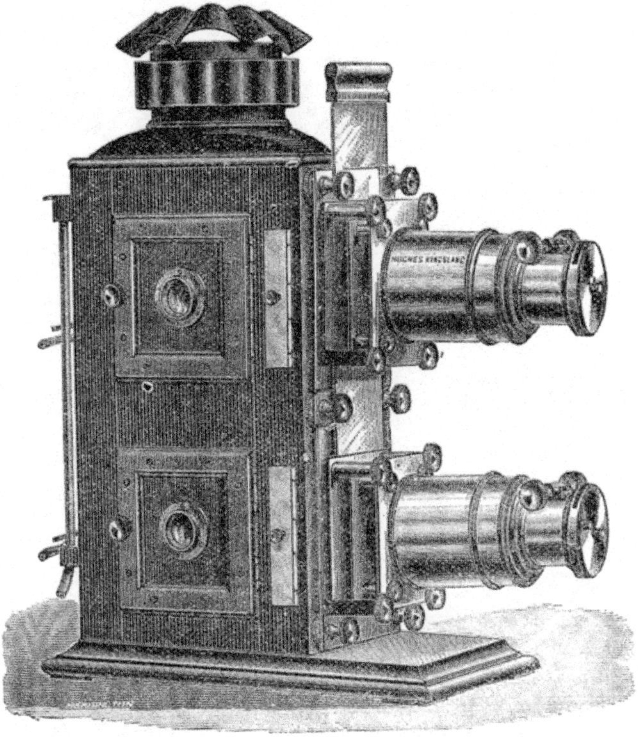

No. 159.

159.—The Excelsior Star (W. C. Hughes, Kingsland).

Details.—Body of mahogany, bound together by brass screws for hot climates; inner linings of iron, rivetted together; 4in. achromatic condenser, mounted in brass screw cells, with outlets for condensed steam; four doors, two on either side of lantern, blue glass sight holes, with brass flange mountings, solid plinth foot, for rigidness when stationed ready for use; fronts entirely of solid brass. The two adjusting plates for arranging the optical systems have ornamental brass fittings, milled screw heads and bolts. The stages of polished mahogany, having openings top, bottom and sides for the introduction of effects. Double combination achromatic front lenses of large diameter, in brass screw cells; double pinion and rackwork for working either side, brass sliding tubes that will draw out, giving various focal lengths, improved sliding trays, with springs to ensure steadiness; also brass upright rods for holding the jets, brass diaphragm for rolling effects. £14 14 0

Extras.—4in. condensers, 42/- Panelled doors and brass side handles for lifting, 50/- Brass supply pipes to jets, 25/-

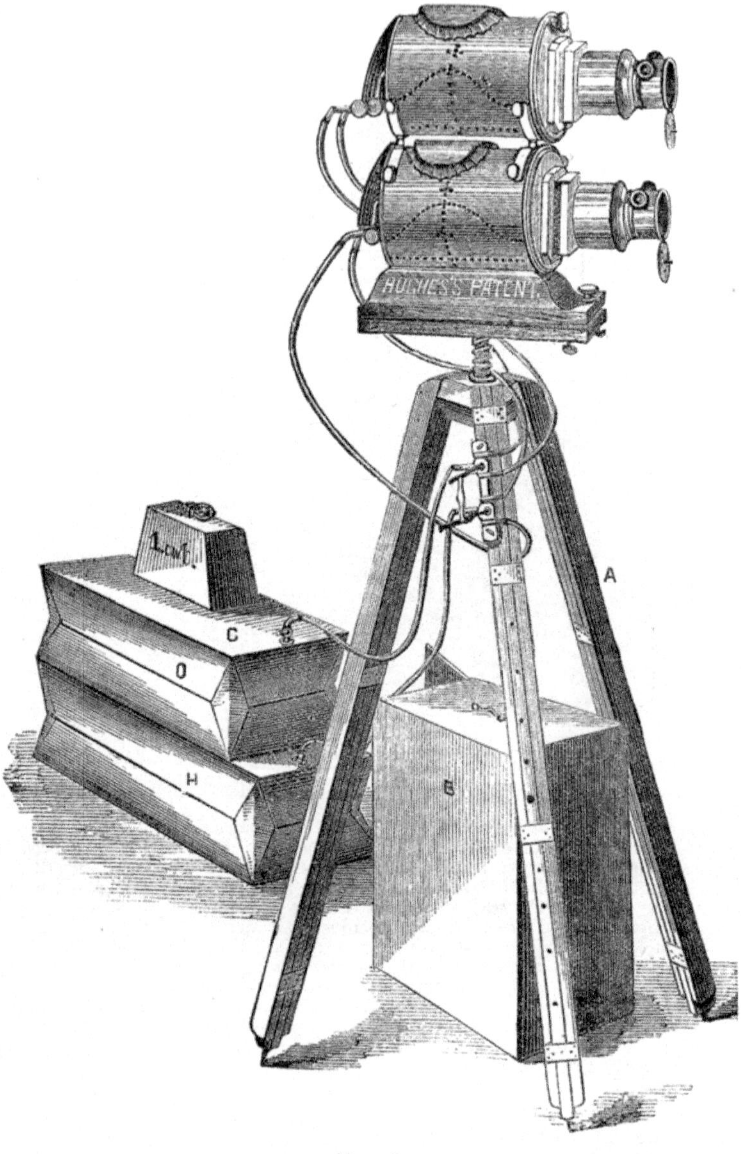

No. 160.

160.—The Bi-cylindrical (W. C. Hughes, Kingsland)

Details.—For the reason that this lantern is quite out of the usual run, a somewhat full description is given. The novel shape, and the extreme portability thereby attained, are the two principal points of difference. It consists of two cylinders, only seven inches in diameter, and the same in length. Each lantern has an inner lining, with outlets intermediate with the outer case to pass the hot air through. The holes that are seen on the body are for keeping up a good draught, thereby causing the lantern to be comparatively cool; each lantern is quite separate from the other, therefore the heat passes off quite distinctly from the upper and in the lower, instead of passing through the entire lantern and then out at the chimney, as in the regular kind. Each lantern carries off its own hot air, and by the internal fittings and arrangements keeps cooler than the others, besides being less cumbersome The illustration shows the lantern mounted on its stand ready for use; at the top front of each lantern is a brass register screw for lowering the front down to a proper angle to get one disc within the other; it does not require much, as the two nozzles are much closer together than in the ordinary kind. There are screws in the brass cradle pieces on which the body of the upper lantern rests, which will also allow the body to shift to any angle, so that the registration of one disc within the other to a nicety is thoroughly attained. As the prices vary, there being four patterns of the same lantern, we quote the prices of each.

No. 1 is made entirely of brass, every part being screwed and cast, and manufactured in a substantial and portable manner, having the hot-air passages, inner linings, chimneys, brass movable cradle-pieces, etc., and mounted on polished mahogany feet, 4in. double achromatic condensers; best achromatic front lenses, double rack and pinion for focussing at either side, spring slide holders and sliding trays, sliding fronts for long-focus lenses, extra foot, and all the arrangements for working them, either as a single side-by-side or a perpendicular lantern; pair of the best "Petite" oxy-hydrogen chamber or pressure jets, with platina points, button handles, etc., twenty feet of tubing and box of limes. The whole packed in polished mahogany cabinet with compartments lined and padded with cloth to prevent injury to the lanterns £30 0 0

No 2 is made of the best Cornwall tin, every part rivetted and hard soldered, inner linings and hot-air passages; 3½in. double achromatic condensers, large diameter double combination lenses, sliding front, with brass flange frontings, and double rack and pinion for focussing either side; spring slide holders and trays; brass movable cradle-pieces; a pair of the best "Petite" safety jets with lever handles, ten feet of tubing and box of limes; each lantern separate, with extra foot, forming perpendicular, single, and side-by-side lanterns. The whole fitted in polished padded case with compartments £12 12 0

Extras.—4in condensers, £2 2/- Brass fronts, 3½in., £2 10/-, 4in., £3 3/-

No. 3 is made of tin, japanned, inner linings and hot-air passages, 3½in. double condensers; double combination portrait lenses with rack and pinion; sliding fronts, spring slide holders; sliding trays; a pair of the second quality "Petite" safety jets. The lanterns do not take to pieces.. £6 6 0

Extras.—4in. condensers, 34/- Ordinary brass sliding tubes and fronts, 30/-

No. 4 is made of japanned tin, without inner linings; brass sliding fronts and achromatic front lenses; 3½in. double condensers; sliding trays; spring slide holders, pair of "Petite" safety jets, second quality £5 18 0

Extras.—4in condensers, 20/-

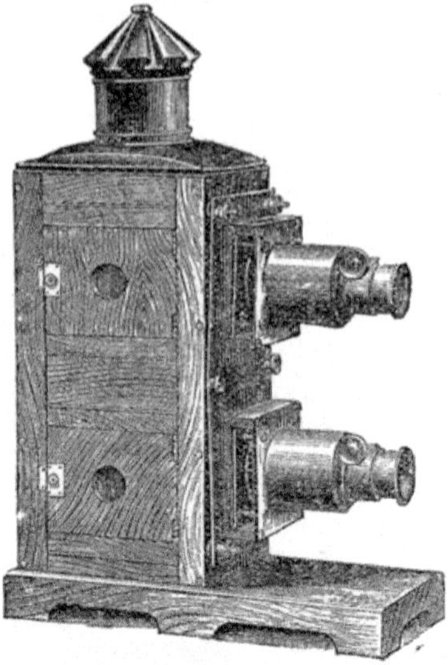

No. 161.

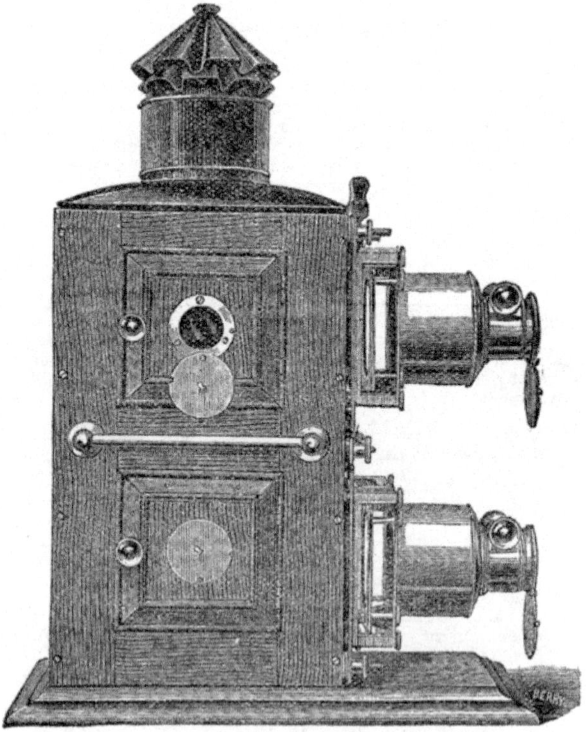

No. 162.

161.—Mason's No. 1 (Mason & Co)

Details.—Polished mahogany body, lined with iron; four doors; 4in. compound condenser; double combination achromatic objectives, with rack adjustment; dissolving tap and pair of safety jets.. .. .. £7 10 0

162.—Mason's No. 2 (Mason & Co.)

Details—Polished mahogany body, lined with iron; panelled doors; 4in. condenser; special achromatic front lenses, solid brass fronts; diaphragm for curtain effect; brass hand-rail, star dissolver and pair of safety jets £11 0 0

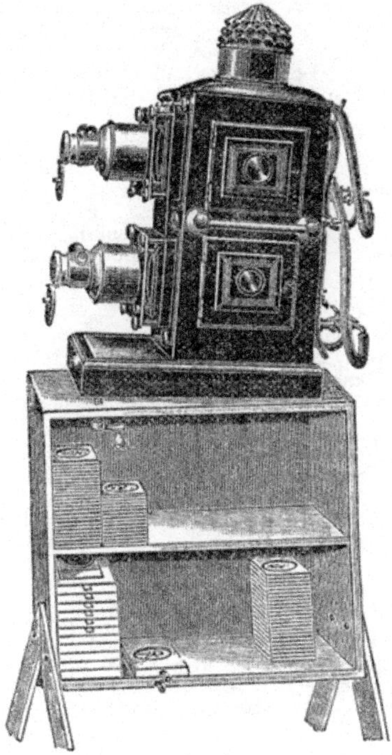

No. 163.

163.—Mason's No. 3 (Mason & Co)

Details.—Same as foregoing, but a superior instrument. The jets are supported by a novel arrangement of sliding tubes, an account of which will be found under Section II, Part I The packing case is supplied with legs (which unscrew), to form a steady and convenient stand, to which the lantern is screwed down. A tilting screw is also provided, dissolving tap, and two jets £18 10 0

Extras.—New triple rack telescopic front with four sets of double combination lenses for showing at different distances, £7 10s., or single achromatic lenses, £6.

Remarks.—A good lantern at a moderate price, and a credit to the makers.

164.—Medland's Iron Body (J. B. MEDLAND).

Details.—Iron body, 4in. compound condensers, mounted in brass cells; brass rackwork fronts, with compound achromatic lenses; brass O.G. sliding tubes to front, adjusting screw, and dome and rose top £5 15 0
Extras.—4in. condensers, 12/6

165.—Medland's Mahogany Body (J. B. MEDLAND).

Details.—Mahogany body, lined with iron; four doors, with sight holes; 4in compound condensers, brass rackwork fronts, with double pinions and compound achromatic lenses; japanned tin stages, relieved with brass fittings and brass O G. sliding tubes, dome and rose top .. £8 10 0
Extras.—Stages and entire front in brass, and brass shutters to front lenses, £3 10s.

166.—Medland's Superior (J B MEDLAND).

Details.—Mahogany body, lined with iron; four doors, with sight holes; 4in. compound condensers, brass rackwork fronts, with double pinions and compound achromatic lenses; stages and front entirely of brass, brass telescopic fronts, with three drawers extending to twelve inches, brass handrail fixed on each side of body, rolling curtain diaphragm.. £14 10

167.—Newton's Japanned Tin (NEWTON & Co.)

Details.—Japanned tin body and fronts; best quality 4in. plano-convex condensers; double combination achromatic lenses of large diameter, mounted in brass rack and pinion focussing fronts, and gas tap for dissolving. In cabinet case, forming table for the lantern .. £10 10 0

168.—Newton's Mahogany (NEWTON & Co.)

Details—Polished mahogany body, lined with tin, and japanned tin fronts; 4in. plano-convex condensers; double combination achromatic lenses of large diameter, mounted in brass rack and pinion focussing fronts, and gas tap for dissolving. In cabinet case, forming table for the lantern £12 12 0

169.—Newton's Mahogany (NEWTON & Co.)

Details.—Polished mahogany body, lined with tin; 4in. plano-convex condensers, polished brass adjusting fronts, double combination achromatic lenses of large diameter, mounted in brass rack and pinion focussing fronts; gas tap for dissolving. In cabinet case, forming table for lantern £16 16 0

170.—Newton's Ebonised and Nickel-plated (NEWTON & Co.)

Details.—Polished ebonised mahogany body, lined with tin, 4in. plano-convex condensers, polished nickel-plated brass adjusting fronts; double combination achromatic lenses of large diameter, mounted in brass, nickel-plated rack and pinion focussing fronts, and gas tap for dissolving. In cabinet case, forming table for the lantern.. .. .. .. £19 19 0

171.—Newton's Combination (NEWTON & Co.)

Details—Two polished mahogany lanterns, lined with tin, and so constructed that they may be used one on top of the other, as a biunial, or side by side, as a pair of lanterns, or as two single lanterns; 4in. plano-convex condensers; polished brass adjusting fronts, double combination achromatic lenses of large diameter, mounted in brass rack and pinion focussing fronts; gas tap for dissolving In cabinet case, forming table for the lanterns.
£18 18 0

172.—Newton's Mahogany (NEWTON & Co.

Details.—Polished mahogany upright double lantern, lined with tin, and fitted with two pairs of best quality 4½in. condensers, polished brass adjusting fronts, and improved double combination achromatic lenses of large diameter and best construction; mounted in brass fronts, with rack and pinion focussing tube, and improved gas tap for dissolving. In cabinet case, forming table for the lantern £19 19 0

173.—Newton's Superior (NEWTON & Co.)

Details.—Polished mahogany body, lined with tin, 4in. plano convex condensers; triple combination achromatic front lenses of superior size and quality; this enables the exhibitor to show a picture of any required size at almost any distance from the screen, thus adapting the lanterns for use in either small or large rooms. The lenses are mounted in polished brass telescopic draw adjusting fronts, with rack and pinion focussing tubes, and gas tap for dissolving. In cabinet case, forming table for the lantern £22 0 0

Remarks.—All Messrs. Newton's lanterns are well made. A stout iron partition between the two lanterns, and a shaft or chimney from the lower lantern up the side which carries the heat away, make the apparatus very cool to work All the lanterns are fitted with good quality long focus lenses, made specially for the purpose. With No. 171 three or four-wick mineral lamps may be used.

174.—The Noakes No. 1 (DAVID NOAKES & SON).

Details.—Polished mahogany body, lined with tin, with brass sight holes and mountings, 4in compound condenser; double combination achromatic rack fronts, with arrangement for depressing and raising fronts Complete, with trays and stout travelling case £7 10 0

175.—The Noakes No. 2 (DAVID NOAKES & SON).

Details.—Same as No. 174, but with bright brass fronts and plates, rolling curtain diaphragm, and the Noakes improved adjustable stages. In stout travelling case £10 10 0

176.—The Noakes No. 3 (DAVID NOAKES & SON)

Details—Polished mahogany body, lined on an improved principle, two moulded doors, two mock doors, windows with cover shutters, entire brass telescopic fronts and stages, brass mediæval handles to doors, mock handles to mock doors, rolling curtain plate, with stop. The stages are light tight, with either a ⅜in slide or an inch tank inserted, and are open all round for chemical experiments. The lenses consist of 4in best quality compound condensers, steam wayed and bayonet jointed; two sets of double combination achro-objectives, with 2in large back lens, 4in. and 6in. foci, mounted in the Noakes improved front mount, by which the lenses, 4in. foci, can be instantaneously changed for the 6in., without anything to unscrew (see Section II, Part 2), double bye-passed dissolver and back tubes, adjustable registering stops, two best safety or chamber jets, with lime turning arrangement. Complete, in extra strong travelling case, into which the lantern packs with dissolver and jets in position, without taking them off £25 0 0

Remarks.—A very fine lantern, upon which much thought and attention have been bestowed. The description of the triple described elsewhere applies largely to this one as well

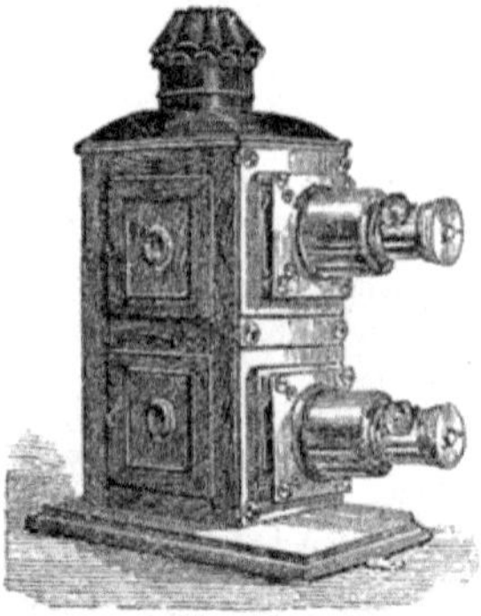

No. 177.

177.—The School (Ottway & Son).

Details—Polished mahogany body, lined with stout tin, metal fronts and stages, black japanned, with brass mounts and adjustments; double combination achromatic focus lenses, with rack and pinion adjustment; 3½in. condensers. Packed in case to form stand £9 9 0

Extras.—4in condensers, 20/-

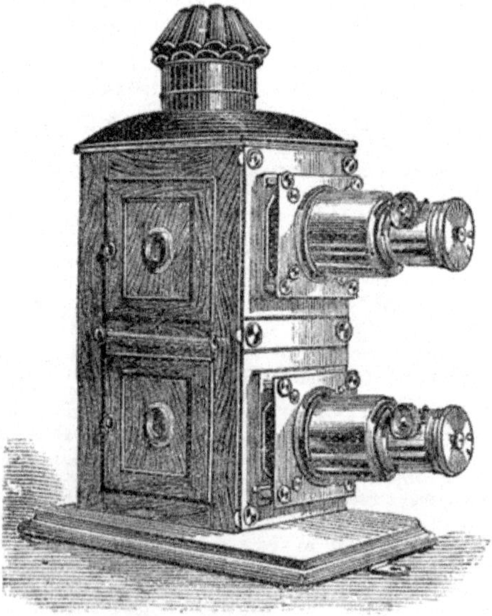

No. 178.

178.—The Lecturer's (Ottway & Son).

Details.—Body of polished mahogany, lined with tin, 4in. condensers; achromatic focus lenses, the fronts and stages entirely of brass, with large sliding tube for long focus; the slide opening or stage is formed with brass pillars, giving access for effects on all sides; sliding diaphragm for rising curtain effect, and shutters for flashing on effects, and six way dissolving tap. Packed in case, forming table for lantern £14 0 0

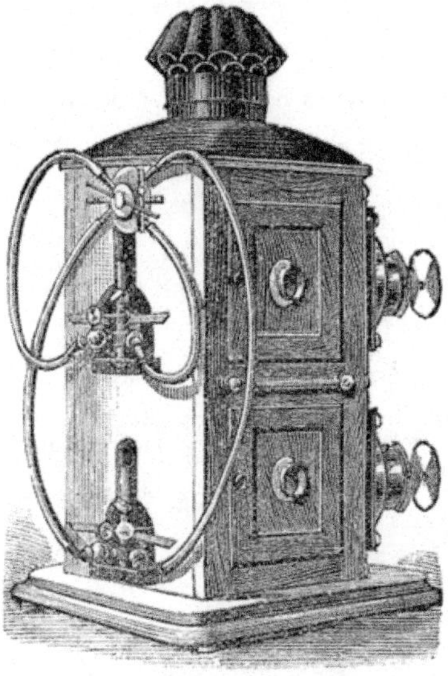

No. 179.

179.—The Lecturer's Standard (OTTWAY & SON).

Details.—Polished mahogany body, with brass rail handles, rosewood doors, with raised panels, the body lined with tin to keep perfectly cool; $4\frac{1}{4}$in. condensers, achromatic focus lenses for giving picture of equal size at five different distances; the fronts and stages entirely of brass, highly finished, with three telescopic draws to suit the various foci; sliding curtain diaphragm, shutters, six way dissolving tap. Packed in case to form stand, and having two movable shelves on which to arrange the slides £21 0 0

Remarks.—A very fine instrument. It can be still further improved by the addition of rackwork to curtain slide and the new method of tinting by doors or grooves in body of the lantern (For details of these see Section II., Part 4)

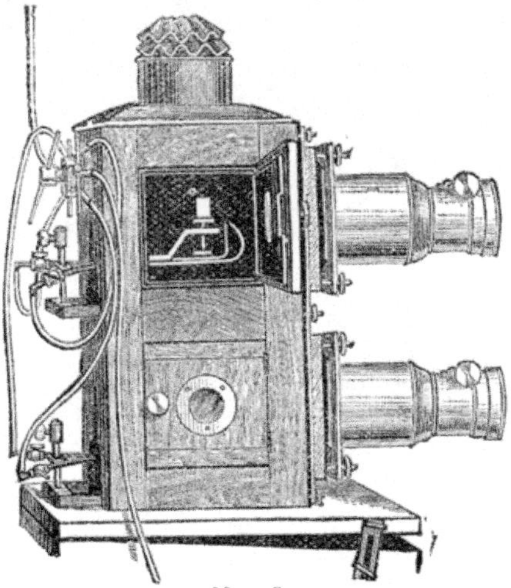

No. 180.

180.—Pexton's Mahogany Body (Chatham Pexton).

Details.— Mahogany body, lined with tin ; rose top chimney ; sliding jet trays ; spring slide holders ; 4in. double condensers ; achromatic rack and pinion fronts £7 10 0

Extras.—Polished brass tube and slide holder instead of black japanned tin, 20/- Complete brass fronts and sliding diaphragm for curtain effects, 50/- Massive brass fronts, open all round, and sliding diaphragm, £4. Three-draw telescopic brass fronts, 70/-

181.—The Chatham (Chatham Pexton).

Details.—An illustration and description of the Chatham Triple appears further on, to which we refer our readers, the biunial form being exactly the same in arrangement and finish. Two best jets, either safety or mixed ; six-way dissolver, with two bye-pass taps ; extra lenses, and same case as the triple £30 0 0

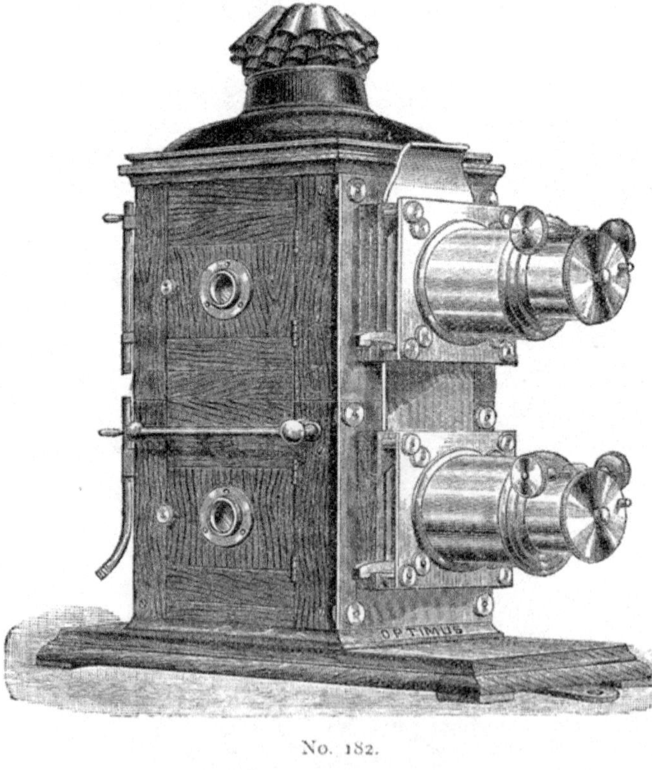

No. 182.

182.—The Optimus (PERKEN, SON & RAYMENT).

Details.—Seasoned mahogany body, well polished, lined tin, with hinged doors fitted with tinted glasses in brass cells; 4in. compound condensers in brass cells; compound achromatic front lenses, with rack adjustment for focussing with double pinion £5 12 0

Extras.—Stages and fronts of polished brass, £4 5/- Brass handrail, three-draw telescopic fronts, and the necessary sets of front lenses, giving a focal range varying from 16 to 120 feet distance from the screen, £4 15/-

Remarks —We have no hesitation in saying that a better lantern for such a low sum is not to be obtained. It is well made, of good sound mahogany, and handsome in appearance.

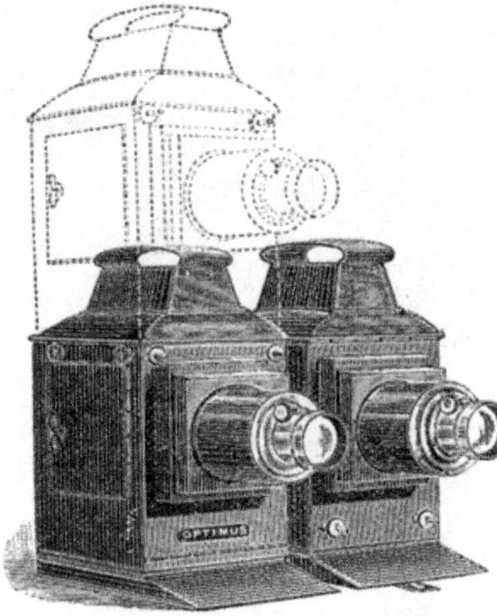

No. 183.

183.—The Optimus Combination (PERKEN, SON & RAYMENT).

Details.—Japanned metal body; 4in. compound condensers; best double achromatic portrait combination objectives; the grooves for limelight trays are made to the same gauge as the lamps made by the same firm, without lamps or jets £5 0 0

Extras —Three-wick lamps, 14/- each

Remarks.—The dotted lines in the illustration show that this lantern is so constructed as to work one above the other in biunial form. The lanterns can, of course, also be used side by side, and, as each is complete in itself, they may be used in two distinct places at the same time, either with oil or gas. It must be a good all-round apparatus to possess, and should especially commend itself to those who hire out lanterns, as it adapts itself readily to the various calls made upon them by the general public. It can in reality be used in four different ways—as a biunial, side by side, and two distinct lanterns.

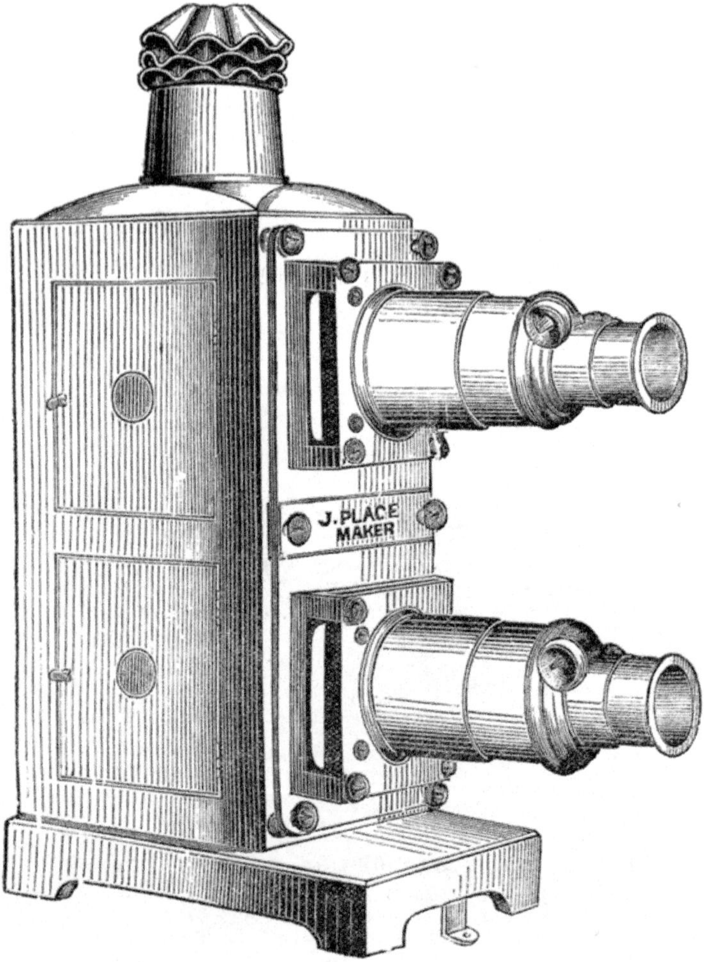

No. 184.

84.—Place's (J. PLACE).

Details.—Polished mahogany body; brass and mahogany fronts; brass telescopic tubes, and rack and pinion achromatic fronts, with lenses suitable for showing at different distances; 4in condensers, mounted in brass cells. Complete in case £10 10 0

Remarks.—Considering that the brass telescopic fronts and lenses for different ranges of focus from the screen are included in the price, together with handsome mahogany and entire well-polished brass fronts, this is a first-class lantern at a very moderate price.

185.—Blakeborough's (SOLGRAPH CO)

Details—Japanned tin body, brass rackwork fronts, achromatic front lenses, O.G. sliding tubes, 3½in. condensers.. .. .. .. £5 5 0

Extras.—4in. condensers, 7/6.

186.—Blakeborough's Mahogany (SOLGRAPH CO)

Details.—Same as No. 184, but has mahogany body, lined with tin £6 17 6

Extras.—4in. condensers, 20/-

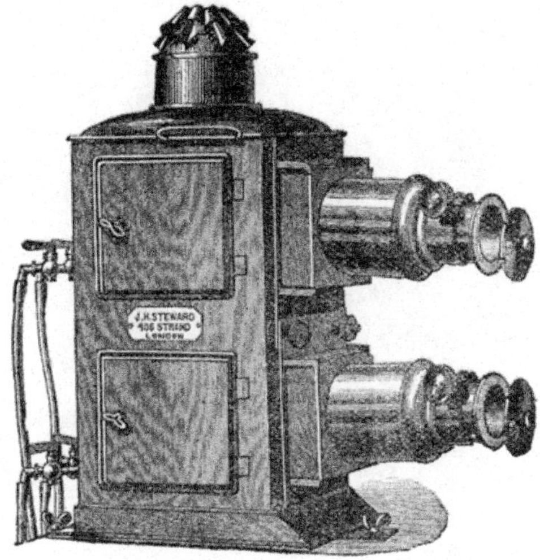

No. 187.

187.—The Luke (J. H. STEWARD).

Details.—Body made of stout japanned tin, which is very firm, although light; adjusting plates, with brass milled head screws and clamps, improved stages open at top and bottom for insertion of the rolling curtain shutter, and effects of rising or descending, etc., brass sliding body tube, rackwork adjustment to brass fronts, carrying achromatic lenses of long and short focus; 4in. condensers. Packed in case, forming stand, size about 18in x 17in x 8½in £9 9 0

Remarks.—A very portable biunial. The height is only 17½ inches, and the width in proportion.

188.—Steward's Mahogany Body (J H STEWARD).

Details—Mahogany body, lined with japanned tin; spring stage, adjusting front tube, and rack and pinion adjustment to achromatic front lenses, with adjustments to japanned tin stages, etc, for centring the two discs. Packed in case complete, 3½in condensers £11 11 0

4in.	,,	.. ..	12 12 0
4½in.	,,	.. ..	15 15 0

189.—Steward's Superior Mahogany (J. H. Steward).

Details.—Mahogany body, lined with tin; entire brass fronts, consisting of adjusting plates, stages with opening at the top, and body tube having one side adjustment, rack and pinion focussing tubes to achromatic double combination front lenses, 4in double condensers. In case, price .. £15 15 0

Extras.—Larger size front lenses, brass shutters and four doors to body, £3 3/-

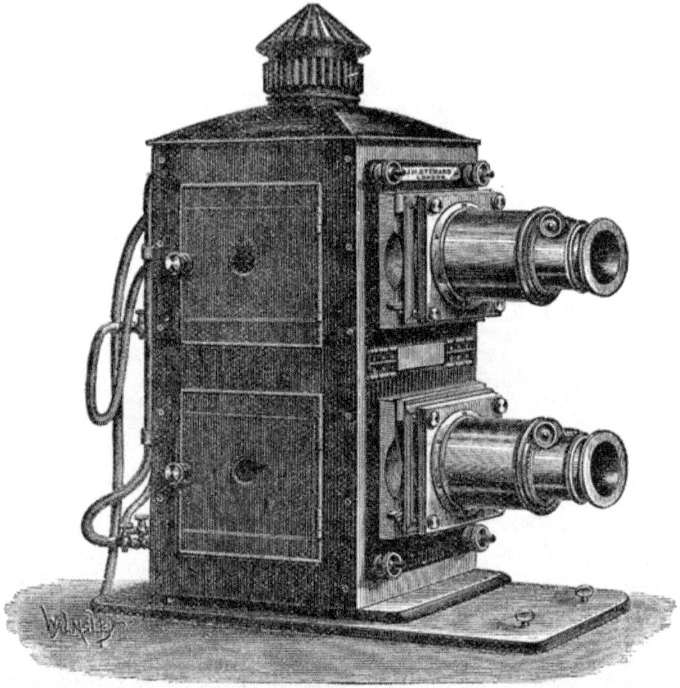

No. 190.

190.—Steward's Standard (J. H Steward).

Details.—Body of mahogany, lined to be perfectly cool, and is fitted with specially worked 4¼in condensers, and showing the square Standard photographs sharp and distinct to the edge; the adjusting plates and the stages are brass, and so arranged that the slides can be held in a vertical as well as horizontal position for producing effects; the body tubes are also brass, and have three draws, adjusting like a telescope, and thus getting the required distance for the lens most suitable to produce the necessary size disc at the most convenient distance for the apparatus to be from the screen; the focussing tubes are fitted with a rack and pinion adjustment, and the front lenses are achromatic, and of three long and short foci Price, packed in stout travelling case £23 0 0

Extras.—Rolling curtain shutter with guides, fittings and stops, 21/- Brass bar handles and fittings, 22/6.

Remarks.—A lantern which, whilst the price is well within the power of many people, is yet a high class apparatus, capable of the best work.

191.—The Portable Standard (J H. STEWARD).

Details.—Much the same as 189, but with extra small mahogany body, ebonised edges and mouldings, and made extremely portable. In travelling case, 20in. x 22in. x 11½in £23 10 0

Extras.—Same as 189.

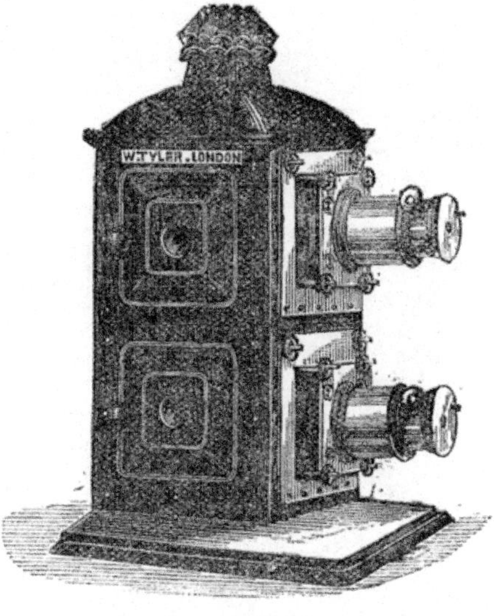

No. 192.

192.—The Exhibition (W. TYLER).

Details.—Body of polished mahogany, lined with iron; four doors, with brass-mounted sight holes, 4in compound condensers; double combination front achromatic portrait lenses, with brass rackwork fronts .. £6 6 0

Extras.—Box, with lock and key, 12/6

193.—Taylor's (T. S. & W. TAYLOR).

Details.—Polished mahogany body, with glazed doors; adjustable brass stages and fronts; platinum tipped oxy-hydrogen jets, with full adjustments; six-way dissolving cock and rolling curtain; 4½in. diameter double condensers; best English portrait lenses, 2in. diameter, with rack adjustment. The whole packed in a substantial iron-bound stained travelling case, 22in. × 22in. × 14in., with inclining top forming the stand for exhibition; the partitions adaptable as shelves for slides, complete set of tools, pulleys, &c., fitted to removable door, drawer, and space for 20ft. screen.. £30 0 0

Remarks.—A first-class lantern, which has received high praise when shown at the Leicester Photo Society meetings, we believe.

194.—Watson's Tin (W. Watson & Sons).

Details.—Japanned tin body, 4in. double condensers, brass rackwork fronts, sliding tubes. Black travelling case.. .. .. .. £7 10 0

195.—Watson's Mahogany (W. Watson & Sons).

Details.—Mahogany body, metal lined, 4in. double condensers, brass fronts and lengthening tubes, achromatic front lenses, screw adjustments for centring discs, doors on both sides of lantern. In travelling case £10 10 0

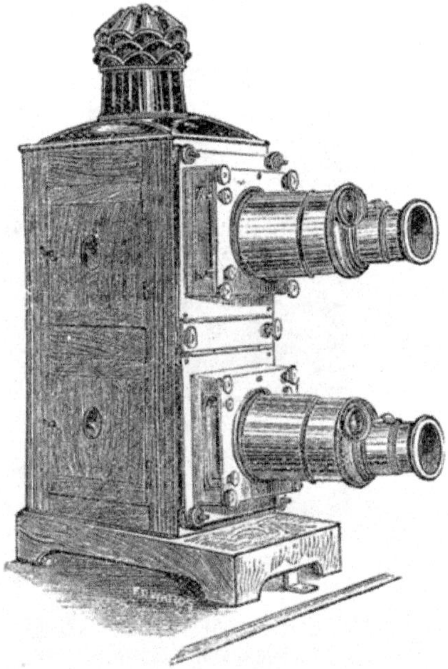

No. 196.

196.—Watson's Superior (W. Watson & Sons).

Details.—Mahogany body, metal lined, doors to open on both sides, fitted with dark tinted glasses in centre, 4in. condensers, achromatic front lenses, with rack and pinion, the entire front of polished brass. In case £14 14 0

197.—Watson's High Class (W. Watson & Sons).

Details.—Mahogany body, metal lined, with four doors, in which are tinted sight holes, front and stages entirely of brass, telescopic draw tubes giving range from 6 to 16 inches between the condensers and front; 4½in. condensers; best achromatic front lenses, three different foci, for use at varying distances from the screen, rackwork adjustment for focussing with double milled heads. Brass supply tubes; brass rail handles on each side; stops to each of the slide holders, with screw adjustment, diaphragm or curtain slide In strong travelling case £25 0 0

Remarks.—A sound, substantial and well-finished biunial. The material and workmanship alike first class. We noticed the telescopic fronts to be particularly rigid. With 4in. condensers instead of 4½ the price is £3 less.

198.–Wilkinson's No. 1 (Wilkinson & Co)

Details.—Body of polished mahogany, lined with tin, japanned tin fronts; 3½in. best condensers and achromatic focus lenses, with rack adjustment. In case, forming table for lantern £9 0 0

Extras.—4in. condensers, 30/-

199.—Wilkinson's No. 2.

Details.—Body of polished mahogany, lined with tin; best quality 4in. condensers; achromatic focus lenses, the fronts and stages entirely of brass, with large sliding tube for long focus, the slide opening or stage gives access for effects on all sides; sliding diaphragm for rising curtain effect, and flashing shutters. Packed in case, forming table for lantern .. £13 10 0

Remarks.—A very useful all-round biunial, and handsome too.

200.—Wilkinson's No. 3.

Details.—Body of polished mahogany, lined with tin; brass rail handles. rosewood doors, with raised panels; best 4in condensers; achromatic focus lenses for giving picture of equal size at five different distances; the fronts and stages entirely of brass, with three telescopic draws to suit the various foci, sliding diaphragm; shutters to lenses Packed in polished pine case to form stand, having movable shelves on which to arrange the slides £21 0 0

Remarks—Messrs. Wilkinson & Co., being very old exhibitors, know what kind of lantern to be useful and attractive. The one above quite hits the mark.

201.—Wood's (E. G. Wood).

Details.—Mahogany body, divided into two chambers, with provision for carrying the heat away from the lower lantern without diffusing into the upper one; 4in. compound condensers; all brass fronts; double achromatic objectives, lengthening tubes; adjustments to secure coincidence of discs; rolling diaphragm shutter. Strong cabinet case, with adjustable case and metal quadrants, forming table or stand £15 15 0

Extras.—4½in. condensers, and extra large objectives, £4 4/-

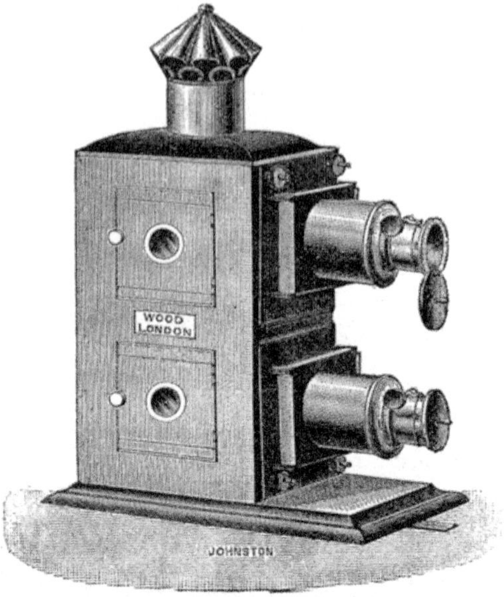

No. 201.

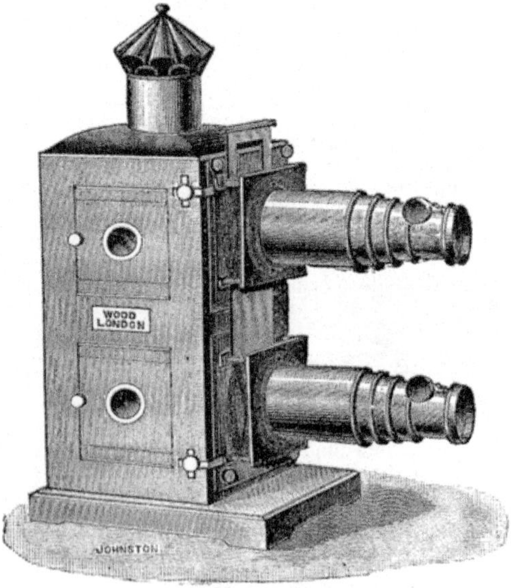

No. 202.

202.—The Lecturer's (E. G Wood).

Details.—Mahogany body, with 4½in. condensers and triple achromatic objectives, forming six different magnifying powers, enabling pictures to be produced at distances between 10ft. and 120ft. from the screen, brass telescope fronts, closing to 8in. and opening out to 20in.; rolling curtain diaphragm. Fitted into substantial cabinet £27 0 0

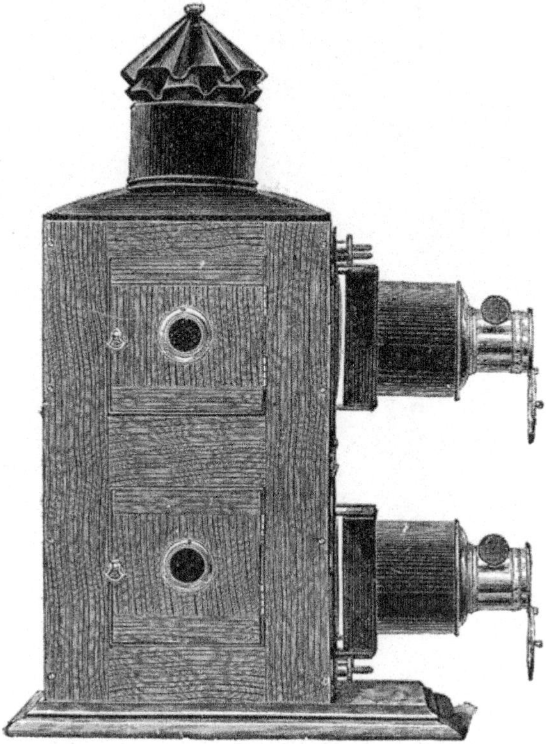

No. 203.

203.—The No. 1 (Wholesale Firm).

Details.—Polished mahogany body with moulded foot, four doors with brass-bound sight-holes, lined with japanned tin; japanned tinned dome and rose top, tin stages (open at top for chemical experiments), with brass condenser cells, front sliding tubes with brass O.G, and fitted with a rolling curtain effect; 4in. plano-convex compound condensers in brass cells; double combination achromatic front lenses mounted in brass (with large size lenses to the back combination), rack and pinion adjustments, and sliding shutters or flashers £6 6 0

Extras.—Stained and French polished case, lock and key, handles, 14/6

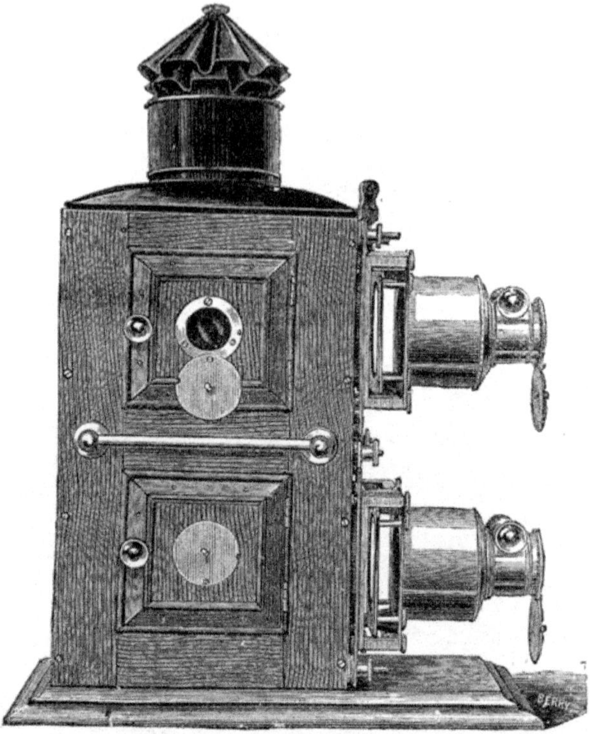

No. 204.

204.—The No. 2 (Wholesale Firm).

Details.—Polished mahogany body, moulded foot, four panelled doors with brass-bound sight holes, sliding shutters and brass rail handles on each side, lined with japanned tin; brass trays and double supports and binding screws for fixing the jets firmly; japanned tin dome and rose top, stages and front tubes made entirely of brass; brass rolling curtain effect, 4in. plano-convex compound condensers mounted in brass cells, double combination achromatic front lenses in brass mounts, with large size lenses to the back combination, and with double pinions to the rack adjustments, and fitted with sliding shutters or flashers, having grooves to carry shapes of coloured glass for tinting slides £10 10 0

Extras.—Ebonised body, and all the brass work nickel-plated, 15/- Stained and French polished cabinet case, with lock and key, and handles, 14/6.

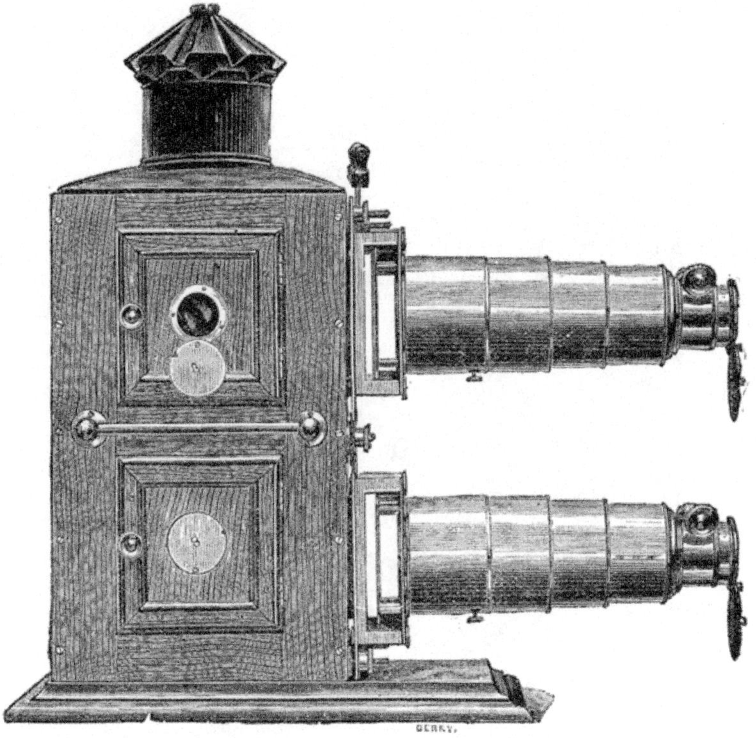

No. 205.

205.—The No. 3 (Wholesale Firm).

Details.—Polished mahogany body, lined with japanned tin, with moulded foot, four panelled doors, brass-bound sight-holes, with sliding shutters, brass rail handles on each side; brass trays, with double supports and binding screws for fixing the jets firmly, brass stages and front tubes, with three brass telescopic draws extending to 12 inches; brass rolling curtain effect; 4in. plano-convex and Meniscus compound condensers in brass cells; brass fronts with double pinions to the rack adjustment, and fitted with sliding flashers or shutters, having grooves to carry shapes of coloured glass for tinting slides; four sets of single achromatic lenses, mounted in brass cells (6in., 8in., 10in., and 12in. foci), made to screw together, or to be used separately £14 5 0

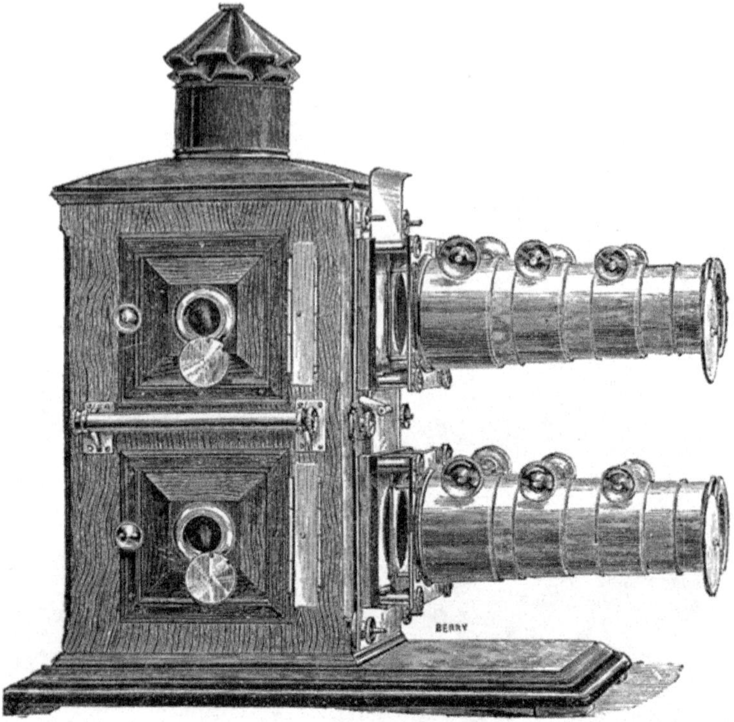

No. 206.

206.—The Triple Rack (Wholesale Firm).

Details.—Body of Spanish mahogany, with four rosewood panelled and frame moulded doors, each corner bevelled and inlaid with rosewood, the back framed and tongued together, panelled with rosewood moulding, the cap and base being of the same; the foot framed and tongued together, and boldly moulded. The four doors have brass-bound sight-holes, with sliding shutters, and the body has brass rail handles on each side. Openings on each side of each lantern, just behind the condensers, bound with brass-hinged shutters for inserting a long painted slide (which is sent with each lantern), used for tinting plain photographic slides from daylight to moon-light. Lined throughout with Russian iron, with Russian iron dome and rose top, and brass trays with double supports and binding screws for fixing the jets firmly. The stages are made entirely of brass, with brass registering side pieces. Patent triple-rack telescopic front tubes, as described in Section II., Part 2. Brass rolling curtain effect worked by rack and pinion; 4in. diameter Plano and Meniscus compound condensers mounted in brass cells; double combination achromatic front lenses (four sets, eight in all), mounted in brass cylinders, with sliding shutters or flashers (having grooves to carry shapes of coloured glass for tinting slides) of the following foci—4in., 6in., 8in., and 10in. back focus. All the cylinders are the same size in diameter, and each one slides into the front of the triple rack as required. Each cylinder has its focal length engraved on it. A set of six coloured glasses in a leather case, for tinting slides, is sent with each lantern. Strong stained and French-polished dovetailed cabinet case, with two locks and keys and handles, having at top a hinged board forming a canting table, which is canted by means of two powerful screws. The inside of the case is fitted with an extra foot, into which the lantern body is placed, having four rollers running on iron rails, enabling the lantern to be withdrawn and replaced easily. The case has also fittings for all the different parts connected with the lantern, including a drawer with fittings for carrying all the brass cylinders and condensers £54 0 0

Extras.—4½in. condensers, 15/-

Triple or Triunial Limelight Lanterns.

In the majority of cases the triple lantern is merely an extension of the biunia That is, the details are the same, but a third, or top, lantern is added. Most of the remarks on page 65 (Biunials) equally apply to this form.

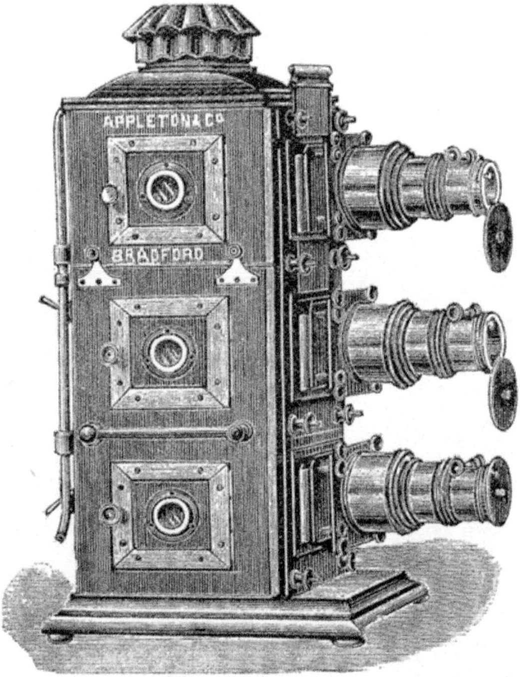

No. 207.

207.—Appleton's (Appleton & Co.)

Details.—Mahogany body with six doors, brass flanged sight holes, entire brass fronts and stages, 4in. condensers, sets of achromatic object lenses of 8, 10, and 12in. focus, mounted in brass cells, with rack and pinion adjustment and flashing shutters, curtain diaphragm. The top lantern takes off and fits into extra base board, with extra dome top. Three blow through jets, and two dissolvers, brass supply tubes, double set of lengthening tubes for lenses. The case has a shelf and rising top, to form stand .. £30 0 0

Remarks.—This is certainly a well-finished apparatus at a cheap price

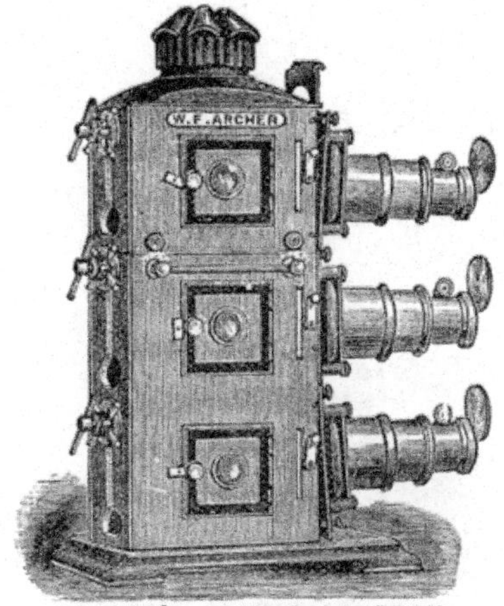

ARCHER'S TRIPLE LANTERN.
No. 208.

208.—Archer's No. 1 (ARCHER & SONS).

Details.—Polished mahogany body, lined with iron, well ventilated; six panelled doors, three each side, brass-bound sight holes, brass rail handles; japanned fronts, with one-draw-out brass tubes, and Archer's improved slide holders, screws for registering the fronts; 4in. condensers, mounted in brass; double achromatic object lenses; rack and pinions. The top lantern is made to take a three or four-wick oil lamp, and lifts off to form a separate lantern, so that it can be used as a single, double, or triple lantern at will. An extra foot and dome are supplied, so that it can be used as a single oil lantern £15 0 0

209.—Archer's No. 2 (ARCHER & SONS).

Details.—Same as No. 207, but with solid brass fronts and stages, one-draw telescopic tubes, and brass diaphragm to produce the rolling curtain effects, Dallmeyer's pattern new front lenses, flashing shutters, brass supply pipes, &c. £20 0 0

210.—Archer's No. 3 (ARCHER & SONS).

Details.—Same as No. 208, but with three-draw telescopic tubes, extending to twelve inches (for showing across large halls when necessary); four sets of achromatic lenses, for showing long or short distances from the screen £25 0 0

211.—Archer's Grand (ARCHER & SONS).

Details.—Same as 209, but with combination long and short range triple lens condensers, giving the finest results possible; Archer's attachments for the true registration of effect slides, fixing screws to the brass trays, highly finished brass top, Archer's quick-action screws for ease in tilting the lantern, &c., &c. £35 0 0

Extras.—To Nos. 207 to 210, cabinet case, forming stand, 20/-

Remarks.—All these lanterns are well made, and appear excellent value for the money. No. 210 is a specially fine instrument, fit for use by the most fastidious of operators.

212.—Armstrong's (Thos. Armstrong and Brother).

Details.—Polished mahogany body, six panelled doors, with brass-rimmed sight-holes; all brass fronts and stages, three-draw telescopic front tubes and three sets of lenses, rackwork diaphragm for curtain effect, three jets, dissolver. The lantern can be used either as a single, biunial, or triple £45 0 0

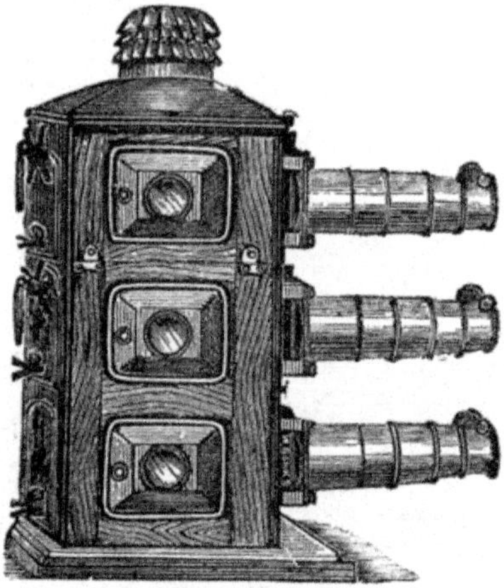

No. 213.

213.—Banks (W. Banks).

Details.—Polished mahogany body (or walnut), six doors with sight holes, ornamented top; 4in. condensers; fronts and stages all brass; telescopic front tubes; achromatic front lenses, with rack and pinion adjustment; curtain diaphragm; two sets of jets, blow through and mixed, three dissolvers; top lantern takes off. Panelled box, which forms stand £55 0 0

Remarks.—It will be noted that with this triple two sets of jets are included.

214.—Catlin's No. 1 (W. Catlin).

Details.—Polished mahogany body, three doors, fronts and stages entirely of polished brass; 4in. condensers, achromatic double combination objectives, with rack and pinion focussing, and flashing shutters; diaphragm for curtain effect; top lantern removable; fitted with one single and one double dissolving tap. Cabinet case £25 0 0

Extras.—Base board and fittings for using top lantern separately, with extra dome and rose top, 25/- Doors on both sides, 21/- Three brass adapter tubes for long focus lenses, 18/-

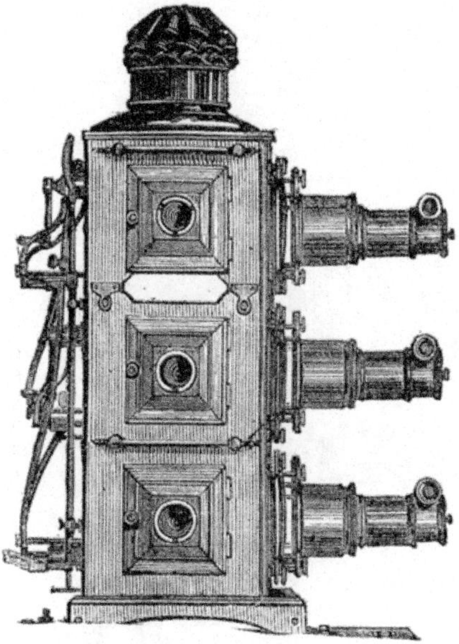

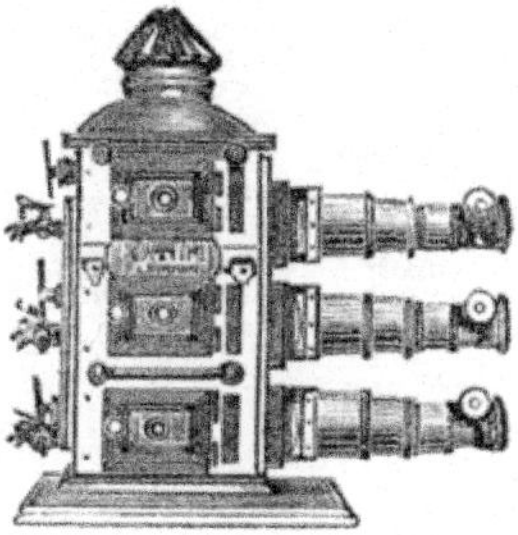

No. 215.

215.—Catlin's No. 2 (W. CATLIN).

Details.—Body of polished Spanish mahogany, with six rosewood panelled doors and mouldings, fitted with brass cells and coloured glasses, two pairs brass rail handles to lift it by, fronts and stages entirely of highly-finished brass, three sliding telescopic adjustments for focussing to various distances, twelve achromatic focus lenses, 4in. condensers, sliding brass diaphragm for curtain and statuary effect, top lantern removes and fits on to second base boards, three best oxy-hydrogen jets, fitted with mitre wheel (for turning lime), platinum points, &c., the jets, fixed on to turned steel pins, fitted in stout iron plates, sliding in dovetails; brass supply tubes to carry the gas to the dissolvers Strong case, which also forms stand.. £40 0 0

Extras.—4½in. condensers, and fitted with adjustable metal masks, producing coincidence of the three discs, and correcting the irregularity of matting in slides; also the new tinting effect, by means of which a graduated tint slide is passed through the lantern between the condensers and light, £5.

216.—The C.P.S. No. 1 (CUBLEY & PRESTON).

Details.—Polished mahogany body, lined with tin, six doors with sight holes; 4in. compound condensers mounted in brass, and brass rackwork fronts, with double pinions and compound achromatic lenses; japanned tin stages, relieved with brass fittings and brass O.G. sliding tubes, dome and rose top; the top lantern made to detach £15 0 0

217.—The C.P.S. No. 2 (CUBLEY & PRESTON).

Details.—The same as 215, but the stages and entire front made in brass, with brass-bound sight holes to the doors, and brass rail handles fixed on each side of body, and brass shutters to the rack fronts .. £23 5 0

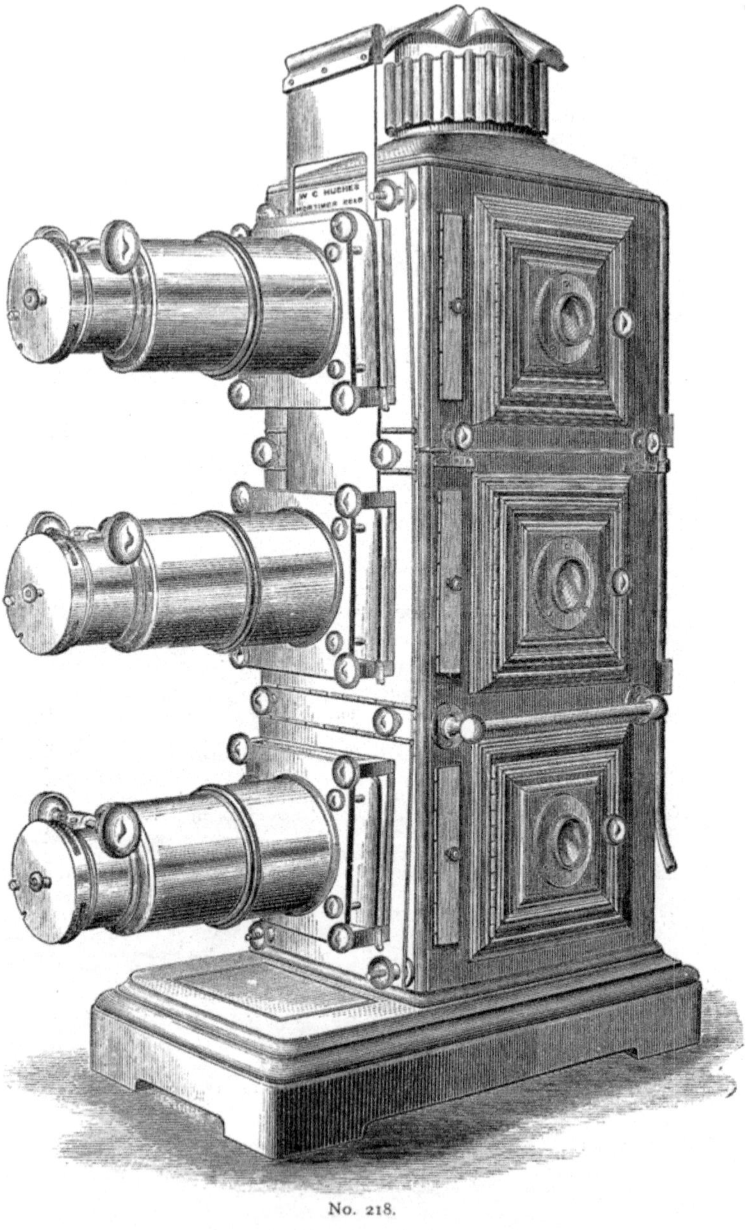

No. 218.

218.—The Grand (W. C. HUGHES, Kingsland).

Details.—Framework made from well-seasoned mahogany and walnut wood; six panel doors, three on each side, with walnut and ebonised mouldings; the back arranged in three parts, the openings being fashioned in a useful way, with additional mouldings to receive the jets and dissolvers, the doors have inlaid brass flange mouldings, with coloured glasses; the linings are of japanned iron, and the sliding trays for fastening the jets are of solid brass ground into solid runners, with additional brass tray to prevent falling limes; brass handles for lifting purposes, fronts entirely of brass with mahogany blocks, held together with brass bolts, &c., the brass plates carrying the condensers and front optical systems are so adjusted that there is perfect coincidence of discs, and are kept stationary by brass clamping collars; these adjustments are arranged so that the rolling curtain, or diaphragm, which is also made of brass, can be used under any conditions at any angle or distance from the screen, having a special arrangement of stops and rests, which is free to be removed in a moment without depreciating the picture on screen; the stages are open on all sides for purpose of effects, and in order to be able to use extraordinarily large framed and peculiar effects, the top lantern has a special arrangement doing away with the top bars and leaving a clear space, so that any frame can be pushed right down from the top, which is useful and desirable at times; the slide holders are actuated by special springs, and secure each slide firmly; telescopic front tubes, with three draws, there are brass runners for slides, which can be removed; a new form of register stop to keep the slides central, and will turn up when larger slides or panoramic frames are used; the condensers are Mr Hughes's special new arrangement, and are the finest in the world for long or short focus front lenses in order that less light may be lost than hitherto, mounted in elegant brass screw cells. Three sets of condensers, together with an extra four-and-a-half, same principle, should a special large size picture be used, or for realising a continual proscenium on the screen; three extra half-condensers are given in case of breakage, the front lenses, which screw into flanges of the telescopic draw tubes, are Mr. Hughes's new lantern objectives. See particulars in Section II, Part 2. Regulated to focus by rack and pinion, with double milled head attachments; screwed on to the front of this is a large deep collar box diaphragm for inserting stops or coloured glass in front, including brass flashing shutter; four sets of these lenses adapted to this large rack mount, making twelve in all, ranging in focus, viz., 4, 5, 6 and 7 inches, also an extra set of close combinations for very short distances when it is necessary to show behind the screen; the jets are special high pressure, and only sold with this lantern, are elegantly finished in solid brass; they will require extra large bags, and the hydrogen bag must be, at least, a third larger than the oxygen, the top lantern can be removed to form a single lantern, or for other optical attachments, while the lower forms a biunial. Improved telescopic box, forming stand for operator, with arrangements for lowering, resting, and elevating lantern.

£105 0 0

Remarks.—This is probably the highest-priced triple in our book, but it is one upon which the resources of the optician and manufacturer have alike been concentrated It is got up, to quote a hackneyed phrase, "regardless of expense." Brass and mahogany in conjunction, lavishly used, make the Grand Triple something to look at, something to ponder over. And as in outside appearance so in working capabilities, the main object being to produce the best possible results in the most reliable and at the same time the most easily manipulated manner. The description probably speaks sufficiently for itself, and we can add only a general opinion that a finer apparatus will not be found.

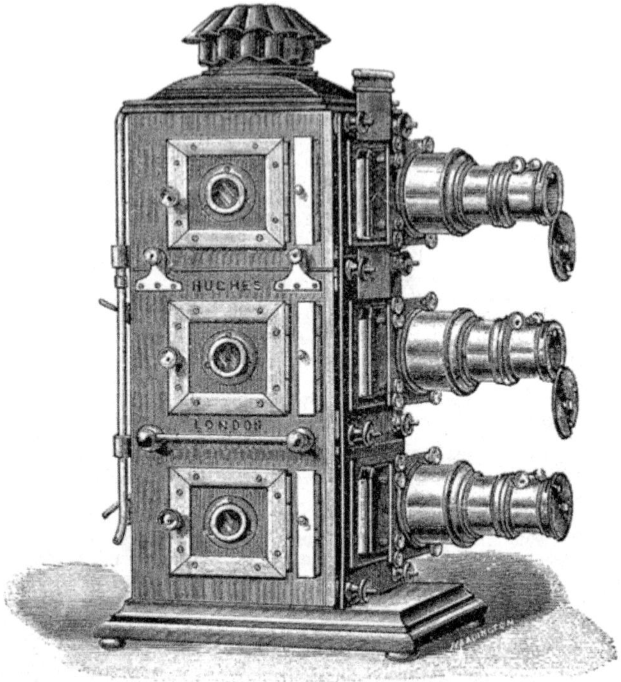

No. 219

219.—The 1886 Pattern (W. C. Hughes, Kingsland).

Details.—Polished mahogany body, lined with japanned iron, walnut panelling, six doors; brass jet trays, on brass ground runners, three pairs of double condensers; objectives fitted in brass tubes, which slide in the large rack; rack and pinion adjustment; fronts entirely of brass; rolling curtain diaphragm; three Malden dissolving taps, brass supply pipes, three jets; brass narrow doors for tinting effects; top lantern can be detached, three mixed jets, with lime burners. Cabinet case £42 0 0

Extras — Sets of longer focus objectives, 30/- each.

Remarks.—An extremely good instrument at a reduced price. It possesses all the advantages of the Grand.

220.—The Bijou (W. C. Hughes, Kingsland).

Details.—Nearly exactly the same as No. 218, but the whole apparatus is extremely small and portable, occupying no more space than a biunial £42 0 0

Remarks.—Mr. Hughes calls this "the smallest triple in the world" It is certainly very ingeniously constructed to attain the end of extreme portability.

221.—The Excelsior Co-operative (W C. Hughes, Kingsland).

Details.—Body of japanned tin, with brass draw tubes, plinth foot, iron 3½in. double condensers, with double combination rack and pinion, achromatic front lenses, sliding fronts, adjusting screws, &c.; the top lantern can be removed when required for use as a single lantern £8 8 0

Extras.—4in. condensers, £2 2/- Polished mahogany bodies, £2 2/- Solid brass stages and fronts, with brass draw tubes, £6 6/- Box, 18/6

Remarks.—As its name implies, a cheap lantern, but nevertheless sound and useful.

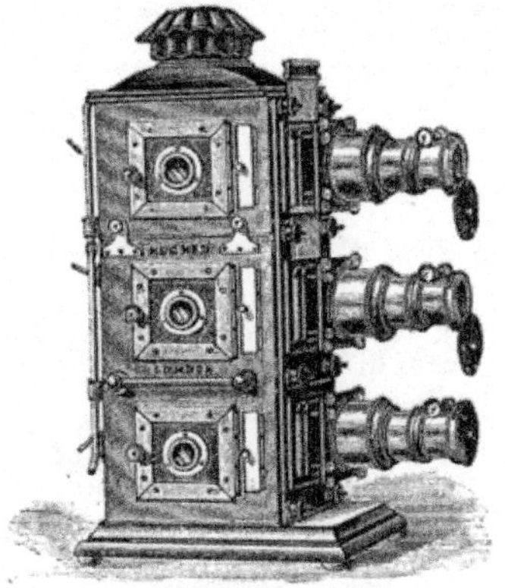

No. 220.

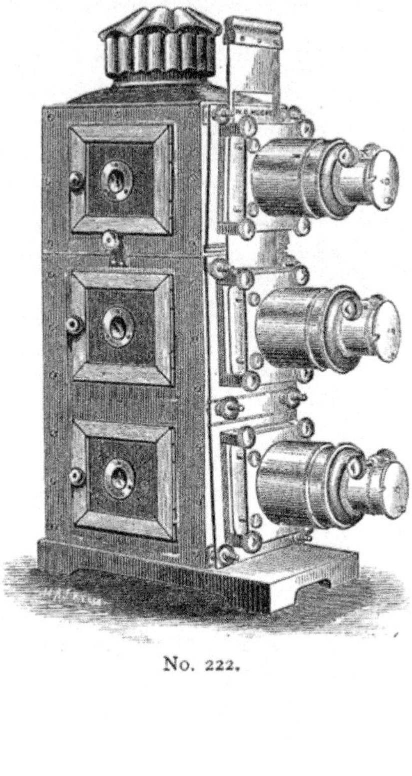

No. 222.

No. 222.—The Excelsor (W. C. Hughes, Kingsland).

Details.—Polished mahogany body, lined with iron, solid brass fronts, and diaphragm for rolling curtain effects, 4in. double condensers in brass screw cells, large sliding tubes for long focus lenses, three sets of front lenses, double combination, consisting in each set of four distinct lenses, mounted in solid rackwork mounts, with pinion and double-milled heads, and shutters for flashing, or three portrait combination lenses, with racks and pinions, three chamber or high pressure jets, or three blow through jets, two Malden dissolving taps, brass back, supply pipes. Box or case £21 10 0

Extras.—Telescopic draw tubes, £5 10/-

223.—The Dockwra (W. C. Hughes, Kingsland).

Details—Polished mahogany body with mouldings and inlaid panels of walnut; very massive brass fronts; centre lantern is arranged to drop, to secure coincidence of discs when shown from an elevated position; lantern stands 3ft. 6in high on a base 24in × 13½in.; large diameter long focus objectives The lining of Russian iron is kept entirely independent of the wood, with clear air space for ventilation. The back fittings, together with the dissolvers, are connected with brass. Special chamber or mixed jets.

Remarks—With so exact an illustration a description is almost superfluous. This is the lantern which gained the medal at the recent Crystal Palace Exhibition. It was designed by Mr. Collin Dockwra, and made by Mr. Hughes. A feature of it is the size of objective, and the plain brass tubes instead of the usual telescopic. Prices may be obtained on application.

224.—Medland's No. 1 (J B. Medland).

Details.—Polished mahogany body, lined with tin; six doors, with sight-holes, japanned tin stages, relieved with brass fittings; 4in. compound condensers, mounted in brass, and brass rackwork fronts with double pinions and compound achromatic lenses and brass O G sliding tubes; dome and rose top, the top lantern made to detach £13 10 0

225.—Medland's No. 2 (J B Medland).

Details.—Same as 223, but having brass stages and front, with brass bound sight-holes to the doors and brass rail handles fixed on each side of the body; brass shutters to the rack fronts, and diaphragm for rolling curtain effect £21 0 0

226.—Medland's No. 3 (J. B. Medland).

Details.—Same as 224, but with the addition of telescopic fronts, with three drawers extending to 12in., and brass shutters to the rack fronts £25 0 0

227.—Newton's (Newton & Co.)

Details.—Polished mahogany body; three pairs of best quality 4in condensers, and double combination achromatic front lenses of large diameter, mounted in polished brass single-draw telescopic adjusting fronts, with rack-and-pinion focussing tubes, on French-polished mahogany bodies, lined with tin to prevent heating. The brass fronts are secured to the mahogany bodies by brass bolts and nuts, so that they may be removed; with one double and one single dissolving tap. In cabinet case.. £25 0 0

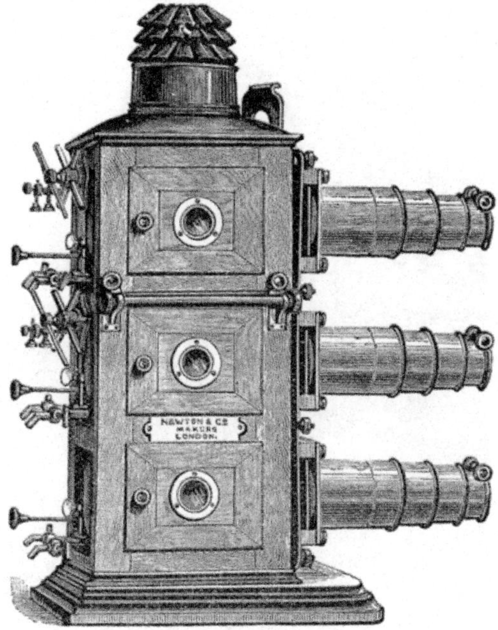

No. 228

228.—Newton s (Newton & Co)

Details —Polished mahogany body, with six doors; three pairs 4in. condensers, and triple combination achromatic front lenses of large size, enabling the exhibitor to show a picture of either 6ft., 12ft., or 18ft. diameter at the same distance from the screen; the diameter of the picture can be increased to 25ft. or 30ft ; the lenses are mounted in polished brass telescopic three-draw adjusting fronts, with rack-and-pinion focussing tubes, on French-polished mahogany bodies, lined with tin to prevent heating; the brass work is secured to the mahogany bodies by brass bolts and nuts, that it may be removed at pleasure, three oxy-hydrogen safety jets, with one double and one single dissolving gas tap and tubes, etc. In cabinet case £42 0 0

Extras.—To either 227 or 228, diaphragm shutter for rolling curtain effect, 21/-

229.—Noakes's No. 1 (David Noakes & Son).

Details.—Polished mahogany body, lined with tin, windows in doors, mounted in brass flanges, brass handles to doors, 4in. compound condensers, double combination achromatic fronts, with rack and pinion, jet trays, and stout travelling case £13 0 0

230.—Noakes's No. 2.

Details —Same as 229, but with brass fronts and plates, rolling curtain diaphragm, and the Noakes improved adjustable stages. In stout travelling case £19 0 0

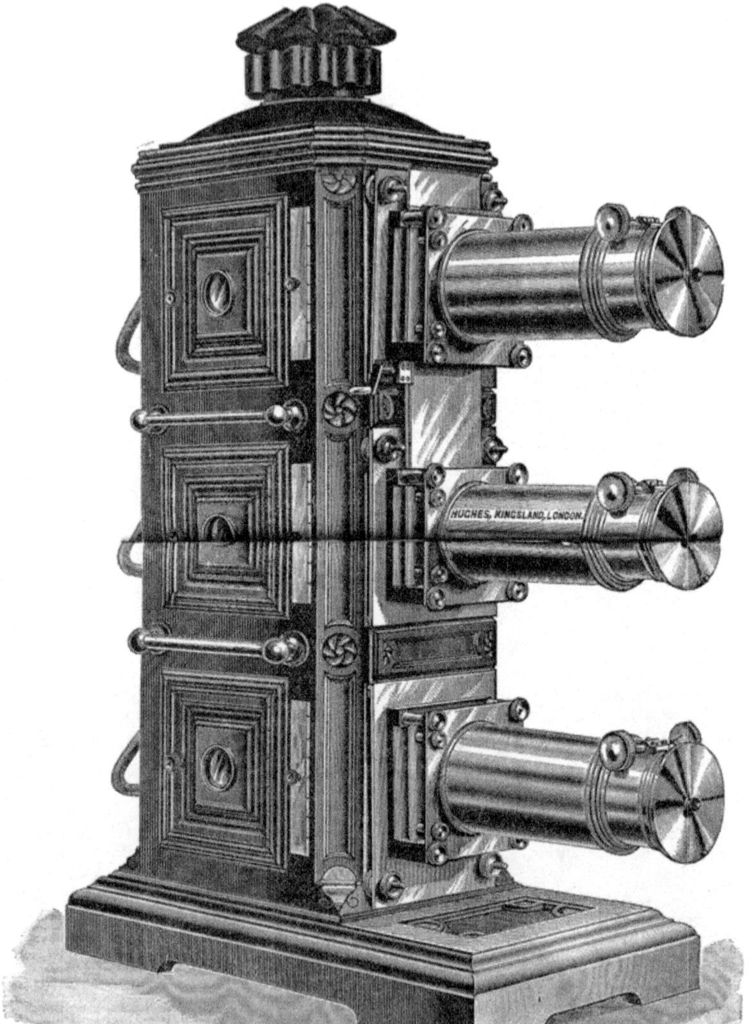

No. 223.

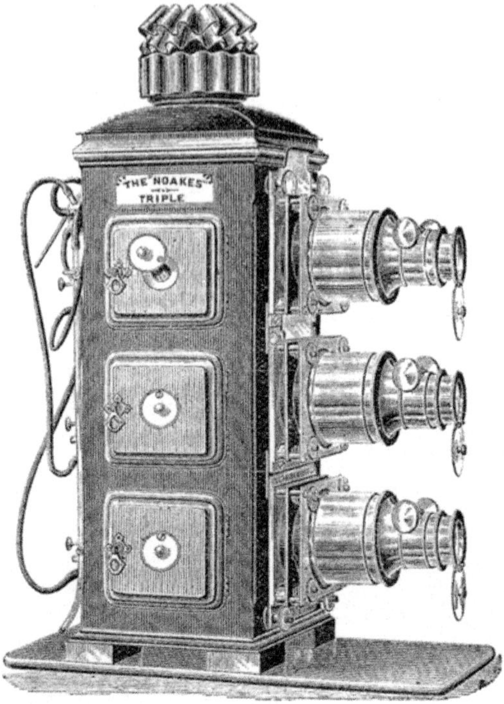

No. 231.

231—The Noakes (David Noakes & Son).

Details.—Mahogany body, moulded doors, ornamental handles, tinted glass sight holes, with brass covering shutters; fronts entirely of brass with telescopic tubes; rolling curtain diaphragms, adjustable register stops; stages are fitted with the Noakes adjustment, and are light tight, yet open all round; arrangement for making slides all one size; improved jet clamp, the Noakes triple dissolver (see "Dissolvers"), 4in compound condensers The front objectives are mounted in an improved mount, by which one set is very readily changed for another (see "Lenses"), three sets, with large back combination of 4, 6, and 8in. focus, giving a 20ft disc at about 20, 30, and 50ft respectively. Three best jets of mixed gas. In stout travelling case £50 0 0

Extras —Case, forming stand and cupboard with shelf when in use, but which when packed will contain all the accessories, 25/-

Remarks.—The features of this apparatus we may briefly describe. Messrs. Noakes believe in registering each lantern by a special system rather than by the use of special apparatus This they do by their own method, which they treat as a secret. So that once done, the registration of effect slides for the future is perfectly reliable. Much, in fact all, of the charm of dissolving effects vanishes if the slides do not correctly register Another feature is the easy changing of front lens already referred to. The fronts really are very light-tight—in fact, one gentleman said that the only fault he had to find was that he could not read his lecture by the rays of light from the stages, as he had been accustomed to The stages also work most noiselessly. The lantern is very neat in appearance, and is altogether a first-class instrument, with which the very best work can be done.

232.—Ottway's No. 1 (Ottway & Son).

Details.—Body of polished mahogany, 4in condensers, the fronts and stages entirely of polished brass, achromatic double combination focus lenses, rack and pinion adjustment fly-shutters, curtain diaphragm; the top lantern can be detached to use separately, fitted with one single and one double dissolving tap. In cabinet case £25 0 0

Extras—Base-board and fittings for using top lantern separately for special effects, or for microscope, &c, with extra dome and rose top, 25/- Three doors on *each* side, 21/- Three brass adapter tubes for long focus, 18/-

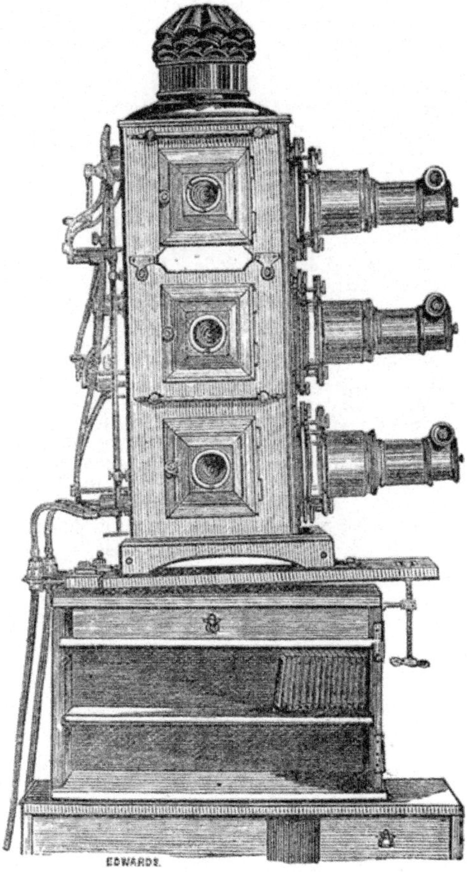

No. 233.

233.—Ottway's High Class (Ottway & Son).

Details—Polished Spanish mahogany body; six rosewood panelled doors and mouldings, fitted with brass cells and coloured glasses; two pairs brass rail handles to lift by; fronts and stages entirely of finished brass; three sliding telescopic adjustments for focussing to various distances; four sets achromatic focus lenses; 4in. condensers, sliding brass diaphragm for curtain effects; top lantern removes and fits on to second base boards; three best oxy-hydrogen jets, fitted with mitre wheels (for turning lime), platinum points, &c. The jets fix on to turned steel pins, fitted in stout iron plates, sliding in dovetails; dissolvers, brass supply tubes. Packed in strong case, which also forms stand.. .. .. £40 0 0

Extras—4½in. condensers, the lantern also fitted with adjustable metal masks, producing an exact coincidence of the three discs, and correcting the irregularity of matting in slides; also the new tinting effect, by means of which a graduated tint slide is passed through the lantern between the condensers and light, producing the most natural sunset and night effect with any ordinary slide £5 0 0

Remarks.—Certainly a high-class triple at a moderate price. The tinter openings in the body of lantern are a great improvement. Further details of these, as also notes on the objectives, rack curtain shutter and jet supports, will be found under their respective headings.

234.—The Chatham (Chatham Pexton).

Details.—Spanish mahogany body, lined with iron, six panelled doors, fitted with brass rings, containing coloured glasses; brass rods and trays, black japanned rose top, light excluders at back; 4in. double condensers, mounted in brass cells and fitted with bayonet catches to prevent their falling out when travelling; the fronts of solid brass; telescopic draw tubes, extending out 14 inches from fronts, three sets of achromatic lenses, enabling exhibitor to show pictures at five different distances from the screen; double pinion knobs and flashing shutters to each front, stops for registering slides and effects, sliding diaphragm for rolling curtain effects, three best high pressure jets or safety jets, with cog-wheel lime turners and platinum tips; one six-way and one four-way dissolver with by-pass taps, brass tubes at back, all connected up with best red rubber tubing ready for use, four massive handles, two each side In specially-constructed padded case, which forms stand, and has a tilting board to regulate the lantern to any angle, top lantern removes, and is fitted to separate base board of polished mahogany, for use with microscope, &c., &c., and enables the other part to be used as a biunial £60 0 0

Extras.—4½in. condensers, and extra large front lenses, tubes, &c., £6.

Remarks.—Mr. Pexton, as an old hand, should know what kind of lantern to turn out. The Chatham shows that he evidently does know.

235.—The Optimus (Perken, Son & Rayment).

Details.—Polished mahogany body, lined tin, with hinged doors, fitted with brass cells and tinted glasses; 4in. compound condensers in brass cell; compound achromatic front lenses, with rack adjustment for focussing, top lantern can be removed and used separately £9 18 0

Extras.—Brass stages, fronts and hand-rails of finished brass, £6 12/6. Brass telescopic draw tubes, and three sets achromatic front lenses, giving focal range from 16 to 120 feet, £6 15/-

Remarks.—A brief description only is required, the illustration supplying the other details. Though possessing no very special features in style of design, it does possess one, viz., extremely moderate price for so good an instrument.

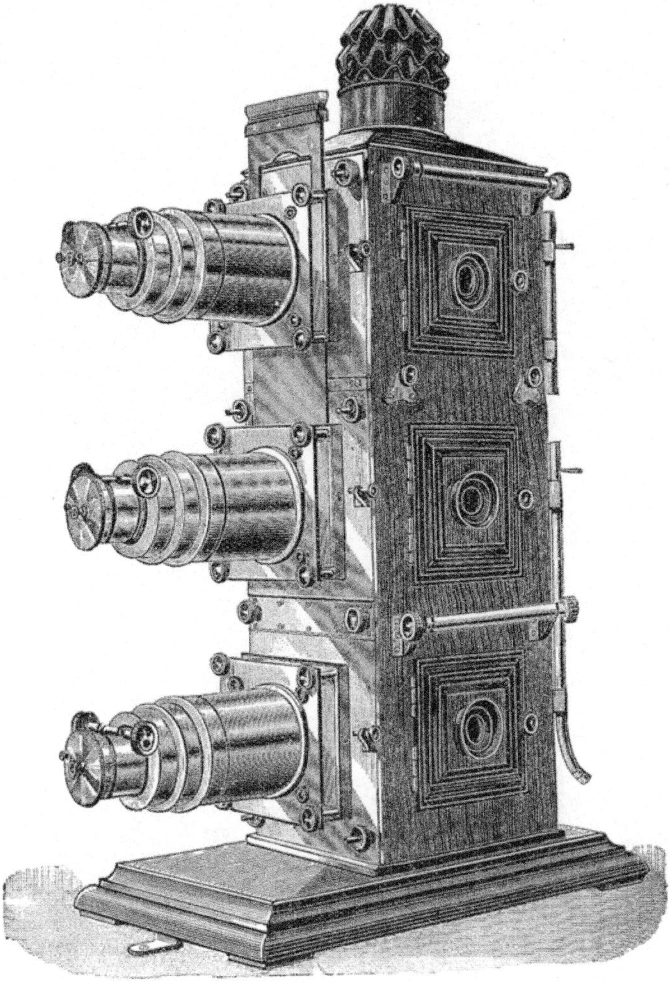

No. 234.

236.—The Bridgman No. 1 (J. H. Steward).

Details—Polished mahogany body, lined on the best principle to avoid heat; three doors on each side, with tinted sight holes protected by brass flange covers, the adjusting plates, stages, and entire fronts of brass; telescopic draw tubes extending from three to twelve inches; sets of achromatic front lenses for varying distances (six); 4¼in. condensers; rack and pinion to front lenses for adjustment; brass flashing shutters; brass

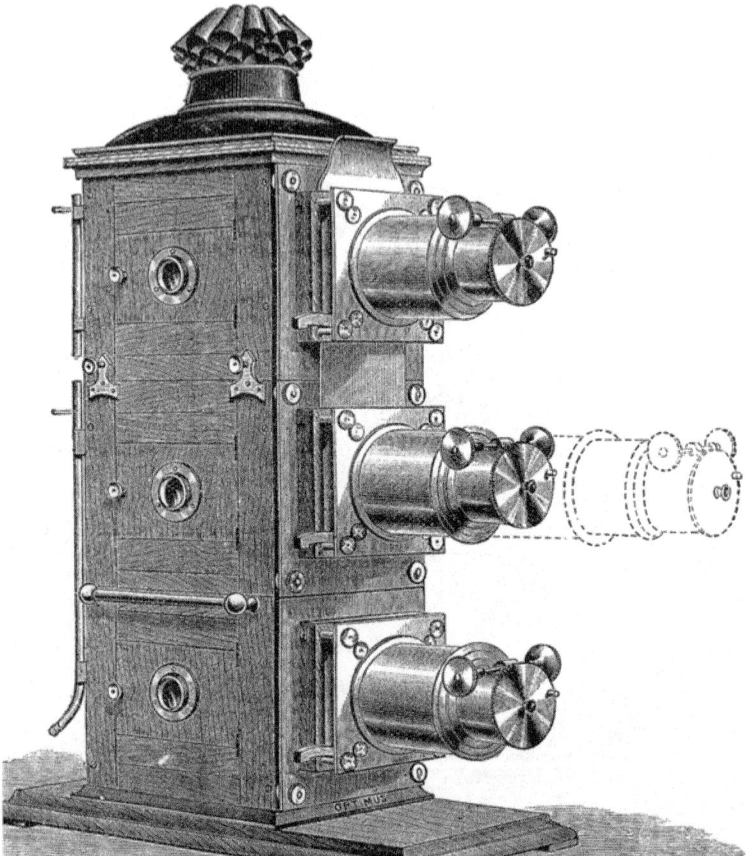

No. 235.

rail handles, the mahogany baseboard is fitted with two hinges, which allow of the apparatus being placed at the right angle. Iron-bound case, forming stand. Combination jets, with lime-turning arrangements, brass diaphragm for curtain effect, register stops for stages; metal runners for slides; two dissolvers. £55 0 0

Remarks.—We are sorry that we could not obtain an illustration of this excellent apparatus, as plain descriptions must of necessity be somewhat stereotyped. It is, however, an all-round first-class triple, with a most business-like look about it.

237.—The Bridgman No. 2 (J. H. Steward).

Details—Same as 233, but with 4in. condensers, the mahogany body being proportionately smaller £50 10 0

238.—The Registered Bridgman (J. H. Steward).

Details—Mahogany body, lined with tin; brass adjusting plates; stages, with openings at top as well as sides, and one draw front tube; 5in. condensing lenses, of three foci; rose top chimney. In case .. £25 0 0

Extras—Base board and fittings for using top lantern of triple as an independent lantern for accessory apparatus or special effects, and extra dome and rose top, 25/- Stop-cocks and brass fittings for shutting off gas from top lantern, 21/- Three brass lengthening tubes for long focus lenses, telescopic, each 11/6. Register stops for slides, 4/6 Improved Universal gas dissolver and fittings, 31/6. Gas dissolver for top lantern, 25/- Brass shutters to fronts, and extra large front lenses, of three foci, each 14/- Three doors *each* side, 21/- Extra packings and fittings to case, and tubings, 15/- Mechanical rolling curtain shutter, 21/-

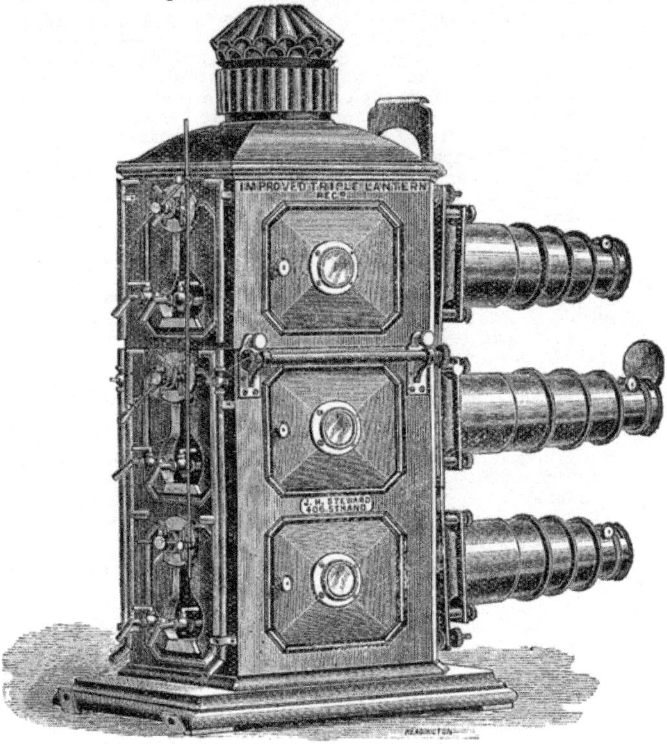

No. 239.

239.—The Malden (J. H. Steward).

Details.—Handsome mahogany body, three panelled rosewood doors on each side, and ebonised mouldings to same. The back is cut and moulded to receive the jets and dissolvers. The doors have brass cells, with tinted glasses let in, and brass bar handles are fitted for lifting the apparatus by. The fronts entirely of brass, excepting the mahogany block containing the condenser, which is bolted to the stage plates by brass pins and screws. The plates are made to adjust up or down, so that the angle may be correct, to allow of all three discs coinciding, and when in position clamping flanges prevent them shifting. The curtain and statuary shutter is made of brass, and is kept in proper position by stops, etc. It has also an adjustment, so that it can be used at various angles and distances from the screen. The stage is open all round for the admission of effect slides, and the slide runners resting on the bottom pillars are removable at will. The spring plate is actuated by spiral springs, and holds the sides firmly in position. Registering stops are fitted on reverse side to stage to stop the slides when central, and will turn up when long panoramic slides are used. Telescopic focussing tubes of three brass sliding draws, beside the outer brass body tube, and extend from four to eleven inches. The jacket containing the front lens is moved by rack and pinion, and the front is arranged to receive tinted glasses just behind the flashing shutters. 4¼in. compound condensers; the front lenses are a new combination, for medium and short focus use; an achromatic lens is used in front, and a correcting lens at the back of the combination, and for extreme long focus, or for experimental work, the front lens is removed, and one, two, or three achromatics used in interchanging cells fitted in place of the corrector at back. In this way any focus can be obtained from four inches up to twelve inches, or, in other words, a perfect disc can be produced without changing the position of the apparatus, of diameters varying from six feet to twenty-five feet. The top lantern is made to remove and fit on to a separate base board, so as to form an independent lantern for special effects or for optical attachments, and then the lower two form a perfect biunial lantern; the body is lined to secure the best possible ventilation, and the dome is made with a flush ledge to the opening so as to economise space when the crinkle top is removed and put in the case; this also enables the packing to be more secure. The case is iron-bound at the corner and edges, plates are let in the top and sides so as to receive the hinged plates of base board of lantern, forming an adjustable stand for the apparatus; the jets are of the improved combination or interchanging form, and have a power of balance adjustment and clamping, so that they may be raised and lowered and moved backwards and forwards in the most accurate manner, making it an impossibility for the light to move after once it has been centred, the dissolving arrangement can either be Mr. Malden's or the Chadwick-Steward —if the former, it consists of a universal dissolver, with cross handles so as to work from either side, and a four-way dissolver, with cross handle for the effect lantern £65 0 0

Extras.—The Chadwick-Steward dissolving system, for particulars of which see "Dissolvers," £5.

Remarks.—We have given an extended description because a lantern used by such an experienced man as Mr. B. J. Malden deserves it. We will content ourselves now, however, with quoting from a letter of his, in which he says "that it has been in constant use for nine years, has travelled about 40,000 miles, and appeared in public over 1,500 times."

240.—Steward's Portable (J. H. Steward).

Details.—This is much the same as 236, but has been designed to occupy as small a space as possible, consistent with perfect ventilation, and convenience in manipulation. It has several points of novelty, viz., the lenses are made to pull out from the front of rackwork tube, so that fresh ones can be rapidly fitted during the entertainment, and thus reduce or increase the size of an effect, figure or statue at will. There are brass doors on each side between the ordinary doors and the brass front, so that a long

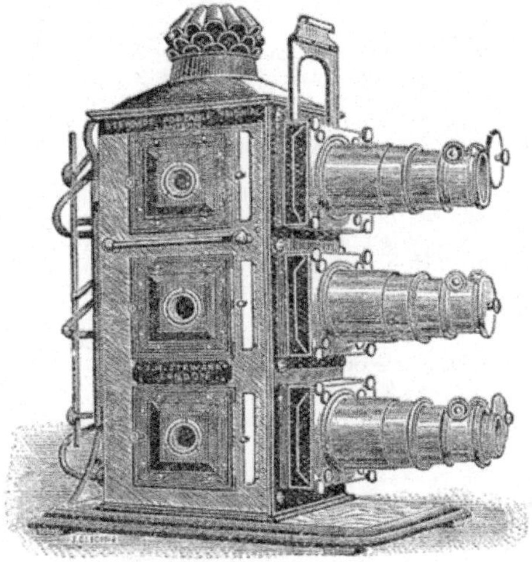

No. 240.

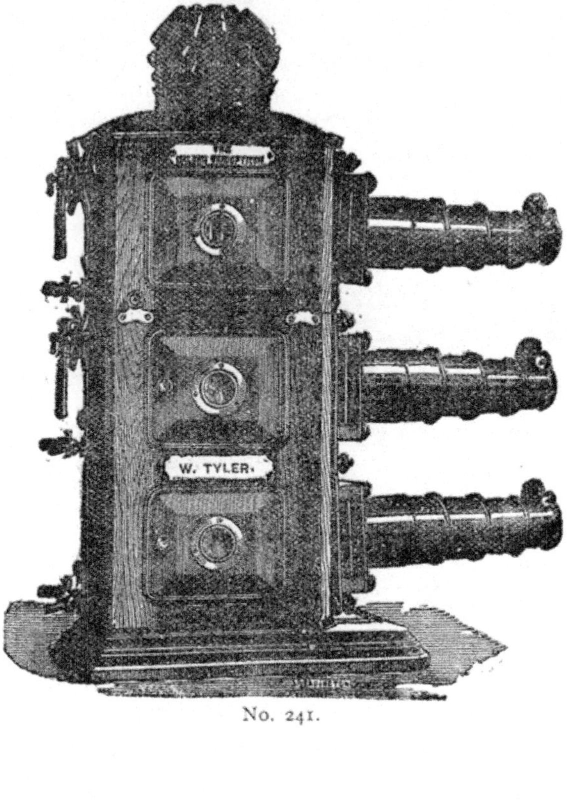

No. 241.

slide with various tints painted on it can be inserted between the light and the condenser to produce on the screen, when showing a plain photograph or other suitable slide, gradations of tone from daylight to sunset and moonlight. As this colour is out of focus of the lenses, the tints are harmonious and pleasing. The jets are fitted with improved bevelled wheel vertically adjusting lime support, and can be rotated from outside. The condensers are 4in., and the front lenses are after the style of best portrait lenses, and give by combination the same size disc at three useful distances from the screen. The adjusting plates, stages, and tubes (two-draw telescopic) are all superior finished bright brass Three gas dissolvers (four way), with connecting rods, so that any one or all three can be raised or lowered at will, independently or together. Cabinet case, with lock and key £52 10 0

Remarks.—A neat and handy little triple, yet possessing most of the advantages of the larger ones.

241.—Tyler's No. 1 (W. Tyler).

Details.—Polished mahogany body, lined with tin; six doors, with sight holes; 4in. compound condensers, mounted in brass; japanned tin stages, relieved with brass fittings; brass rackwork fronts with double pinions, compound achromatic front lenses; japanned sliding tubes; dome and rose top; top lantern detachable £12 10 0

Extras.—Stages and entire front made in brass, brass bound sight holes to doors, brass rail handles on each side of body, brass flashing shutters, £6. Three draw brass telescopic fronts, extending to 12in., £4 10/- Brass shutter for rolling curtain effect, 10/- Sliding trays, brass, bronzed, with two uprights and two brass clamping screws to keep the jets in place and perfectly rigid, can be supplied in lieu of ordinary trays, at 3/6 each. Brass lengthening tubes, for screwing into nozzle of lantern, 2in., 4/6, 4in., 5/6; 6in, 6/6, 8in., 8/6. Achromatic lenses mounted in brass cell, any focus, 7/6. Compound achromatic lenses, 12/6. Cabinet cases, 15/-

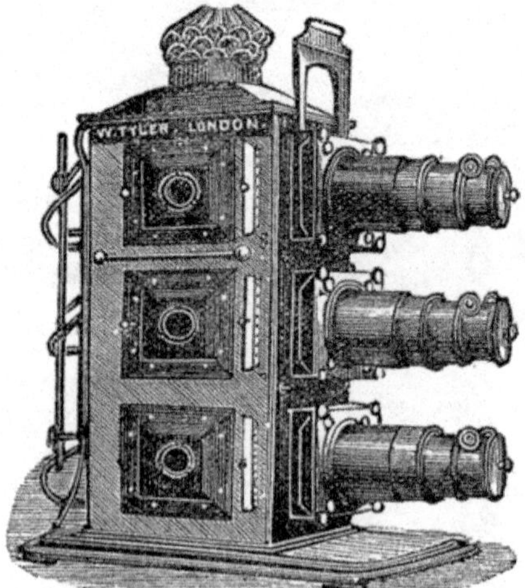

No. 242.

242.—Tyler's High Class (W. Tyler)

Details—Mahogany body, six rosewood panelled doors, also six small brass doors for using the new tinting slides behind condenser and for better ventilation; entire brass fronts; telescopic draw tubes, achromatic front lenses, with double racks and pinions and extra sets of lenses (for showing the same size picture at nine different distances); 4½in. best condensers; three Malden dissolving taps with bye passes (or the new triple dissolving tap); Malden rolling curtain effect, fitted with rack-work motion, interchangeable jets, which can be used either for blow-through or mixed gases. The top lantern can be detached so that they can be used either as single, biunial, or triple lanterns. Cabinet case £48 0 0

Remarks.—Not only a handsome instrument, but a very powerful one. Our readers will note the size of condensers.

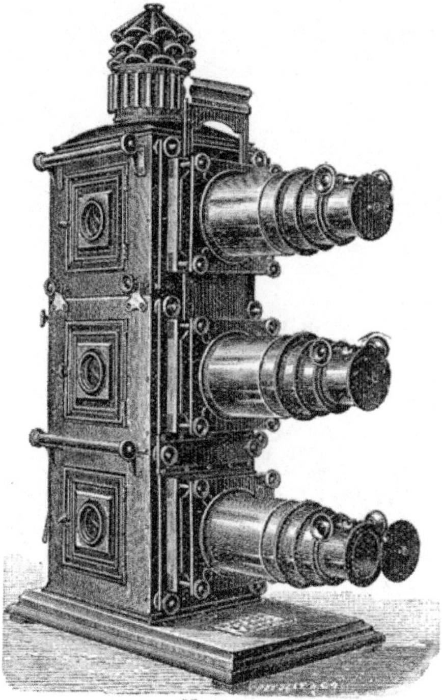

No. 243

243.—Watson's No. 1 (W. Watson & Sons).

Details.—Mahogany body, metal lined, 4in. condensers; achromatic front lenses; brass fronts and lengthening tubes; screw adjustments for centring discs; doors on both sides; top lantern removable. In travelling case £15 15 0

244.—Watson's No. 2 (W. WATSON & SONS).

Details.—Polished mahogany body, six doors, each fitted with sight holes of darkly-tinted glass; whole of the front of polished brass, 4in. double condensers, achromatic front lenses, with rack and pinion adjustment; lengthening tubes of brass; top lantern removable. Stout travelling case £22 0 0

245.—Watson's No. 3 (W. WATSON & SONS).

Details.—Body of polished mahogany, lined throughout with metal; front, stages, &c., entirely of brass; 4in condensers; three sets achromatic front lenses of different focal length, brass telescopic draw tubes; rack and double pinions to front lenses; six doors, each fitted with brass flanged sight holes; stops and screws to slide holders; diaphragm shutter in brass, for rolling curtain effects; brass supply tubes on back of lantern; brass rail handles for lifting purposes; top lantern removable, extra stand being supplied to use it as a single. Strong travelling case £35 0 0

Extras.—4½in. condensers, £5.

Remarks.—We noticed a double arrangement of metal in the dome and rose top, which ought to make this handsome triunial very cool to work. We can also speak most highly of the rigidity of the telescopic tubes; they are above the average in this respect.

246.—Wilkinson's High-class (WILKINSON & Co.)

Details.—Polished Spanish mahogany body with six rosewood panelled doors and mouldings, fitted with brass cells and coloured glasses, two pairs brass rail handles to lift it by; fronts and stages entirely of finished brass, three sliding adjustments for focussing to various distances; twelve achromatic focus lenses; best 4in. condensers; top lantern removes and fits on to second base board; three best oxy-hydrogen jets of improved and very solid construction, fitted with mitre wheels (for turning lime); platinum points, etc; the jets fix on to turned steel pins, fitted in stout iron plates, sliding in dovetails; one dissolver only, fixed on brass plate which hinges forward and clamps, so as to come under the eye of the operator; sliding brass diaphragm for curtain effects. Packed in strong case, which also forms stand £45 0 0

Extras.—The new panoramic tinter, with six small brass doors to admit it, etc.; new manipulating slide box and drawers containing lenses, screw elevator, etc. Price, packed in two strong cases, with double locks, which form together a stand high enough without any table or other contrivance being necessary.. .. .. £15 0 0

Remarks.—The above is a really first-rate production, though coming rom a town out of the ordinary way. The firm are celebrated for their slide-painting, and are also well known about the district as exhibitors. We have not seen a more convenient stand than the one (included in the extras) they issue. With drawer, shelves, and a long tilting screw, it must be comfort itself to work. Perhaps the feature attracting our attention the most was the small but most useful one of having the dissolver at the side instead of the back. This makes the lantern much easier to work. The dissolver is fixed on to a hinged brass which comes forward, and is then fixed firmly by clamps. Altogether the triple is a credit to the firm.

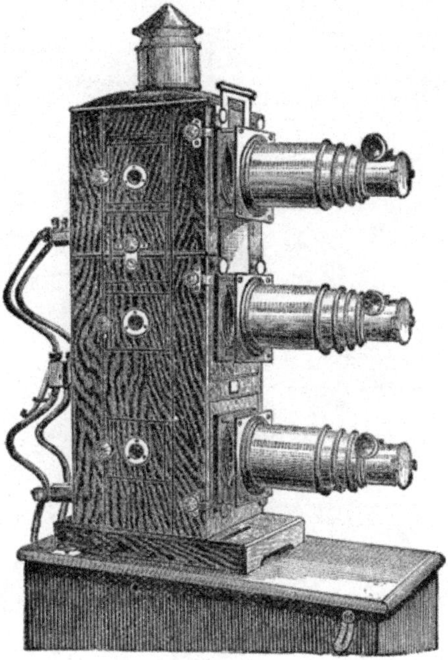

No 247.

247.—The Lecturer's (E. G. Wood).

Details.—Mahogany body, 4½in condensers; triple achromatic objectives, forming six different powers, enabling pictures to be shown at distances between 10 and 120 feet; brass telescopic fronts, brass diaphragm for rolling curtain effect, top lantern removable. Packed in cabinet .. £35 0 0

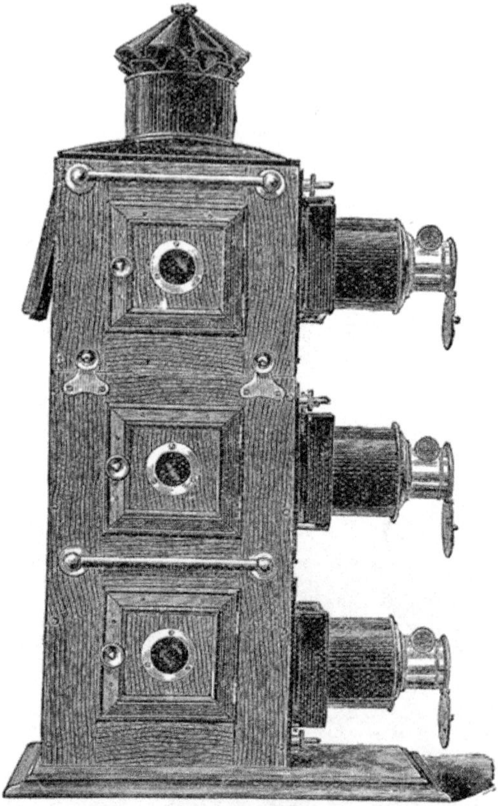

No. 248.

248.—The No. 1. (Wholesale Firm).

Details.—Polished mahogany body with moulded foot and six panelled doors, with brass-bound sight holes, brass rail handles on each side of biunial portion, and also to the top lantern, lined with japanned tin, brass trays, with double supports and binding screws for fixing the jets firmly; japanned tin dome and rose top and japanned tin stages (open at top for chemical experiments); 4in plano-convex compound condensers mounted in brass cells and double combination achromatic front lenses mounted in brass (with large size lenses to the back combination), with rack and pinion adjustment, and sliding shutters or flashers. The top lantern is made to detach and an extra moulded foot is supplied, so that the lantern can be used as a single lantern, a biunial, or a triunial. The top lantern is arranged so as to take a three or four-wick Russian iron lamp to burn paraffin oil, and an extra dome is supplied so that it can be used as a single oil lantern £12 12 0

Extras.—Stained and polished case, lock and key, and handles, 22/6.

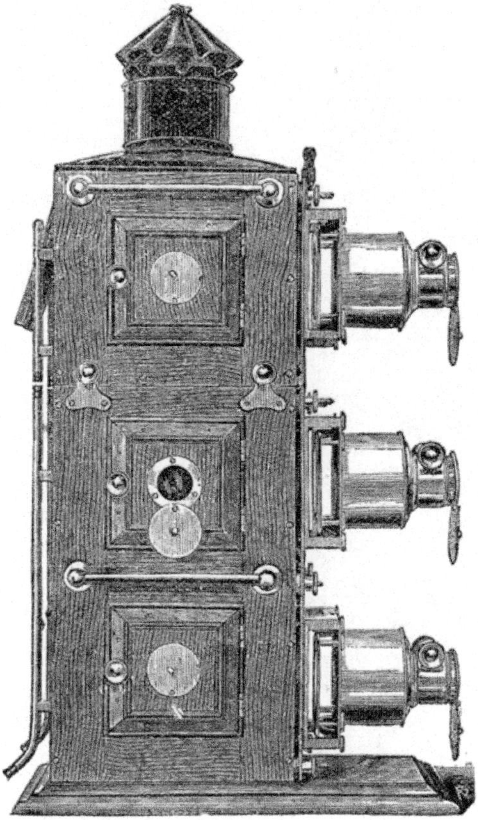

No. 249.

249.—The No. 2 (WHOLESALE FIRM).

Details.—Polished mahogany body with moulded foot; six panelled doors with brass-bound sight-holes, with sliding shutters and brass rail handles on each side to the biunial portion, and also to the top lantern; lined with japanned tin, brass trays, with double supports, and binding screws for fixing the jets firmly; japanned tin dome; rose top and stages and front tubes, made entirely of brass, and with brass rolling curtain effect; 4in. plano-convex compound condensers mounted in brass cells, and double combination achromatic front lenses in brass mounts (with large size lenses to the back combination), and double pinions to the rack adjustment, and fitted with sliding shutters or flashers, having grooves to carry shapes of coloured glass for tinting slides, brass tube supply pipes for connecting the gases to the dissolving taps The top lantern is made to detach, and an extra moulded foot is supplied so that the lantern can be used as a single lantern, a biunial, or a triunial The top lantern is arranged so as to take a three or four-wick Russian iron lamp to burn paraffin oil, and an extra dome is supplied so that the lantern can be used as a single oil lantern £18 0 0

Extras.—Case, same as 248.

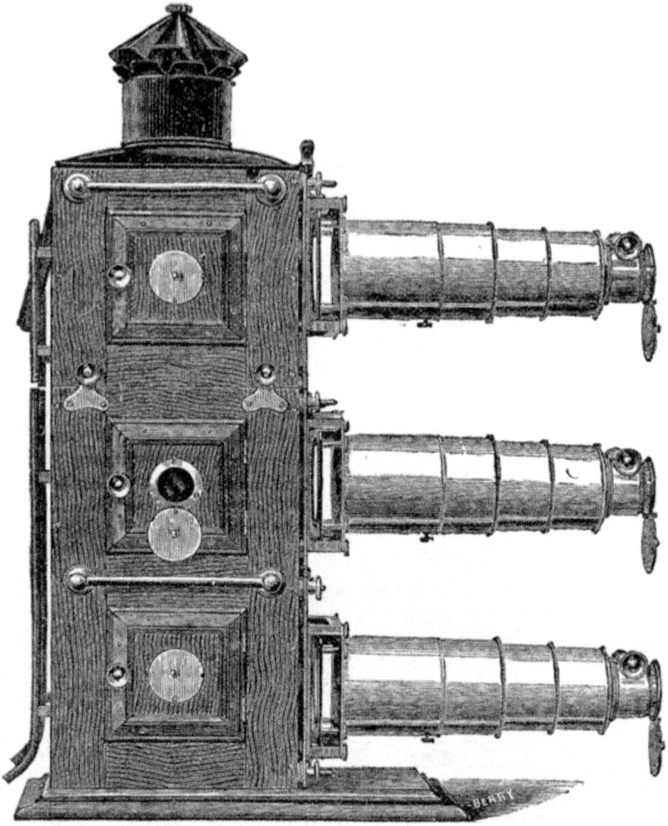

No 250

250.—The No 3 (Wholesale Firm)

Details.—Same as 246, but the front tubes have three telescopic draws extending to 12 inches, made entirely of brass, and with brass rolling curtain effect; 4in. plano-convex and Meniscus condensers in brass cells; double pinion rack fronts fitted with sliding shutters or flashers, having grooves to carry shapes of coloured glass for tinting slides; four sets of best single achromatic lenses, mounted in brass cells (6in., 8in., 10in., and 12in. foci) made to screw together or to be used separately.

Extras.—Case, same as 248. Rack fronts, with long and short focus double combination achromatic lenses, as described amongst the objectives, in place of the rack fronts and single achromatic lenses, £7 17/-

251.—The Triple Rack (Wholesale Firm).

Details.—This triple is exactly the same pattern as the biunial of same (No. 206), but of course with the addition of a third lantern on top. This extra lantern is made to detach, and an extra base-board is supplied, so that it can be used as a single lantern. Extra Russian iron dome and rose top, for use

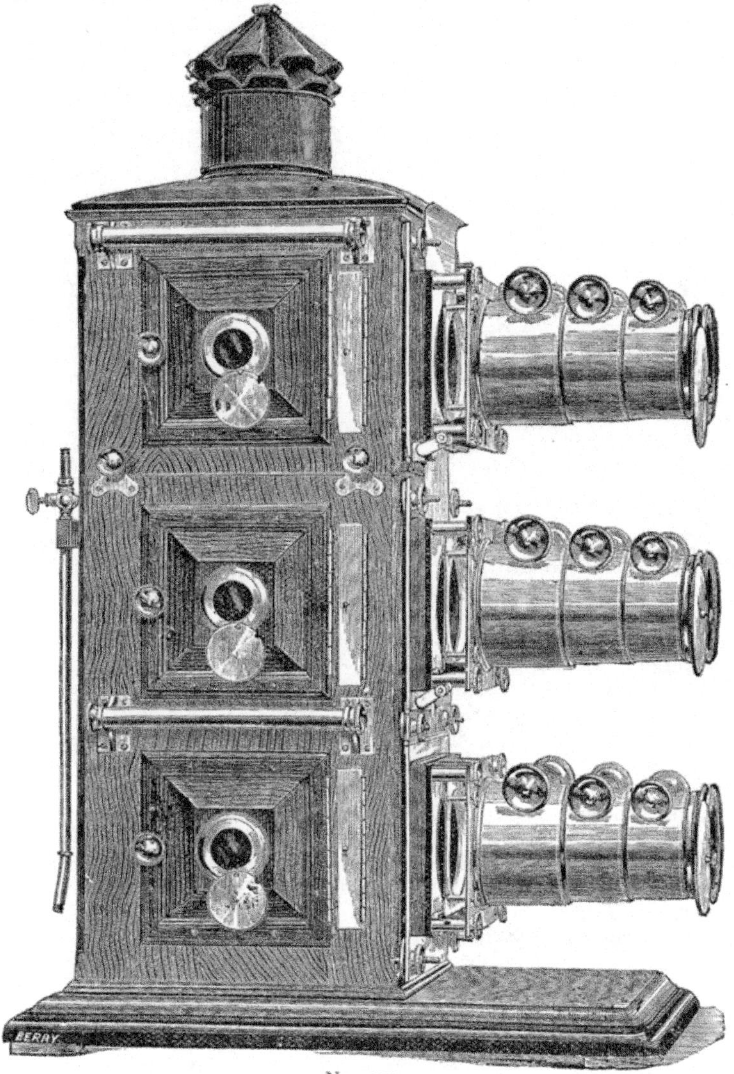

No. 251.

when the biunial portion only is required. Top lantern has a pair of handsome brass rail handles, in addition to those on the biunial portion, and the back of the lantern is fitted with brass tube supply pipes, for connecting the gases with the dissolving taps. The three sets of four double combination achromatic lenses are the same size and focus as those sent with the biunial, and are mounted in the same way with sliding shutters or flashers. Strong cabinet case, with locks and keys and handles £82 10 0

PART III.

Special and Scientific Lanterns.

OF special and scientific lanterns there are not a great variety, the demand for them not being great, as nearly everything which the average exhibitor wishes can be done with the lanterns of ordinary construction. For scenic effects on a large scale, the limelight lends itself admirably, and it is, as a consequence, largely utilised by those whose business it is to cater for the public by spectacular display. The light which plays around the fairies in theatrical entertainments is cast by the lantern, and the changes of tint in the light effected by the use of colour screens; the illumination of fountains, waterfalls, &c., such as proved so attractive a feature of the Kensington Exhibition illuminations, is obtained by throwing the light—either white or tinted—from a limelight lantern on to the water to be illuminated; ghostly figures in theatrical representation are likewise obtained by the use of the lantern, and many other effects of a like nature, but, of course, the use of a lantern for such purposes is of a rougher character than that ordinarily met with, and the rough handling likely to be undergone would do a high-class instrument no good, whilst, as in such cases, the lantern itself not being seen by the audience, its appearance need not be considered at all. Several special lanterns for rough work of this kind will be found described in the following pages, whilst the purposes of the different varieties of scientific lanterns are noted with each; suffice it to say, that by their use a great variety of very interesting experiments can be made and illustrated to a large audience. A glass tank may be inserted in the slide stage, and the various visible changes which take place by the chemical combination of various solutions exhibited on the screen, of course on a highly magnified scale; then, by the union of the microscope with the lantern, microscopic slides and subjects of any kind may be thrown upon the screen—a flea 10 or 15 feet long being, for instance, something not seen every day, and the magnification to such a degree affords ample opportunity for critical study. The—to the unaided eye—invisible forms of pond life—that world in a drop of water—may be seen to the best advantage through the microscopic lantern, and, in like manner, anything which may be placed in the live-box of the microscope, or those sections of animal and vegetable tissues, &c., in which microscopists take so much delight, exhibited in all their wonderful forms and beautiful colours to a large audience in a far clearer manner than when seen through the microscope only. By the use of the reflecting prism, objects which are not of a transparent nature may also be cast magnified and brilliantly illuminated upon the screen—a photographic print, a picture, a coin or medal, a watch showing the works in motion, or any other small article that may be desired—and by a little special modification of apparatus the optical lantern may be used by the scientific lecturer to illustrate to his audience in the clearest and most interesting manner a great variety of the phenomena met with in the various sciences.

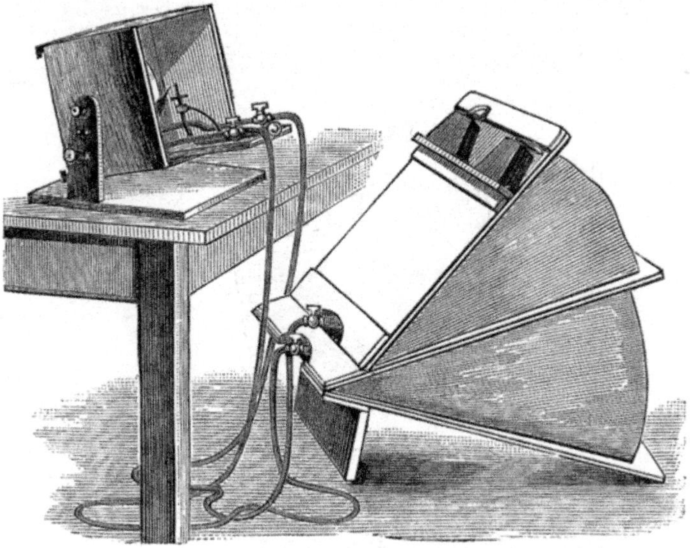

252.

252.—Dale's Light Box (H. Dale & Co., Limited).

Details.—Best form, lined with iron, when in use swings on two screws, a third clamping it, if required, in any position; grooves for coloured glasses; special condenser, which is fitted in panel which slides in a second groove to the coloured glasses; for travelling, the screws are put into extra holes, allowing the box to rest on the base.

Box, without condenser or jet	£0	12	6
" with 4½in. condenser	1	7	0
" " 6in. "	2	0	0
" " 4½in. " but no grooves	1	1	0

Extras.—Special form of mixed jet for use with this box, 15/- Better quality, as in illustration, 22/-

Remarks.—This is the same pattern light-box as used in the majority of theatres. It is very effective for amateur use, when used for *tableaux vivants*, amateur theatricals, fancy dress balls, &c. Having thoroughly tried it ourselves we can testify to its power of light, and convenience in use. Mr. Dale has had a long experience of this special limelight work, and knows, therefore, how to combine ease in manipulation with power and effect. And, further, the prices are very low.

253.—Dale's Scenic Lantern (H. Dale & Co., Limited).

Details.—This is an apparatus for throwing moving water, clouds, advancing or retreating figures, words, or other things, on to a cloth or sheet. The lantern itself resembles an ordinary shaped oxy-hydro lantern lined with iron, but is made much stronger to bear rough usage; fitted with 6in. condensers, very short focus objective, because behind the flats on a stage there is often only 8 to 12 feet distance. The pictures are usually painted 5½in. diameter and blacked in, as, for instance, witches, fairies, or phantom ships, &c., to appear on a dark cloud. The price includes a best jet .. £6 15 0

Remarks.—This apparatus is possibly only adapted for its particular work—a work, too, that but few have to do with. Still it is a lantern somebody, from somewhere, at some time may desire details of, so they are given.

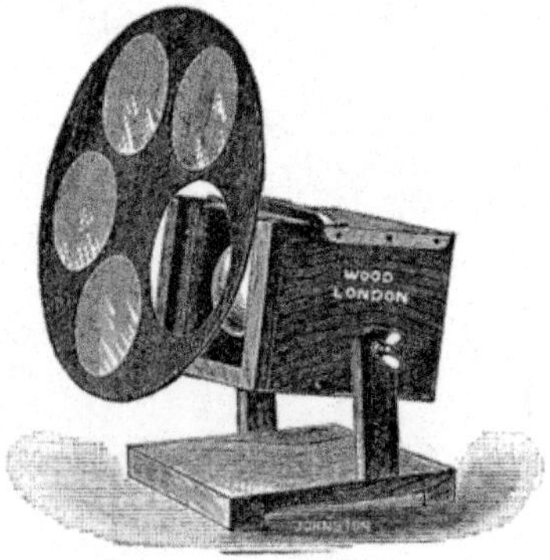

No. 254.

254.—Wood's Illuminating Lantern (E. G. Wood).

Details.—A convenient box with revolving diaphragm or colour disc; 6in. lens; but without jet £4 4 0

Remarks.—The revolving disc saves a deal of trouble, though it adds to the expense somewhat.

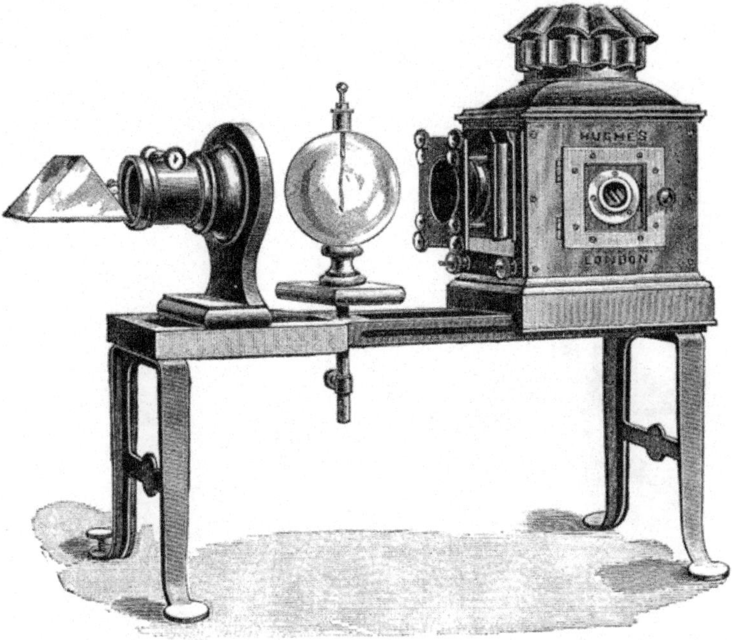

No. 255.

255.—Scientific and Experimental Lantern No. 1 (W. C. HUGHES, Kingsland).

Details.—Single lantern, polished mahogany body panelled and grooved, with inner linings of iron; brass sliding trays of a special character for holding jet, brass front with full open top, that can be removed, 4in. double condensers in brass screw cells, for throwing a parallel beam; long focus objective of large diameter; special chamber jet, with platina point, Pamphengos lamp; polished mahogany platform on bronzed feet; experimental registering table, with fittings, an extra pair of best double condensers of shorter focus, and an extra high-class shorter focus front objective, rack and double pinion for use when large pictures are desired nearer the screen.
£42 0 0

256.—Scientific and Experimental Lantern No. 2 (W. C. HUGHES, Kingsland).

Details.—Mahogany bodied lantern; 4in. double condensers of best make, brass fronts that can be removed; high-class new objective, rack-work and double pinion, best Pamphengos lamp; ordinary chamber jet, or the best blow through oxy-hydrogen jet, stand, registering table, and fittings erecting prism £22 10 0

Remarks.—Lanterns for the projection of chemical, electrical, magnetic, and other scientific apparatus on the screen. The front lens and the body of the lantern are each separately movable along the stage or platform arranged with this object. A registering table runs in the centre of the platform, for supporting the various experimental apparatus which are placed upon it. An erecting prism in front renders the objects on the screen upright, and this prism can be adjusted to any angle. It will thus throw the picture in various directions, and can be mounted on a stand to slide along platform. A further advantage is that mechanical or other slides can be shown in the ordinary way, and the lantern stage in turn entirely dispensed with by an ingenious arrangement when physical apparatus is used. Altogether capital lanterns for science teachers or lecturers.

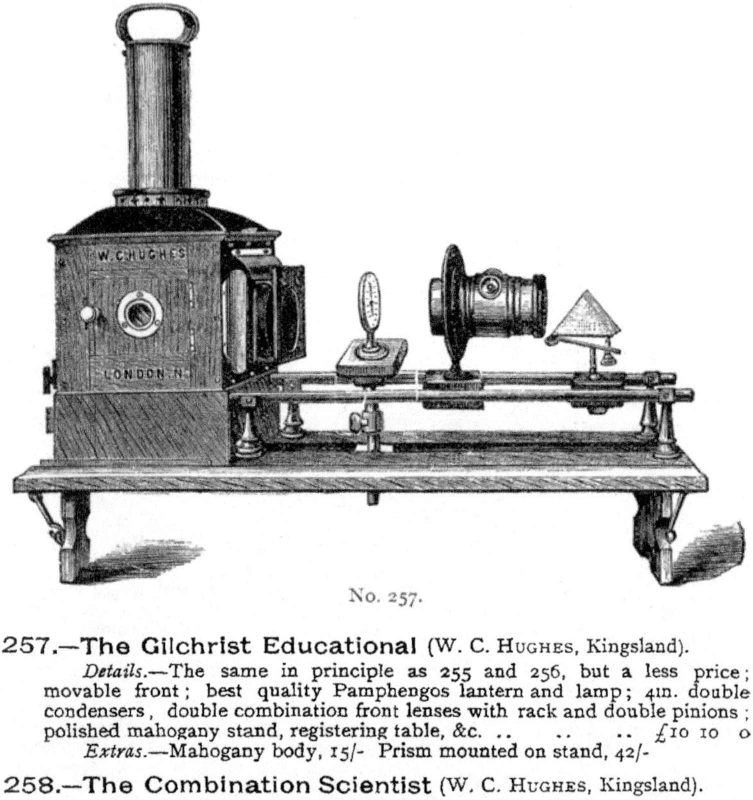

No. 257.

257.—The Gilchrist Educational (W. C. Hughes, Kingsland).

Details.—The same in principle as 255 and 256, but a less price; movable front; best quality Pamphengos lantern and lamp; 4in. double condensers, double combination front lenses with rack and double pinions; polished mahogany stand, registering table, &c. £10 10 0

Extras.—Mahogany body, 15/- Prism mounted on stand, 42/-

258.—The Combination Scientist (W. C. Hughes, Kingsland).

Details — Body of mahogany, hexagon shape, lined with best Russian iron; it revolves on a mahogany base board, three feet square with broad moulding round. This revolving action is obtained by means of metal rollers, and the counterpoise also with roller action is constructed that the balance of points is equalised. There is, therefore, no undue strain owing to weight in front. A spring click registers each front of the body when arrived at its required position. The first side is fitted with the Gilchrist lantern (see 257), upon the condensers of which is a best form of adjustable spectre slide, which is removable without becoming detached. This enables the decomposition of white light and the various spectrum experiments to be shown. The second side is supplied with a microscopic attachment as shown in 259 The remaining side is the ordinary optical lantern of transparencies. The price includes special oxy-hydro jet £78 15 0

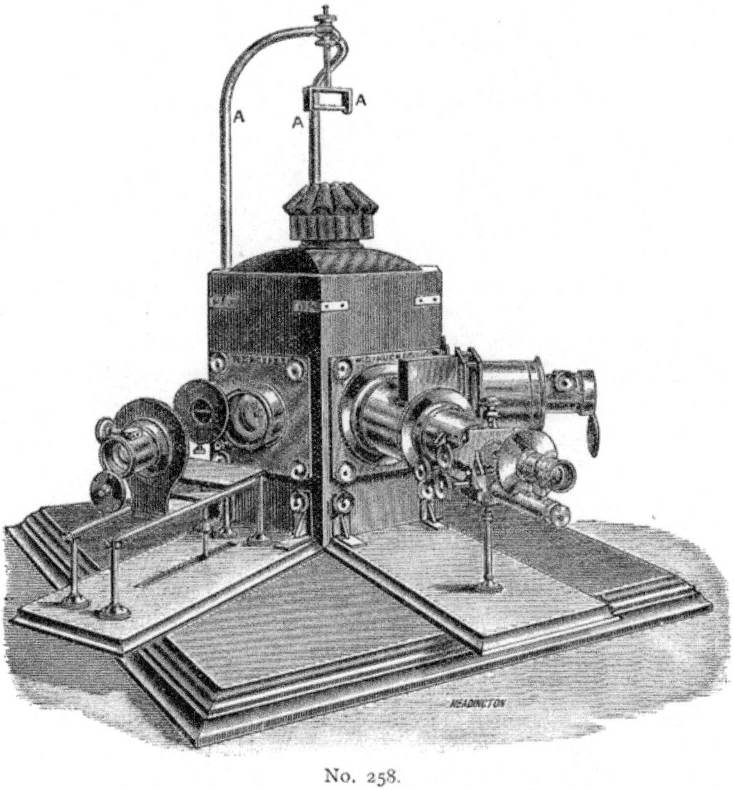

No. 258.

Remarks.—A remarkable novelty, which ought to satisfy the professional lecturer, either for theatre or class room purposes, the great advantage being gained of rapidly throwing and changing upon the screen the general view of an object, the mocroscopic portion of the same enlarged, and if required as in chemistry, for instance, the actual experiment in operation. The workmanship throughout is of the highest order.

259.—The Microscope (W. C. HUGHES, Kingsland).

Details.—Polished mahogany body; special arrangement of triplet condensers, 5in. diameter; solid brass front, bolted through body; best mixed jet with lime turner; alum trough; mechanical stage, having ¾in. vertical and horizontal motion, by which any portion of the object can be brought into the field by means of milled heads, without the operator placing his finger on the object rack; adjustment for the sub-stage condenser, with the standard 1½in. fitting; coarse rack and fine screw focussing adjustments, adapter with R.M.S. standard screw; one sub-stage condenser; amplifying lens. In case without objective £35 10 0

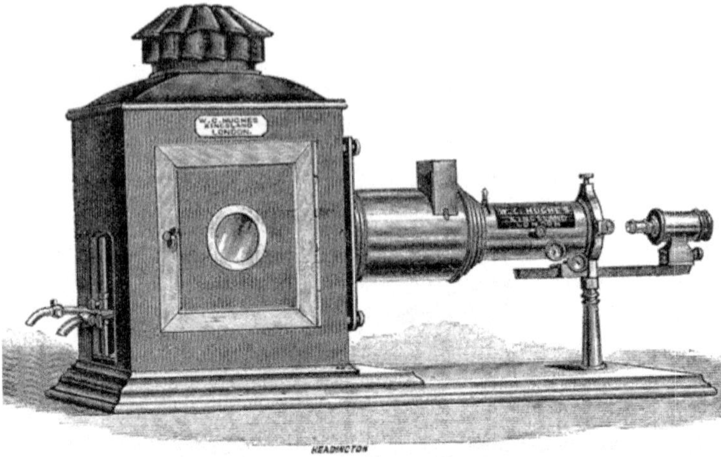

No. 259.

Remarks.—To some, the lantern microscope possesses a peculiar fascination; the foregoing is a convenient and very powerful lantern and attachment for the purpose.

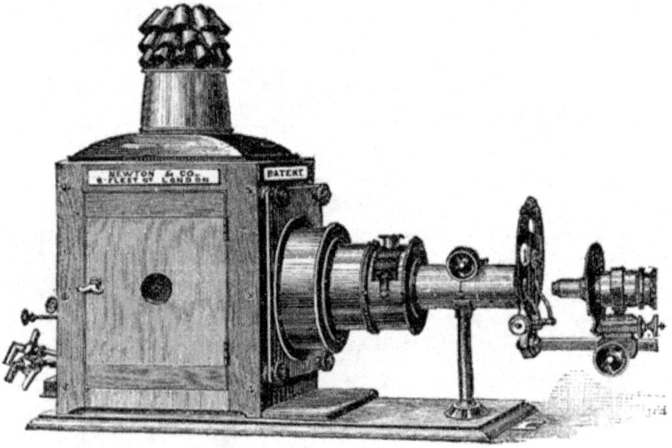

No. 260.

260.—Wright & Newton's Microscope (Newton & Co.)

Details.—Mahogany body, lined with tin; special triple wide angle primary 5in. condensers; solid brass front bolted through body, top and chimney, and lime tray; best mixed jet with lime turning movement. Two cases £25 0 0

Extras.—To use this lantern in the ordinary way an additional brass front is required. With 4in. double condenser, achromatic front lens with rack and pinion focussing adjustment, £5 10/-

Remarks.—Mr. Lewis Wright has, in conjunction with Messrs Newton, produced this excellent apparatus. We quote from the firm's list the following notes as to its powers :—"Ample light is obtained for the magnification of ordinarily transparent subjects to 1,250 diameters, which will display in a clear and beautiful manner all the parts of insects, the minute details of anatomical sections, vegetable tissue, &c. A flea may be shown on the screen, 15 feet long, quite as sharply, and almost as brilliantly as a magic lantern slide; the proboscis of a blow fly is easily displayed with the various powers from 8 to 14 feet long; and all the details of an insect's eye are shown in section clearly. The circulation of the blood is easily displayed in the veins of a living frog, and pond life is, of course, shown without any difficulty. Geological sections are admirably shown either by ordinary or polarised light."

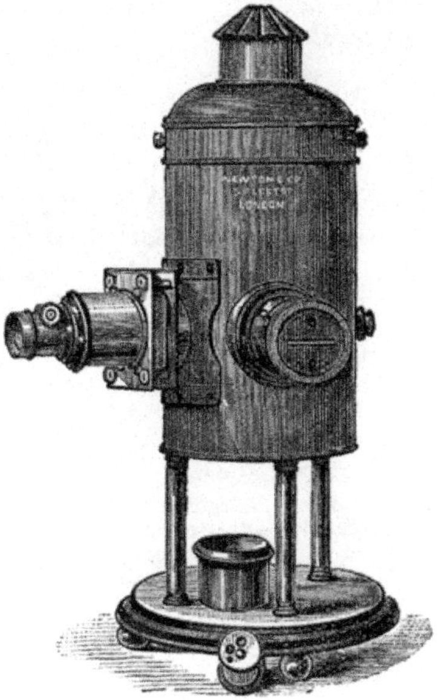

No. 261.

261.—Newton's Electric Scientific (Newton & Co.)

Details.—Cylindrical bronzed brass body on brass pillars and mahogany stand, with two pairs of best quality 4in. condensers; socket for condenser for parallel beam work, revolving diaphragm; adjusting slit front, brass optical front fitted with improved double achromatic long focus front lens with rackwork adjustment for focussing, for showing diagrams, &c., and gas jet fitted inside body. The lantern is double-fronted, and so arranged as to exhibit both spectra and diagram on the screen at the same time without moving the lantern £15 15 0

262.—Newton's Experimental (Newton & Co.)

Details.—Polished mahogany body and base-board, lined with tin, the entire front made of polished brass, bolted through the mahogany body, fitted with double plano-convex condenser, and double combination achromatic front lens of large diameter, with rack and pinion adjustment for focussing, and Newton's four-wick "refulgent" lamp. In polished case, with lock and key £9 9 0

Remarks.—A feature of this lantern is that all the fittings and screws are of standard sizes throughout, so that all ordinary spectroscopic, polariscopic, and other apparatus will fit without alteration It will exhibit diagrams, &c., and work all experiments as well as any oil-lit lantern made. If preferred, it can be fitted with long focus limelight lens with top and tray (without jet) in place of the lamp and oil-light lens at the same price.

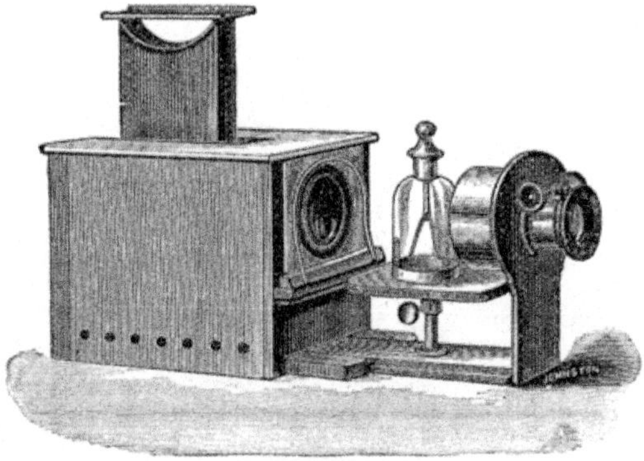

No. 263.

263.—Wood's Projection (E. G. Wood).

Details.—Mahogany body; 4in. double condensers; achromatic objective with rack and pinion focussing adjustment £6 6 0

Remarks.—Designed to supply science teachers and lecturers with a lantern that will exhibit ordinary slides, and also small pieces of scientific apparatus, at a moderate cost.

Section II.

Apparatus pertaining to Lanterns.

PART I.

Illuminants.

AS will have been gleaned by the preceding chapters, the principal illuminants used in the optical lantern are mineral oil—petroleum—and the limelight. Amongst the oil lights petroleum has a practical monopoly, the other oils being used in little else than the cheap toy lanterns; and in the use of mineral oil but little real difference exists, the principal aim being to combine the lights of several wicks into as intense and concentrated a flame as possible, this being usually done by placing the long wicks parallel with and sloping towards each other, so that by a proper regulation of the ventilation the flames converge and unite in one brilliant light, the principal exception to this arrangement being the "Euphaneron," in which the parallel arrangement of wicks is departed from and the fourwick chambers are arranged in the form of a W.

Of the limelight there are four systems, the oxy-calcium or spirit jet, the blow-through, the mixed, and the ether-oxygen or eth oxo light, all of which depend for their illuminating properties on the incandescence of a piece of lime by the impact upon its surface of a jet of oxygen passed through or mixed with a hydrogen or other flame to maintain combustion, for oxygen by itself will not maintain a flame. It will thus be seen that for all these lights the making of oxygen—or the purchase of it ready made—is the first necessity, and this we will treat of fully in a future chapter. Of the four systems in use the first is, perhaps, the simplest and safest, although the light given by it by no means approaches that of the others, and, consequently, it is not by any means largely used. The oxy-calcium or spirit jet utilises spirits of wine as the flame maintainer, a reservoir containing spirit being attached to the jet in such a manner as to keep up a constant supply of spirit in a small lamp placed just in front of the lime support. This spirit lamp being lit, the oxygen is turned on, and is thrown by the jet through the flame on to the lime, which is lit up with great brilliance. The blow-through jet is precisely similar in its action to the oxy-calcium, save that in place of the spirit flame the flame of a jet of hydrogen is used. As, however, the ordinary coal gas of commerce is a hydrogen compound, and answers the purpose well enough, besides being always ready and handy for use without the trouble of making and the expense of the necessary apparatus, the household gas supply is usually requisitioned by a nipple being unscrewed and the hydrogen tap of the jet connected with it. As will be seen by reference to the illustrations in the following pages this hydrogen pipe is carried to a point just in front of the lime, and the oxygen tube is carried out in front of it and brought round at an angle, so as to cross the lines of flame and blow the oxygen through the hydrogen flame as it emerges from the nozzle. This system gives a light of about 200 candle power, or about one-third greater intensity than that obtained by the spirit jet. It is, moreover, perfectly safe, and although, as will presently be described, the other and more powerful forms of limelight are dangerous in the hands of inexperienced persons, there need be no occasion for apprehension by the most timid persons when this form is being used. As a consequence this is perhaps the most largely patronised system of using the limelight, as it will give full illumination of a 15ft. disc, which is quite as large as the average exhibitor requires. The next system of light, the oxy-hydrogen or mixed jet, is far more powerful, giving a light the intensity of which is at least double that of the last described, from which it differs in that the two gases are mixed in a small chamber an inch or two below the nozzle, through which the mixed gases issue on to the lime as before. The mixture thus formed is highly inflammable and explosive, and danger is present in its use from the fact that, should the pressure upon the gas-bags—a matter which will be fully dealt with in a later chapter—be removed for a second, the mixed gas and flame are sucked back into the bag, and a tremendous explosion caused. There are, it is true, several safety valves in the market, which have as their object the stoppage of the passage of the gas should a suck-back occur, and which act more or less effectually, but for all that the use of the mixed jet requires the greatest care, even in the hands of experienced persons. With this jet, too, a greater pressure is required than with the blow-through, and the use of gas-bags for both hydrogen—or coal gas—and oxygen is virtually a necessity. By its use, however, the best results are obtainable, and for large exhibitions of a high order, where 20 to 30 feet screens are used, the mixed jet is a *sine quâ non*.

The ethoxo limelight utilises the mixed jet in precisely the same way as the last system described, but in place of hydrogen, oxygenated ether is used, a current of oxygen being passed through a "vapouriser" which contains fibre, saturated with ether, the vapour of which, mingling with the oxygen, is conveyed with it through the hydrogen tube of the jet to the mixing chamber, where it unites with a stream of pure oxygen in precisely the same manner as with the oxy-hydrogen system. It gives a beautiful light, fully equal, if not superior, in power to the oxy-hydrogen, when properly manipulated, and entails the use of only one gas, so is useful where coal gas is not readily obtainable, but as, unless the properties and peculiarities of the various ether mixtures used be fully understood, accidents are liable to occur from causes connected with the vapouriser, in addition to the ordinary dangers of the mixed jet, it is not a system which can be recommended for general adoption, and, indeed, is but little used.

As many people do not care for the additional trouble and care necessitated by the use of the mixed jet, except when circumstances compel its use, jets may be purchased provided with interchangeable nozzles, so that the blow-through system may be used for all ordinary displays, and the mixed gases for more imposing exhibitions, without the expense of entirely new jets, an arrangement which will be found very useful.

Whilst speaking on the subject of nipples it may be well to mention that the success of the light depends to a very great extent on the clearness of the aperture from which the gas is emitted, and, as a consequence, the greatest care should be exercised to keep the pipes and nozzles perfectly clean and free from the slightest obstruction or corrosion, to which end the cleaning plug is a useful addition, and if the nipple be made of platinum, gold, or other practically non-oxidisable metal it will be a decided improvement.

The limes which are used for lantern purposes are merely pieces of unslaked lime turned into cylindrical shape and pierced with a central longitudinal hole that they may be placed upon the supporting pins provided for the purpose, though some prefer shaping the limes to a broad disc upon the face of which the jet of heated gas impinges. Moisture or damp will, of course, spoil them, so that it is essential they be kept sealed up in bottles or cases to protect them from the air. Limes can be easily made by shaping a piece in the lathe, or by cutting it out to shape, and if lime is unobtainable, chalk, although not quite so suitable, may be used instead. Limes, however, can, as a rule, be purchased so readily and cheaply that few trouble to make them themselves. The best are made from hard lime, which is a necessity with the oxy-hydrogen jet, though soft limes will do well enough for the oxy-calcium light. The incandescence of the lime is caused by the intense heat of the jet of flame thrown against it, and this same heat very soon burns away the lime at its point of impact, with the result that a small cavity is formed in the lime, which will, if the jet continue to play into it, soon cause a diminution of the light, and also prove a source of danger to the condenser by causing the flame to strike off at a tangent, in which case a very likely consequence would be the cracking of the glass nearest it. On this account the lime has frequently to be moved so as to bring a fresh portion of the surface opposite the jet, and to do this there are several devices for turning the holder on which the lime rests by means of a milled screw on the outside of the lantern, the best systems giving the lime an upward spiral movement, which results in almost every part of the cylinder being utilised by being successively exposed to the action of the flame, this being a great improvement on the double motion system in which one screw raises the lime whilst another rotates it.

As will be seen by reference to the illustrations of the various jets described in the following pages, the two tubes conveying respectively the oxygen and the hydrogen to the nozzle are provided with taps, so that the quantity of each gas can be regulated to a nicety, and as a turning of the wrong tap will give at times somewhat startling results it is essential that some means be adopted of readily distinguishing them. The experienced operator will not, of course, need this, but others will find it useful, and with this end in view, whilst some makers only mark the respective taps O. and H., others make them of different shape, the handle of one being of the usual T shape and that of the other a short lever ; and again, others fit both taps with lever handles and emboss the words OXYGEN and HYDROGEN plainly upon them, the lever handle, too, being a better guide to the direction the turn should be made than the T handle. With these few remarks we leave the reader to the study of the various commercial patterns and special varieties contained in the following pages.

Oil Lamps.

Having described the Sciopticon as the first of the mineral oil lanterns, it follows that the special lamp therein used must also receive first attention.

264.—The Sciopticon (Sciopticon Co.)

In the description of the lantern itself, see No. 43, several details of the lamp appear. Mr. Smith, the improver of this cylindrical lamp, constructs it on the principle of having no air to enter, except that required for perfect combustion, which is regulated by means of the length of the chimney, the same being made telescopic for that purpose Fig. 1 shows the lamp ready for use. The glasses closing the ends are circular, and being equally heated all over are not liable to fracture. Severe tests have been applied, extending over a period so long as twelve months, without any sign of a breakage. They are fitted in cells, and are thus practically air-tight. Fig. 2 shows the lamp open for lighting or trimming. It has two wicks of 1½in. width. The reservoir holds enough oil to last about four hours.

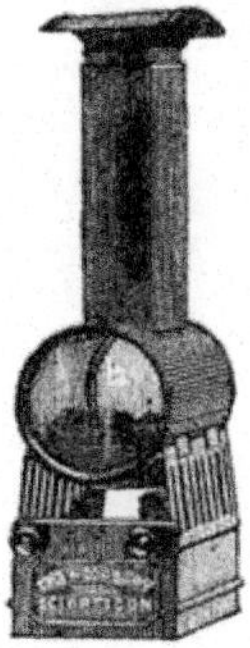

FIG. 1.

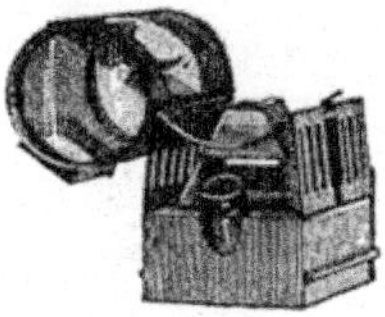

FIG. 2.

No. 264.

Remarks.—We have given this light a good trial, and can assert it to be very powerful. In contrast with some ordinary three and four-wick lamps, it gave a light quite as good, with less heat and chance of smoke. Our experience with it points to the fact that no recent improvements in lamps have yet ousted it from the field.

265.—Archer's Three-wick (ARCHER & SONS).

Details.—This is the lamp they fit to the Excelsior, and is a very good one. It has J jointed iron chimney, toughened glass plates; reflector with blue glass sight-holes £0 15 0

266.—The Photinus (ARCHER & SONS).

Details.—A four-wick lamp with new combustion chamber, new pattern reflector, &c. £2 2 0

267.—The Pamphengos (W. C. HUGHES, Kingsland).

Details—A lamp which has made the lantern in which it is used of world-wide reputation. The burner is composed of four wicks placed parallel with each other and axially in the lantern. They are controlled by pinions projecting behind. The air supply is adjusted in such a manner as to cause the flames from these to be deflected towards the centre, thus forming a rather small flame of exceeding great intensity. The chimney is of metal and somewhat tall, so as to ensure perfect combustion; while the portion immediately surrounding the flame is formed of metal, having a large aperture, both in front and behind, to permit the passage of the light to the condensers on the one hand, and the reflector on the other. These apertures are protected by small curved panes of glass, annealed by special process so as not to be liable to breakage by heat. With full-sized wicks, 2in wide, including metal chimney and all the recent improvements, according to patent and registration £2 2 0

268.—The Triplexicon (W. C. HUGHES, Kingsland).

Details—A lamp that was Mr. Hughes's best until the "Pamphengos" superseded it, but still it gives a capital light, and is equal to the best of three-wick lamps. With full sized curved wicks, two inches wide, including conoidal glass plates and metal chimney £1 5

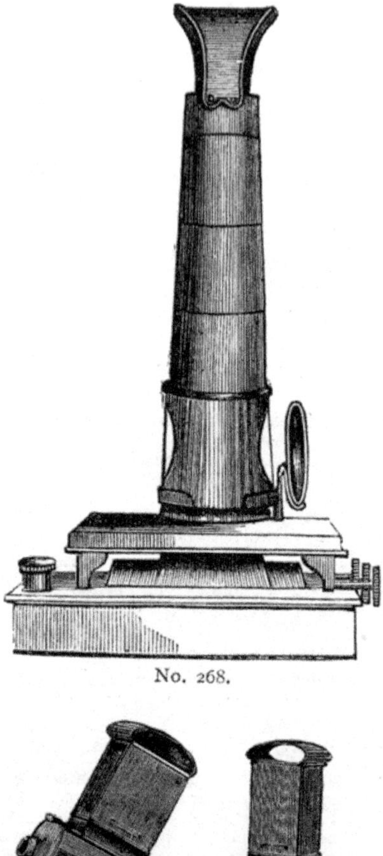

No. 268.

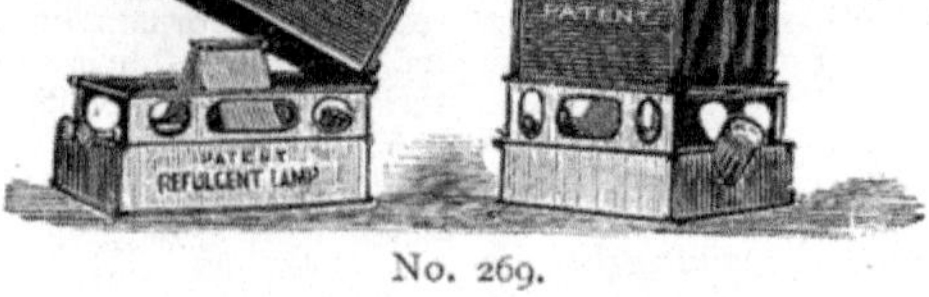

No. 269.

269.—Newton's Refulgent (Newton & Co)

Details —These are made with three or four wicks, each about two inches wide. The end of the hood is formed to carry a square of bent crown glass. These are fitted loosely on the lamp to avoid breakage, and we might note here rather a peculiar coincidence, that Messrs. Newton provide largely for ventilation, a point which in the "Sciopticon" is guarded against. In both cases the reason is to avoid breakage of the glasses, and we must say, as both principles appear to carry it out, we could not place one before the other. These lamps of Messrs. Newton give a very white flame of great power.

Three-wick	£0	16	6
Four-wick	1	0	0

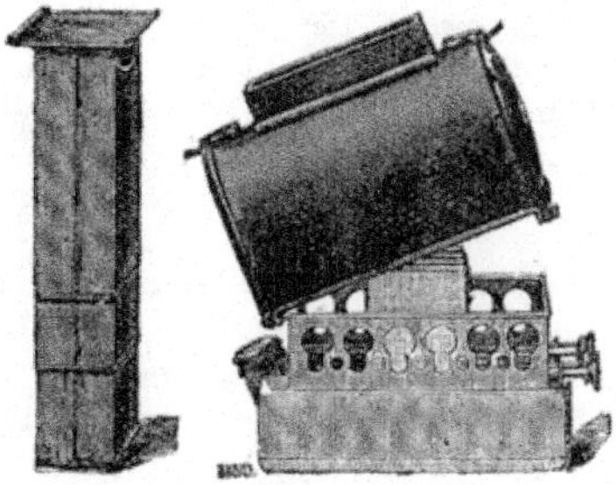

No. 270.

270.—Three and Four Wick Lamps.

The general prices of these are as follows:—

Three-wick, Russian iron, 1½in. wicks ..	£0	13	6
Four-wick " " " " ..	0	16	6
Three wick, better quality, 2in. wicks ..	0	19	6
Four-wick " " " " ..	1	3	6

271.—The Five-wick Parisian (Theobald & Co.)

Details.—Russian iron body, lamp with five wicks, flat chimney 3¼in diameter, with draught shutter. This lamp gives an excellent light £1 10 0

272.—The Euphaneron (E. G. Wood).

Details—The cistern of the lamp forms its base, and is a rectangular tin vessel about 6×4×1in. At the front end is an orifice, closed by a brass cap, for pouring in the oil. The wicks are contained in four rectangular tubes about 2½in. long and 1½in. wide, sloping together as they rise. The four wicks are arranged parallel to each other, but in the form of the letter **W.** The usual milled heads at back adjust the height of each. Round the top of the cistern, and rising 1¼in. above it, is a metal frame on which the combustion chamber is hinged and rests. This chamber is of iron, with a perforated metal base for the supply of air to the wicks. A glass plate, mounted in a copper frame in the front, and a reflector at the back, both of which are hinged for convenience, close the chamber. The chimney is telescopic, with three draws £1 4 0

Remarks—This arrangement of wicks produces a first-class light of a pure white quality, and the lamp is a thoroughly capable one.

Gas Jets for Limelight.

As will be seen in the introductory notes to this part, jets are divided mainly into three classes—the oxy-calcium, called by some the oxy-calcium *lamp;* the blow-through safety, or oxy-calcium *jet,* and the mixed or oxy-hydrogen jet. As the use of the word lamp may perchance mislead, and also because the distinguishing titles we have chosen are the popular ones, we have termed them

The Spirit Jet.
The Blow-through Jet.
The Mixed Jet.

Prices do not vary very much, some extra attachment (such as a lime turner) being usually the cause of differences, nor in actual shape and design does much variety appear. It is more in the methods and movements adopted, or the results claimed, that distinctions can be made.

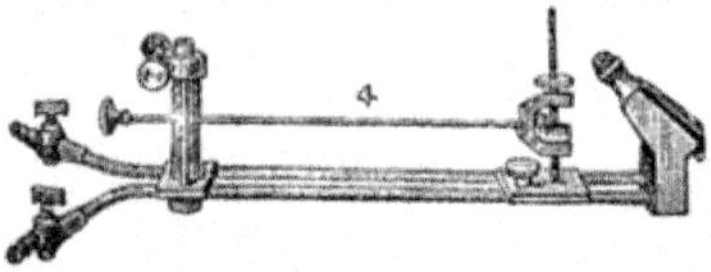

No. 273.

273.—Archer's Blow Through (Archer & Sons).

Cleaning plug at bottom of nozzle, ends of jets turned downwards T taps; single movement lime turner £0 16 0

An extremely good jet at a very reasonable figure.

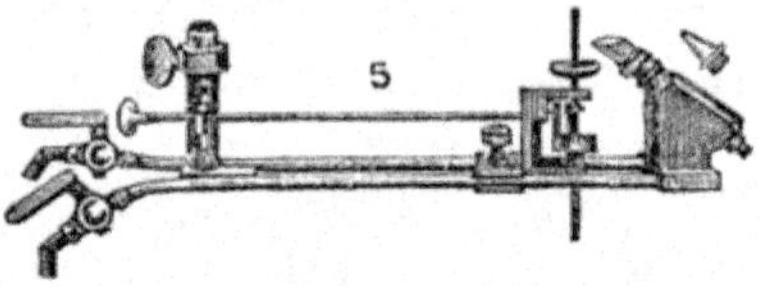

No. 274.

274.—Archer's Interchangeable (Archer & Sons).

Cleaning plug at bottom; handle taps with name of gas engraved, double motion lime turner, interchangeable nozzles for the two forms; all polished brass £1 10 0

275.—Archer's New Jet and Mechanical Stage (Archer & Sons).

The jet used is the same as 274, with a stage and tray all of finished brass; it has two useful features—instead of the usual support, it has a rack-work one, with milled head pinion; this enables the whole jet to be raised or lowered to a nicety and from the outside of lantern. The second feature is another rackwork with pinion, also from the back of the lantern, to move the lime holder nearer to and further from the nozzle, with screws to fix it firmly when once set in position £2 10

Remarks.—This is really one of the best jets we have had the pleasure of trying. Without being cumbersome it gives almost unlimited power over the jet, and that without opening the doors, to plunge the hands into a heated atmosphere, too hot sometimes to conduce to that evenness of temper so compatible with good manners. The cleaning plug referred to was introduced by Messrs. Archer some ten years ago. The illustrations will show its position, and its use is of course to facilitate the cleaning of the nozzle interior.

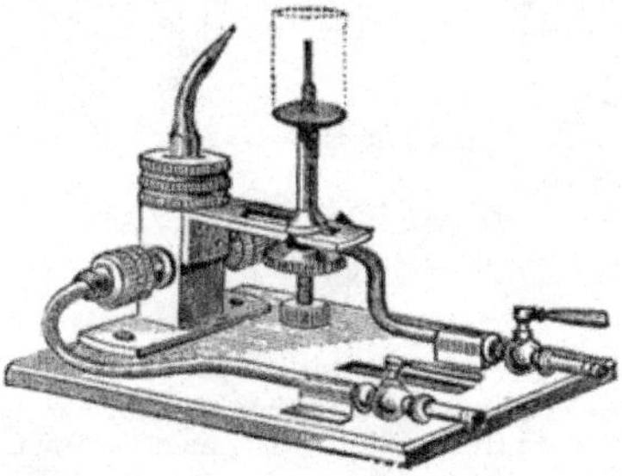

No. 276.

276.—Dale's Mixed

Details.—This is a jet especially adapted for light boxes, screw joints, on mahogany body, with improved revolver and adjustment Very easy to use and simple, at the same time most effective £1 2 0

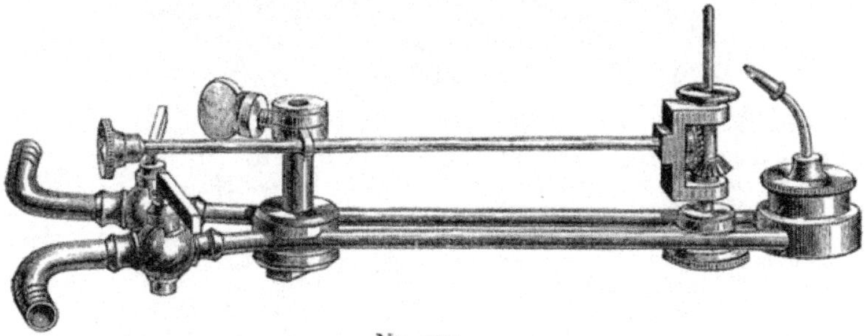

No. 277.

277.—Hughes's Best Mixed (W. C. HUGHES, Kingsland).

Details.—Made in best finished brass, all screwed and cast so that it can be readily taken to pieces; double-motion lime turner; handle taps, ends of tubes turned outwards £1 10 0

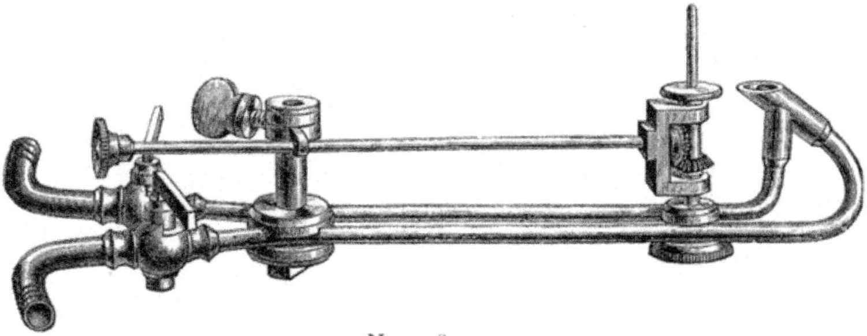

No. 278.

278.—Hughes's Best Blow Through (W. C HUGHES, Kingsland).

Details.—Same in style as No. 277 exactly, but with the different nozzle required £1 10 0

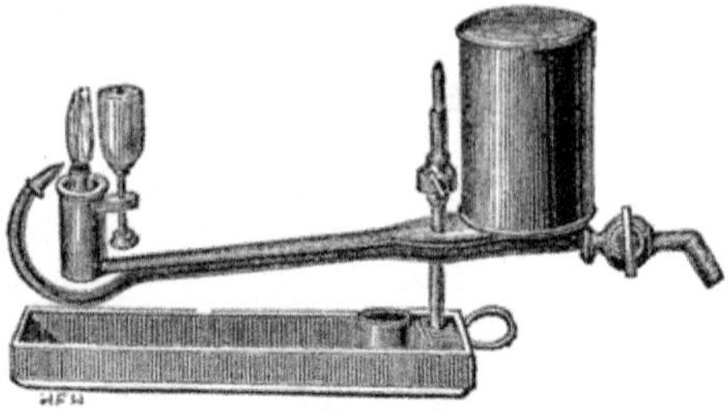

No. 279.

279.—Newton's Spirit (NEWTON & Co.)

Details.—This is constructed so that the spirit reservoir can be placed at an angle, and so avoid the necessity of raising it to the centre of the disc. Platinum nipple £1 1 0

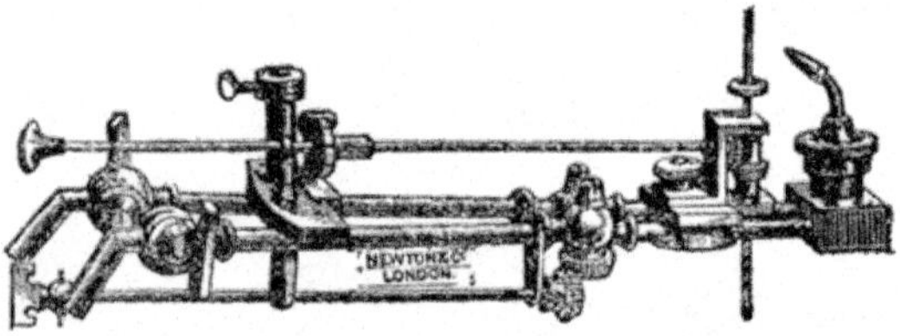

No. 280.

280.—Newton's Best Mixed (NEWTON & Co.)

Details.—Pringle's "cut off" arrangement, see No. 296; check action lime movement, working outside lantern, see No. 291; T taps .. £2 7 6

Remarks.—About as useful and convenient a jet as one could wish to use, the "cut off" arrangement being especially a convenience.

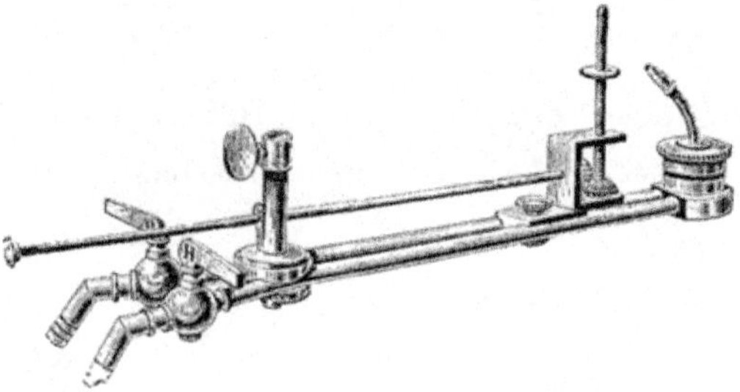

No. 281.

281.—Noakes's Mixed (D. NOAKES & SON).

Details.—Bright brass entirely. A feature of this jet is the chamber. The nozzle is fitted to a ground joint which is gas tight. This obviates the use of a leather washer, which wears out or deteriorates with use. The top part of chamber screws down as a cap. The advantages are (1) that the chamber is easily opened for cleaning, and (2) that the nozzle may be moved a little to either side if required. Double motion lime turner. Handle taps. End of tubes bent outwards £1 10 0

Remarks.—A well finished article, bearing every evidence of care in manufacture.

282.—Noakes's Blow Through (D. NOAKES & SON).

Details.—Entirely of polished brass, double movement lime turner; handle taps; ends of tubes turned outwards. The tubes are kept well apart until the actual lighting point.

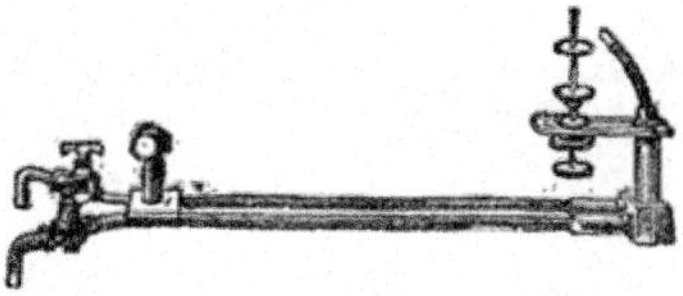

No. 283.

283. Optimus Mixed (PERKEN, SON & RAYMENT).

Details.—Plain substantial jet. T taps £0 18 6

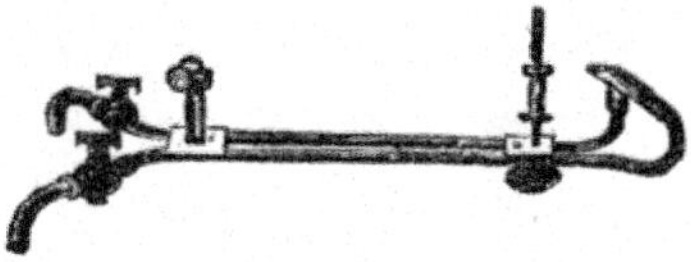

No. 284.

284.—Optimus Blow Through (PERKEN, SON & RAYMENT)

Details.—An ordinary pattern, but at a cheap price £0 12 6

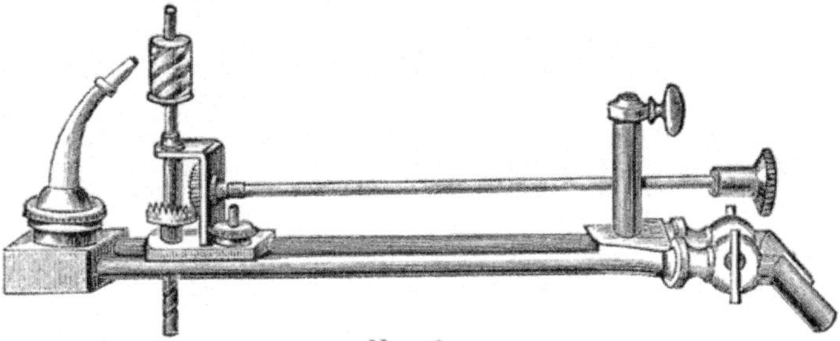

No. 285.

285.—Place's Mixed (J. PLACE).

Details – T taps; lime turner; double motion, with fine rack £0 18 6

Remarks.—A jet we consider extremely cheap. We have not handled one of which the lime turner works more rigidly and yet so smoothly.

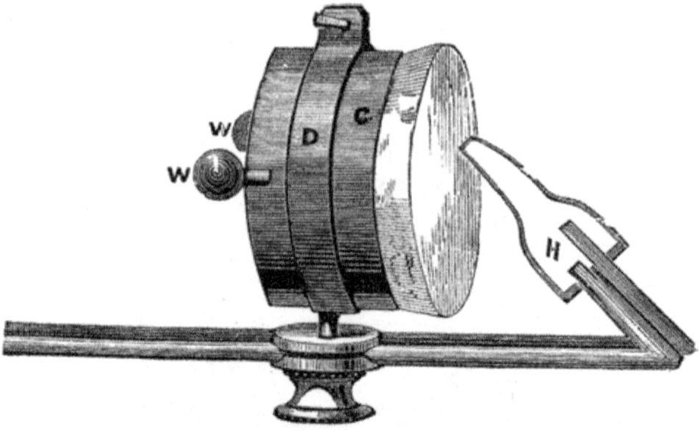

No. 286. Fig. 1.

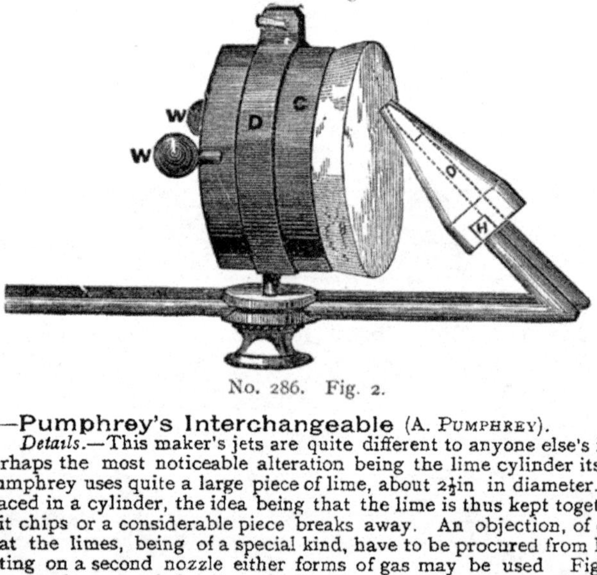

No. 286. Fig. 2.

286.—Pumphrey's Interchangeable (A. PUMPHREY).

Details.—This maker's jets are quite different to anyone else's in design, perhaps the most noticeable alteration being the lime cylinder itself. Mr. Pumphrey uses quite a large piece of lime, about 2½in in diameter. This is placed in a cylinder, the idea being that the lime is thus kept together, even if it chips or a considerable piece breaks away. An objection, of course, is that the limes, being of a special kind, have to be procured from him. By fitting on a second nozzle either forms of gas may be used Fig. 1 shows the nozzle as used for the mixed gases, and fig 2 for the blow-through system £0 18 0

287.—The Sciopticon Mixed (SCIOPTICON COMPANY).

Details.—The neatest jet we have seen for one of full size; T tap for one gas, and handle tap for the other; double motion lime-turner with fine rack £1 15 0

Remarks.—The neatness is largely attained by the bending of the tubes just after the support, not after the taps, as in most others. This makes the taps also well out of the way. Another notable feature is the difference in shape of each tap, which must be a great help in preventing mistakes. Altogether this is a very fine little jet.

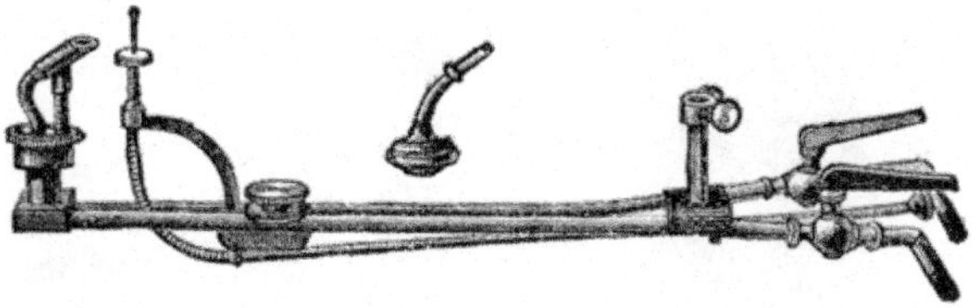

No. 288.

288.—Steward's Combination (J. H. Steward).

Details.—By unscrewing a collar and the substitution of one nozzle for another, this jet gives both powers; platino point; handle taps; spiral spring lime turning, single movement £1 15 0

Remarks—The lime turner in this pattern is well out of the way.

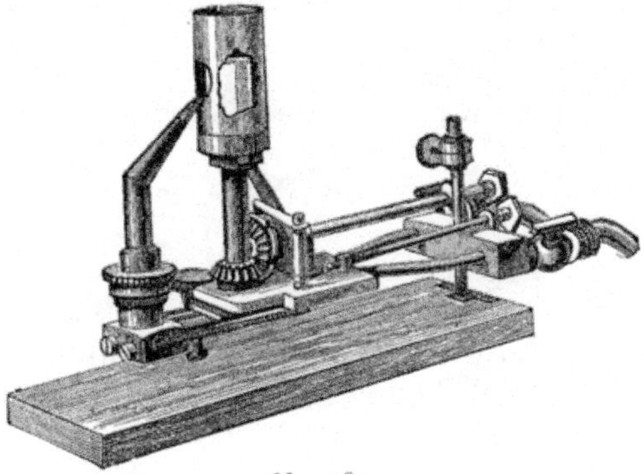

No. 289.

289.—Wood's Mixed (E. G. Wood).

Details.—Facilities for making the axial and conjugate focus adjustments and arrangements for clamping the same. The jet can be kept burning a long while without allowing the flame to impinge upon a part of the lime cylinder already used. Further details of this very excellent jet will be found further on £3 17 6

290.—Wood's Blow Through (E. G. Wood).

Details.—Central oxygen tube; perforated dome. Particulars of some of its features will also be found further on £1 11 6

291.—Wood's Automatic Spirit (E. G. Wood).

Details.—It is provided with a hollow instead of a solid wick; it has the wick chamber inclined to the proper angle at which the jet should impinge on the lime, and a perforated dome is provided, covering the wick cistern. The lamp is instantaneously lighted by the application of a taper or match, but the flame is very little larger than a pea, until the oxygen gas is turned on, when a most brilliant light is produced. When the oxygen is turned off the flame instantly dies down to its original size, and will continue so for any length of time until the oxygen is again supplied to it. A tell-tale is provided for the wick chamber, and an adjustment to the spirit cistern, in order to ensure a proper supply of spirit to the wick chamber when it is necessary to fix the apparatus out of the horizontal position £2 2 0

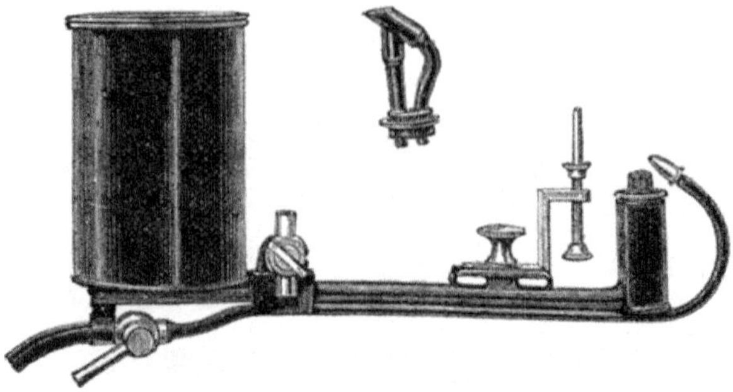

No. 292.

292.—The Ordinary Spirit.

Details.—Fountain reservoir, with arrangement for tilting .. £0 13 6

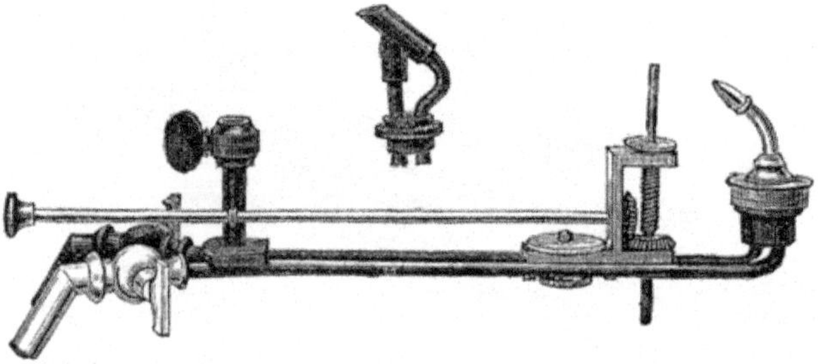

No. 293.

293.—The Interchangeable.

Details.—For both gases under pressure, or for the blow-through form, by simply removing and using the burner as required. There is a platina nipple to the mixed gas jet, and also to the blow-through portion; cog wheel arrangement for turning the lime £1 7 0

Accessories to Illuminants.

There are various accessories, additions to, or parts of jets, which need some amount of attention and description. Some of them are used pretty generally on jets, and would have caused repetition if described along with the jets, whilst others are really accessories or extras.

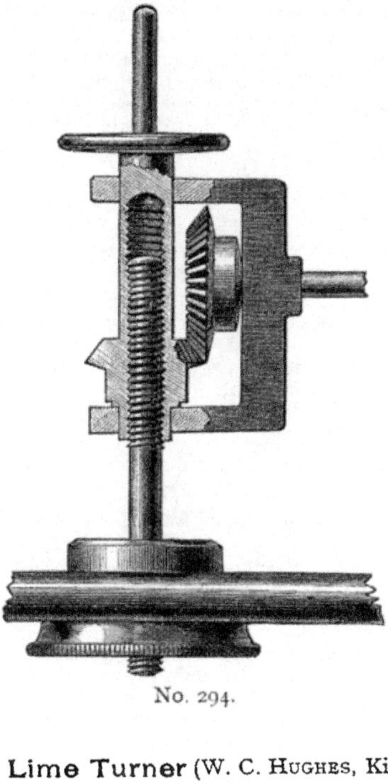

No. 294.

294.–The Star Lime Turner (W. C. HUGHES, Kingsland).

Details.—A cause of annoyance in some forms of lime-turning apparatus is the clogging, by lime dust, of the cog wheels. The object of the lime-turner is to turn and raise the lime-holder so as to obtain perfect incandescence If the cog wheels won't work by reason of accumulated lime dust or dirt, it is better off, than on, the jet. In the above the screw arrangement is protected by a cover. Made entirely of polished brass, with brass spindle £0 12 6

295.—Mason's Jet Holders (Mason & Co)

Details.—To their No. 3 biunial is fitted a novel arrangement to hold the jets, instead of the usual lime tray. The support consists of two brass sliding tubes running through guides fixed to the lantern body. The usual tray rather interferes with the ventilation, but this plan allows the lantern to be open right through Another feature is its superior rigidity.

296.—Pringle's Cut Off Newton & Co.)

Details.—Illustration No 280 shows this capital idea for the convenience of the lecturer. It is perhaps more especially adapted for an occasion when the lantern is being worked by someone inexperienced in limelight work This occasionally has to be done. One who could be trusted to put in the slides as required. Markings, arrangement, talking, and solemn warnings, will perhaps keep that part of the work in order. But when it comes to turning on the gas, it's quite a different matter. With Mr. Pringle's device, the light is first arranged in the usual way by the jet taps, and then the "cut-off" tap is turned. When the time comes for lighting up, this tap is turned, bringing the light at once to its best form. The novice could not meddle with the hydrogen and oxygen in separate taps to attain the essential nicety of quantity. To put it briefly, it is a device for using one tap to turn on the gas instead of two, this one tap at the same time turning on the exact relative quantities for the best light. In the illustration, the cut off tap is shown under the tube ends, and works by a screw with long pinion £1 2 6

297.—Check Action Movement (Newton & Co.)

Details.—This is used in the ordinary way to turn the lime, but by means of a spiral screw the action of turning the milled head outside the lantern not only rotates, but at the same time raises the lime cylinder, so that a spiral series of holes is formed in it, and the light never impinges twice on the same portions of the lime, while one lime will last for a whole exhibition. A check is fitted to the milled head, which renders the action as far as possible automatic £0 10 6

298.—Lime Turning Movements.

Details.—The following, taken from E. G. Wood's book on the lantern, though referring perhaps more specifically to his jet (No 289), will give much useful information. The dissimilar bevel wheels consist of a pair of toothed bevel wheels, one placed horizontally and the other vertically, one working into the other. Their use is to enable the lime cylinder to be revolved by simply turning a rod at the back of the lantern, and thus expose a fresh part of the lime cylinder to the action of the inflamed gases. The horizontal wheel is fixed to a short tube that covers and also moves the screw on which the lime cylinder is placed. The vertical wheel is attached to the rod, before spoken of, that passes out of the back of the lantern. The head on the outside end of this rod is a five-sided polygon, numbered 1, 2, 3, 4, 5, so that this head serves as a sort of index to show to what extent the lime cylinder has been turned on its axis. The horizontal and vertical wheels do not revolve with equal velocity; their diameters are as ·9 to 1, so that when the vertical wheel has made one revolution, the horizontal wheel has only made about 9-10ths of a revolution. The movement of the lime cylinder should not be a continuous, but an intermittent motion, a shift, in fact, from one part of the cylinder to another part of the cylinder. In order to determine the space over which the lime cylinder should move at each shift, a disc with five notches in its edge is fitted on to the rod that carries the vertical bevel wheel. Into these notches the end of a spring falls, and indicates by a click that the vertical wheel has been revolved 1-5th of its circumference, and a fresh portion of the lime cylinder brought opposite to the point of the jet. But as the vertical wheel in five shifts moves over the whole of its circumference, and the lime cylinder only moves over about 9-10ths of its circumference, it follows that there has been a loss in the rotation of the lime cylinder equal to about 1-10th of the circumference. In consequence of this, when five spots of the lime cylinder have been acted on, the sixth and subsequent spots will fall between the spots already made. The lime cylinder should be shifted about every three minutes in order to bring a fresh portion of the face of the lime opposite to the jet. As the diameter of the cylinder will only admit, even with the dissimilar bevel wheels arrangement, of ten shifts in a revolution, it is obvious that an additional motion must be given to the lime cylinder in order to operate upon successive zones of the surface. This is effected by fitting the lime cylinder to the top of a quadruple threaded screw, this screw being moved by the horizontal bevel wheel. By this means the lime cylinder is lowered 2in. during each revolution, and the lime is thus acted upon in a spiral path from bottom to top. The length of the screw admits of six complete revolutions, and this will require about one-and-a-half hours to accomplish. If the exhibition is to be further prolonged, the vertical bevel wheel should be turned in the opposite direction, in order that the lime cylinder may be gradually raised up to the point started from, and the spiral path thus retraced. In order to ensure that the jet flame in this return path shall fall between the portions of the lime cylinder previously acted upon, a reverse action has to be provided. This reverse action will move the lime cylinder through 1-10th of its circumference without moving the notched wheel. This is the first return shift, and the jet flame thus falls between the cavities previously made in the face of the lime cylinder. All the subsequent movements, which will be from notch to notch, will shift the lime about 2-10ths of its circumference as before.

299.—Lime Shield (E. G. Wood).

Details.—An illustration of this useful little addition will be seen at No. 289. It consists of a metal cylinder, about 4in. larger in diameter than the lime cylinder itself, and long enough to cover the cylinder even when it is elevated to its highest position. This shield is supported upon the base-plate of the jet, and does not partake of any of the movements of the lime cylinder. In the side of this shield, opposite to the jet, an elliptical aperture is made sufficiently large to allow all the light that can be taken up by the condensing lenses to pass out, and small enough to prevent the escape from the lime cylinder of any of the deflected streams of flame and heated air. Its advantages are that it prevents splitting or cracking of the lime, and also diminishes almost to nothing the quantity of disintegrated lime thrown off from the lime cylinder, and which speedily in all the old forms of apparatus covers the inside of the lantern with a white, almost impalpable powder. The loose, ragged flame that heretofore has been a serious source of injury to the body of the apparatus is, by the action of the patent shield, utilised in increasing the heating action of the jet upon the lime, and is finally carried directly upwards into the metal ventilator at the top of the apparatus, leaving the body of the lantern all but cold. It can be fitted to most jets.

£5 0 0

300.—Jet Supports (Ottway & Son).

Details.—Instead of trays this firm mounts the jets upon stout metal plates running in dovetailed fittings, which slide to and fro with a smooth and even motion. These are fitted with turned steel uprights, which also fit the upright tube fitting of jet. Towards the lime stage of the jet is a brass forked piece, and fixed into the metal plate is a threaded pin. This works in a fork, whilst above and below are screw nuts which clamp the forked piece when it is in the exact position required. This arrangement gives a very rigid support, and enables the light to be readily adjusted.

301.—Place's Lime Turner (J. Place).

Details.—The advantages claimed for this arrangement are that the elevation and turning of the lime are effected at the same time. The lime is moved in a coarse spiral, and presents a line of surface of 9in. in length to be burnt; also, in the event of the lime cylinder cracking, a good surface may be at once offered to the flame £0 5 0

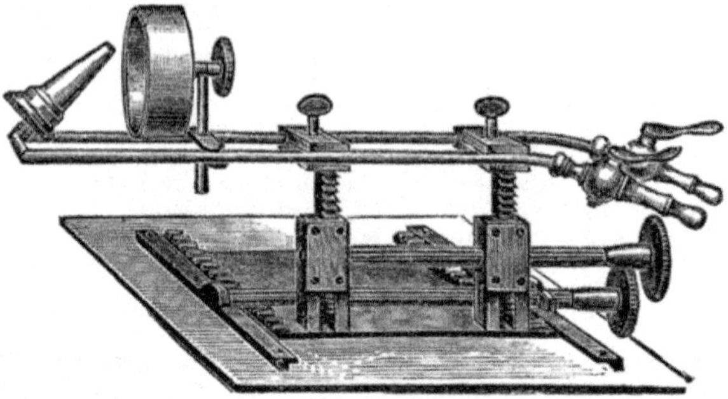

No. 302.

302.—Mechanical Stage (A. Pumphrey).

Details.—An arrangement to more easily effect the exact centring of the light. In addition to the movements of the lime from the nozzle and distance of nozzle from condensers, it has two others worked by rack and pinion; one moves the jet sideways, whilst the other raises or lowers the whole jet. This gives four movements altogether £1 0 0

We have now to deal with the ethoxo or ether-oxygen light. Much difference of opinion exists on the question of the safety of this method. The compiler is known as an advocate for it, but it is only because his experience points to it as the most convenient method of getting the best possible light. He is often told "Oh! you may manage it, but it isn't everybody that can without an accident," or, "You may not have had any bust up yet, but you will some day." But with such vague assertions as these the compiler, of course, cannot argue.

303.—Pumphrey's Vapouriser (A. Pumphrey).

Details.—A represents a bag filled with oxygen gas; this is conveyed by the tube B into the vapouriser, entering at D it passes through one tube of the vapouriser tube into the second, passing out at H. Into the vapouriser is introduced about one and a half pint of sulphuric ether, the oxygen in passing comes into contact with fibre saturated with ether at many surfaces, so as to be fully charged with the ether vapour, each measure of oxygen will take up about nine volumes of ether vapour. If this is connected with the hydrogen tap of the jet, it will burn much like common gas (The jet used must be the mixed jet. At B there is a branch union and a tube, which is carried to the oxygen tap of the jet; the oxygen which comes through this tube will not light, but by turning on a proper proportion of the pure oxygen to that charged with ether vapour the most perfect and powerful light possible is produced.

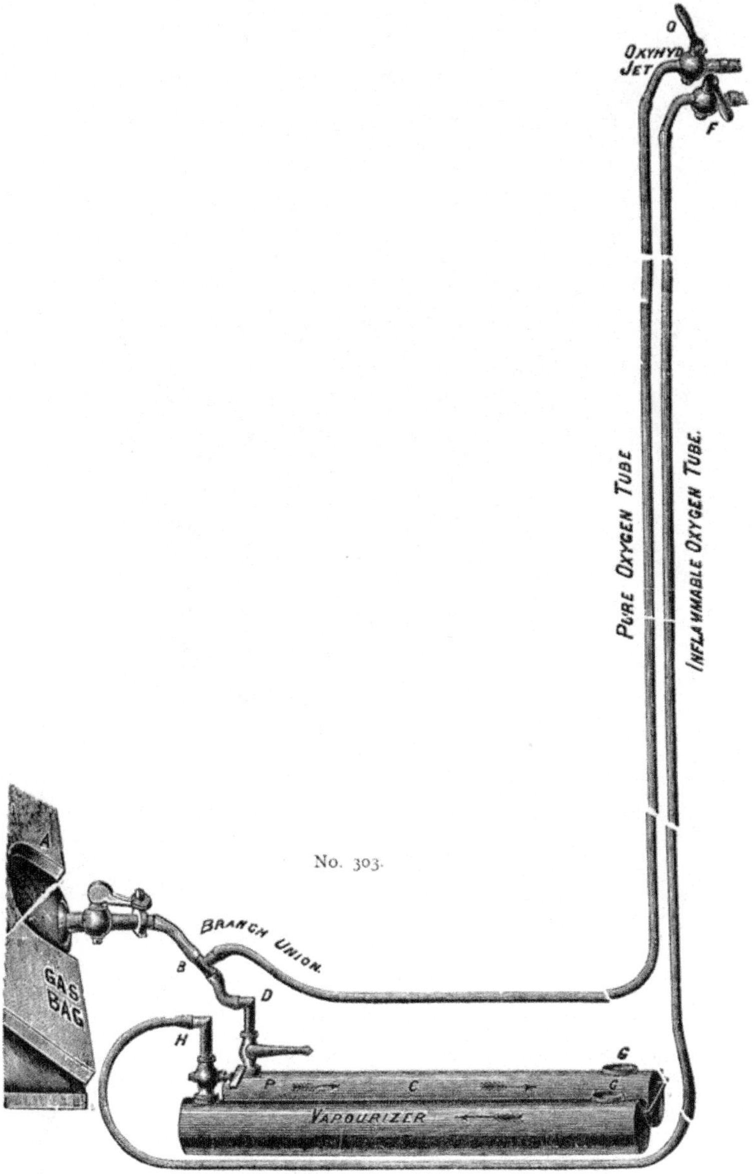
O
OXYHYD
JET
F
PURE OXYGEN TUBE
INFLAMMABLE OXYGEN TUBE.
A
GAS
BAG
BRANCH UNION.
B
D
H
G
P
C
C
VAPOURIZER

No. 303.

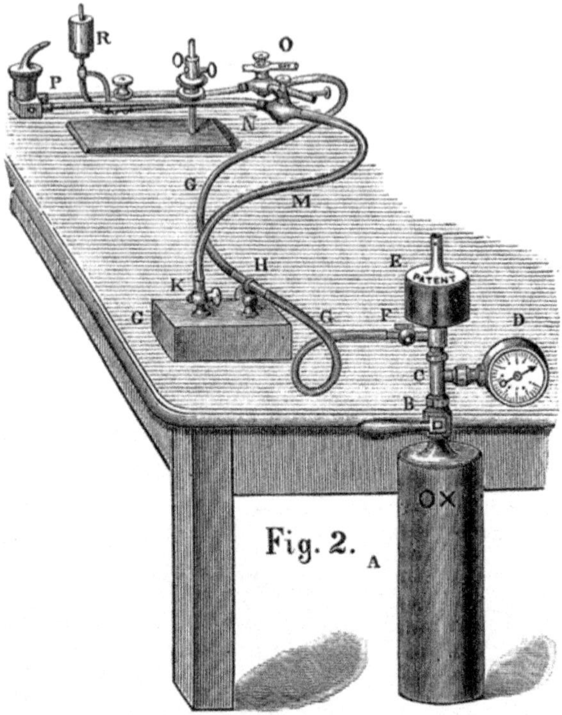

No. 304.

:304.—The Oxy-ether Light (J. H. Steward).

Details.—A copper tank is used, divided into two parts longitudinally, forming an upper and a lower chamber, and having inlet and outlet stop cocks. The lower chamber has a number of compartments in it, so that the oxygen gas may enter at one end, and freely pass along the many compartments, and so get charged with the vapour, then rise into the upper chamber and pass off by the outlet tube to the jet or lantern. Oxygen gas must be made or obtained and used in connection with the ether as described, a portion of same going on pure to the jet to form the oxygen for the limelight. This is accomplished by means of a T tap, one end of the T being the entrance and the other the exit for jet, and the downward pipe conducts the necessary amount of gas through the tank to form the substitute for hydrogen. If the oxygen is used from a bag and pressure supplied by dead weight, say 112 lbs., a light requiring very little attention, equal to the oxy-hydrogen in brilliancy, is obtained, and perfect steadiness and freedom from noise secured. For those who do not care to make their own oxygen gas, the gas cylinders are strongly recommended with the new automatic gas regulators, the light being very superior on account of the extra pressure obtainable The new small size bottles are very portable, and the lime cylinders are not pitted so regularly when using the ether as with the hydrogen, consequently they last longer. The illustration represents the use of gas cylinders. The following are particulars :—B, the valve of gas bottle, C and D, the gas pressure gauge; E, the automatic regulator; F, gas cock for oxygen; G, oxygen supply pipes, H, T shaped oxygen gas cock; K, ether gas as hydrogen for N. tap of jet; R, lime cylinder, P, safety pumice chamber; G, ether oblong tank. The copper tank for ether, with chambers as described, and brass stop cocks and connections £1 1 0

Extras.—Copper bottle for extra supply of ether with screw cap washer, 7/6. Safety pumice chambers for putting immediately behind the ordinary jet to prevent a return of gas, 5/- the pair

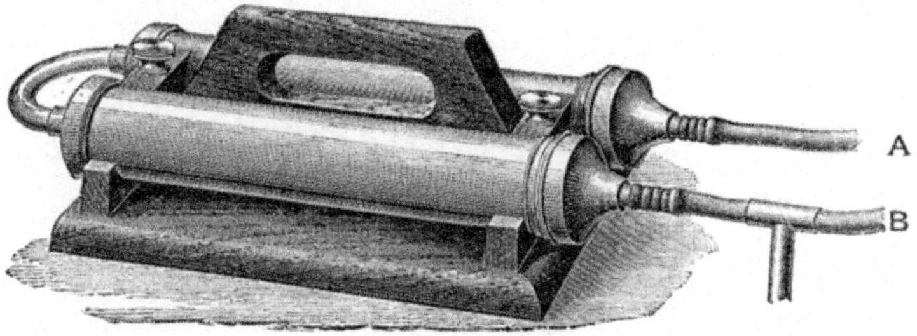

No. 305.

A to ether side jet. B to oxygen side of jet.

305.—Ether Saturator (WHOLESALE FIRM).

Details.—It consists of two brass tubes or bodies, screwed into a bent brass connecting tube, lying side by side on an ebonised wooden stand, which is fitted with a handle for carrying about. The brass tubes or bodies can easily be removed by unscrewing the set screws which fasten them to the board Each tube is fitted at one end with screw down cap, having a nozzle in its centre for the purpose of attaching elastic tubing, and both are fitted with a roll of flannel or coarse cloth, having a spiral wire in the centre to keep them open. Each nozzle has a small screw cap to prevent escape of ether when not in use £1 16 0

PART II.

Lenses, Condensers, &c.

IN the preceding chapters we have really touched upon many matters in connection with lenses and condensers which should have been reserved for the present part, so we must be excused if in any way we are guilty of repetition. As has been before explained, the duty of the condenser is to collect the rays of light given out by the illuminant and throw them evenly and uniformly through the slide into the objective by which the picture is magnified and cast upon the screen. To begin with the condenser, the simplest form, as used on the cheapest-class of lanterns, is a single plano-convex lens, or bull's eye, placed with its flat side to the light. To be of any use, however, a lens of this kind must be of long focus, and as a long focus lens gives a small picture, unless at a long distance from the screen, and there are many other objections to its use, it has in all lanterns of the better class been superseded by the double or compound condenser, which consists of two plano-convex lenses mounted together in a ring or cell with their flat sides outermost, and the crowns of their convex sides almost touching each other. The result of this combination is to give a lens of much shorter focus as well as better optical qualities, and, at the same time, comparative cheapness of manufacture, and this, therefore, has come to be the form of condenser now most commonly used. Another form of double condenser which is fully equal, if not indeed superior, in its optical qualities, although somewhat more expensive, is the Herschel model, in which a double convex and a concavo-convex lens are placed together, the convex side of the latter being next that of the other, and its concave side placed next the light. The condenser, however, which gives the highest possible results is the triple condenser, which may consist of either of the two previously described double combinations with a third and smaller plano-convex placed nearest the light, with its convex form towards the combination and separated from it by some considerable space. The size of the condenser will depend on the size of slide to be used; but for the photographic slides now so generally, we may almost say universally, used the best size will be found to be 4in., or for work at extra long distances 4½in. With a larger condenser much light is wasted, and with a smaller one there is danger of the whole of the picture not being sufficiently illuminated. In the construction of the condenser care should be taken that the lenses are not too tight in their cell, or the heat of the lantern may crack them on account of insufficient allowance for expansion. They should, in fact, be sufficiently free in their cells to be easily revolved by the hand, and—especially when using the oxy-hydrogen light—it is always a wise precaution to take it out and warm it before lighting up. The edge of the cell, too, should be furnished with several small holes between the lenses, so that the moisture which will condense on the inner faces of the lenses may, on heating, be able to escape. The condenser lenses should especially be free from scratches, spots, and air-bubbles, and should be made of the best white flint glass.

The objective, as has been explained before, is a lens or combination of lenses which takes up the picture cast into it by the condenser and throws it in a highly-magnified form upon the screen. The objective is a very important part of the lantern, and, indeed, may well be said to be that part to which most attention should be given, as upon its quality—given, of course, a good slide—depends the fineness and clearness of the picture on the screen. Great care should be taken in selecting it to see that no streaks or dim spots are in it, although an air-bubble in the glass or a scratch is not of so much moment here as upon the condenser. In form the simpler lenses are very similar in construction to the condenser, although, of course, on a smaller scale. We thus have the compound plano-convex and Herschel forms repeated, and in toy lanterns the double convex in place of the plano-convex of the condenser; but all these are what are termed single lenses, or single combinations, all lanterns of any quality at all being fitted with lenses of double combination, the construction of which is based on that of the typical portrait lens of the photographer, and may be described as consisting of a double convex and a plano-concave lens ground so that the concave side of the one exactly fits the side of the convex lens next to it, and the two lenses cemented together with Canada balsam. This is termed the back combination, and is placed at one end of a short tube, at the other end of which a double concave and a double convex lens are fitted, this time with a small air space between them. The use of this combination gives achromatism and a flatness of field and depth of definition unobtainable by a single lens, and, moreover, a shorter focus is obtained than that of either one of the two combinations used. Another advantage about this lens is that by unscrewing the front lens and replacing it by other lenses of different foci the focus of the whole combination can be altered, and full advantage of this is taken by several makers who supply, say, three combinations of different foci, the front and back lenses of which, when interchanged with each other, will give different foci still, so that by this means a great range is obtainable, and the operator is practically independent of space and can get a disc of the size most suitable to his desires in almost any room.

The different methods of mounting and racking out the objective, with the reasons therefor, having already been fully dealt with, we pass on at once to consider the principal commercial varieties.

306.—Dale's Illuminating Condensers.

These are made for the light boxes used in theatres, &c.

4in. diameter ..	10/6 deep focus ..	9/6 shallow focus.
4½in. ,, ..	16/- ,, ..	14/6 ,,
6in. ,, ..	25/- ,, ..	22/6 ,,
7in. ,, ..	40/- ,, ..	32/- ,,

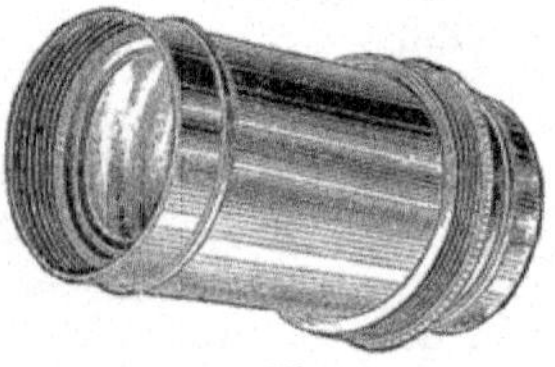

Fig. 1.

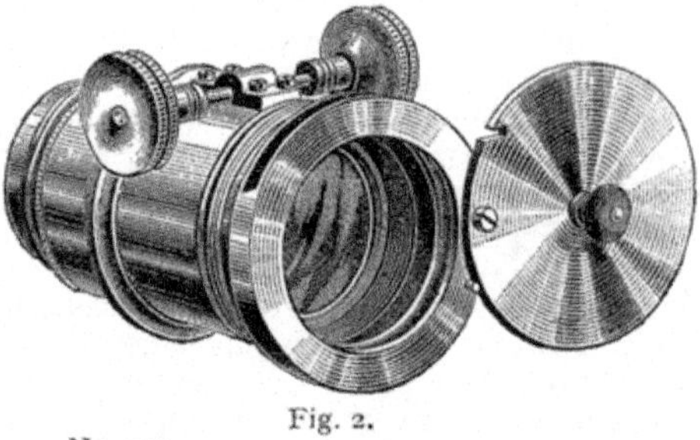

Fig. 2.

No. 307.

307.—Hughes's Objective.

Fig. 2 shows the rack tube complete with flashing shutter, whilst fig. 1 shows the lens separate and ready to fix into the rack mount. The changing from one foci to another is thus easily accomplished. We consider the definition excellent. With ordinary brass mount, rack, and single pinion, and brass flashing shutter £4 10 0

Large size rack tubes, as supplied to their large triples, fig. 2, which will take any given number of lenses of various foci, which are made to screw in as fig. 1 .. .. £5 10 0

Price of lenses separate, with screw, adapted to fit large rack mounts of various foci from 5in. to 12in from back lens, fig. 1 .. .. £3 10 0

308.—Hughes's Long-distance Objective (W. C. Hughes, Kingsland).

Made for the one particular purpose of showing a picture from a great distance. The usual result of work at long distances is to decrease the strength of the light. In the above Mr. Hughes has so striven that in pictures even 100 to 150 feet from the sheet the loss of light is but small. Mounted in brass screw cells, solid brass mounts, rack and double pinion, brass front with flashing shutter, focus from 8 to 12 inches .. £9 9 0

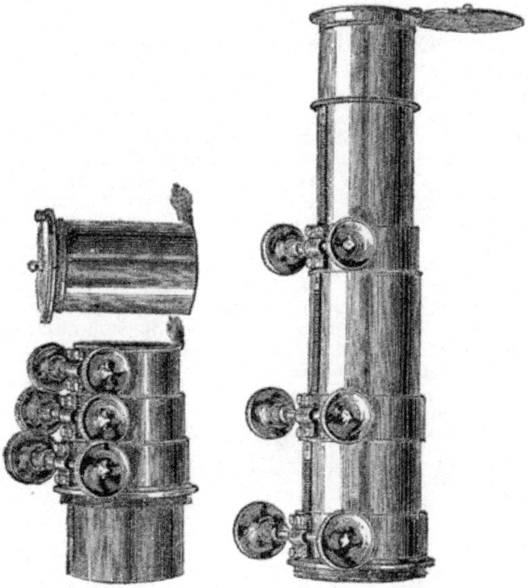

No. 309

309.—Triple-rack Telescopic Objective (Mason & Co.)

No lengthening tube required. Supplied with the following special front lenses, which can instantly be changed as required —

No. 1	4½in	back focus	gives a	15ft.	picture at	32ft.
,, 2	6½in.	,,	,,	15ft.	,,	42ft.
,, 3	9in.	,,	,,	15ft.	,,	52ft.
,, 4	11in	,,	,,	15ft.	,,	65ft.

Set, in polished mahogany box. Single combination £4 0 0
Double ,, ... 5 0 0

310.—Newton's Condensers (Newton & Co)

The following are their styles and prices —

Single	Plano-convex	3½	inch.	diameter	in brass cells	£0	8	6
,,	,,	4	,,	,,	,,	0	10	6
Double	,,	3½	,,	,,	,,	0	17	6
,,	,,	4	,,	,,	,,	1	1	0
,,	,,	4½	,,	,,	,,	1	7	6

311.—Newton's Front Lenses (Newton & Co)

Large diameter double combination achromatic, brass mounting, rack and pinion focussing, adapted for oil light £1 1 0

Long focus double combination achromatic, of large diameter, brass mounting, rack and pinion adjustment, especially made for use with the limelight, as fitted to their biunials £1 10 0

Triple combination achromatic set, consisting of three *large diameter* achromatic lenses, mounted in brass cells, and rackwork front, which can be used singly or combined, enabling the exhibitor to show a picture of either 6ft., 12ft, or 18ft. diameter *at the same distance from the screen*, or of any required size at almost any distance, thus adapting the lanterns for use in either small or large rooms .. .. per set £2 2 0

312.—Ottway's Objectives (J. Ottway & Son).

We quote from their letter :—"The lens we recommend as the best is the double achromatic, *i.e.*, an achromatic lens at one end of a tube of the necessary length, and a Meniscus flint and convex crossed lens at the other; these latter are not balsamed together, and are placed nearest the condenser. We make these in three different sizes—2in., 2½in., and 3in. diameter—for the reason that, as the focal length of the objective is increased or lengthened, the diameter should be enlarged, in order to include all the principal rays from the condenser, and thereby get the utmost brilliancy obtainable upon the screen. We make the 2in diameter 6in. and 7in. foci, the 2½in. diameter 8in. and 10in. foci, and the 3in. diameter 12in. and 14in. foci. These lenses are used without any stop whatever, and for penetration, crispness, and brilliancy are unsurpassed. Our other form of objectives are a set of interchangeable compound achromatics, 6in., 8in. and 12in foci, used singly, and will give 3½in., 4in., 4½in., 5in., and 5½in. foci. Using the combined, for instance, 6in and 8in. together, give 3½in focus, 8in and 12in give 4in., and so on. The lenses are mounted in an inner sliding tube, which can be withdrawn from the rack and pinion nozzle and replaced without any unscrewing, so that the size of picture required can be rapidly determined, and the size of figures, statuary, &c., reduced at will. They are also fitted with two diaphragms, one suited to the short and the other to the long foci."

313.—Ottway's Condensers (J. Ottway & Son).

"The compound plano-convex condenser, consisting of two plano-convex lenses with their convex surfaces inwards, answers very well with short focus objectives with the oil-light, but with the limelight a Meniscus and a double convex lens (the concave face of the Meniscus nearest the luminant) give the best result. Also, they should not be of too short a focus, in order to accommodate the long focus object glasses. The size of condensers depends upon the size of the pictures. If all round pictures are intended to be used, 3in. in diameter, then 3½in. condensers are amply large enough. If square or dome-shaped, then 4in., or even 4½in. is required for some sizes, but on no account should the condensers be larger than necessary, as it would only result in loss of light."

314.—The Noakes Front Lens and Mount (D. Noakes & Son).

The system here adopted is one remarkable for its quick-changing capabilities, the lenses are ground with special curves, and are a double combination with 2in. back lens, the mount is so constructed that lenses of varying foci can be almost instantaneously changed without unscrewing any part of the mount; it is mounted with double milled head on pinion, and flash shutter complete. Noakes and Son have always in stock lens tubes to fit this excellent mount of the following back foci :—

3½in.,	giving	15	feet disc at about	20	feet from	screen.
4in.,	,,	15	,, ,,	30	,,	,,
6in.,	,,	15	,, ,,	40	,,	,,
8in.,	,,	15	,, ,,	50	,,	,,
10in.,	,,	15	,, ,,	60	,,	,,
12in.,	,,	15	,, ,,	80	,,	,,

Price of above mount with 1 lens in tube	..	..	£2 0 0
,, ,, 2 lenses in tubes	..	..	3 0 0
,, ,, 3 ,,	..	..	3 15 0

315.—Taylor's English Portrait Objectives (T. S. & W. Taylor).

Made by themselves at Leicester. Rack and pinion mount. 2in. aperture. A splendid front lens.

6in. focus and under	..	..	£3 5 0
Above 6in. focus ..	..	..	3 0 0

316.—Steward's Achromatic Front Lenses (J. H. Steward).

These are so constructed that several foci can be obtained without deterioration of definition, and the combinations supplied of any special focus to meet the requirements of the exhibitor, according to the distance the apparatus has to be from the screen when in the most suitable position, and yet get the size picture required. For stock lanterns the lenses are fitted for the distances most generally used, to be convenient for the audience—long as well as short. Price, in brass jacket, with rack and pinion adjustment, and brass flashing shutter to front £3 3 0

317.—Theobald's Cheap Front Lens (Theobald & Co.)

Details.—This is a remarkable front lens, being an all brass portrait combination, with rack and pinion focussing adjustment, giving excellent definition, at a wonderful low price £0 10 6

318.—Theobald's Cheap Condenser (Theobald & Co).

Details.—Another very cheap thing. A 4in. compound condenser, in brass cells, with ventilating holes, for £0 8 6

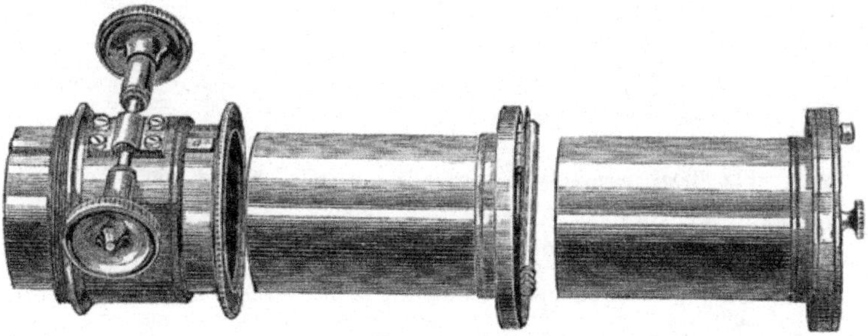

No. 319.

319.—Interchangeable Objectives (Wholesale Firm).

The above fronts are of best make and finish, with double pinions to the rack adjustment, and so constructed that brass cylinders, carrying short and long focus double combination lenses, can be used. Each brass cylinder is fitted with hinged instantaneous shutter or flasher, having grooves to carry shapes of coloured glass for tinting slides, it has the focal length engraved on it, and is fitted with double combination achromatic lenses, with large size lenses to the back combination. They can be fitted to any lantern that has telescopic draws, or to any lantern without telescopic draws by the addition of brass lengthening tubes. Four brass cylinders can be had with each front of the following foci :—

No. 1 cylinder, 4½in. back focus.
No. 2 ,, 6½ ,, ,,
No. 3 ,, 9 ,, ,,
No. 4 ,, 11 ,, ,,

The cylinders can be changed in a moment, without unscrewing the front from the nozzle of lantern. With double pinions to the rack adjustment £0 18 0

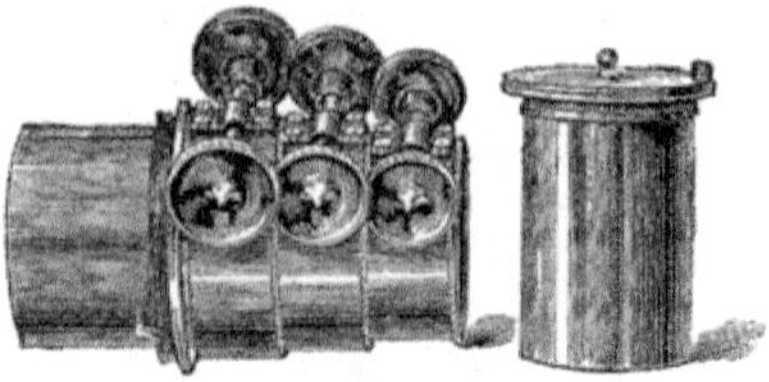

No. 320. Fig 1.

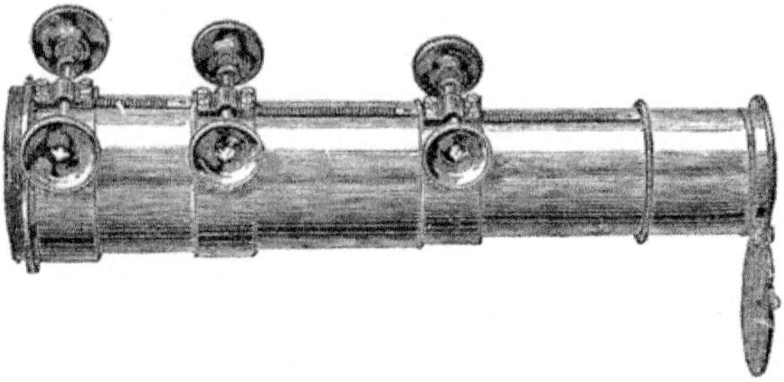

No. 320. Fig 2.

320.—Triple Rack Objective (Wholesale Firm).

The above new rack front is made to carry short and long focus double combination achromatic lenses, for showing pictures at short and long distances from the screen with biunial and triunial lanterns, and with single limelight lanterns, and pairs of limelight lanterns, without any lengthening tubes being required. The brass front has three telescopic draws, with rack and pinion adjustment to each draw, with a range of 7½ inches, being sufficient to allow 11in. focus lenses to be used without any lengthening tubes. The lenses (which are double combination achromatic with large size lenses to the back combination) are mounted in brass cylinders which slide into the smallest telescopic draw of the front, each cylinder being fitted with a sliding shutter or flasher, having grooves to carry shapes of coloured glass for tinting slides. The cylinders are of the following foci, and each one has the focal length engraved on it.

No. 1	cylinder,	4½in.	back focus.
No. 2	,,	6½in.	,,
No. 3	,,	9in.	,,
No. 4	,,	11in.	,,

The cylinders can be changed in a moment, without unscrewing the rack front from the nozzle of the lantern, and the lenses being double combination, give a sharper and better picture than the ordinary single achromatic lenses. The following are a few examples of the different distances from the screen a limelight lantern can be placed to give a 10 feet picture with a 3¼ square photograph with the above lenses.

With the No. 1 cylinder a 10ft. picture can be shown at 21ft. from screen.

,,	2	,,	10	,,	,,	28	,,
,,	3	,,	10	,,	,,	35	,,
,,	4	,,	10	,,	,,	43	,,

A set of six coloured glasses, fitted in a leather case, is sent with each patent rack front. The set consists of four brass cylinders, with sliding shutters or flasher, containing double combination achromatic lenses (with large size lenses to the back combination) of the following foci —$4\frac{1}{2}$in., $6\frac{1}{2}$in., 9in., and 11in. Packed in a fitted mahogany box, with lock and key £6 6

Remarks.—Of course, for a biunial two sets, and for a triple three sets are required. The racks on each draw are for rigidity and ease of focussing. Fig. 1 shows the front closed, and fig. 2 the front extended. Should the above patent triple-rack fronts be required with only one, two, or three cylinders, the allowance will be 18s. per cylinder. Any of the cylinders not ordered at the time as the triple-rack front can always be supplied afterwards, as they are all made to gauge.

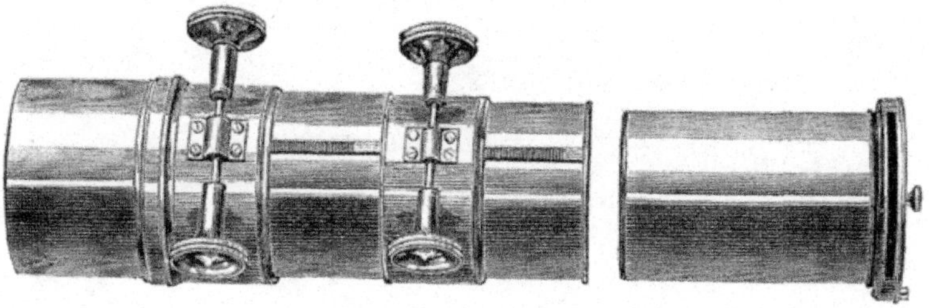

No. 321.

321.—The "Double Rack" Objective (Wholesale Firm).

The above patent telescopic optical lantern objectives, which are similar in construction to No. 320, but have only two racks, are fitted with extra large size double combination achromatic lenses These are of the following diameters and foci, are mounted in brass cylinders, fitted with sliding shutters or flashers, having grooves to carry shapes of coloured glass for tinting slides.

No. 1..	..	$1\frac{7}{8}$in. diameter.	4in. back focus.
,, 2..	..	$2\frac{3}{8}$in ,,	6in. ,, ,,
,, 3..	..	$2\frac{5}{8}$in ,,	8in. ,, ,,
,, 4..	..	$3\frac{3}{16}$in ,,	10in. ,, ,,

Each brass cylinder has its focal length engraved on it, and a set of six coloured glasses, in a leather case, for tinting slides is sent with each. The lenses are the same as in quarter-plate, half-plate, three-quarter-plate, and whole-plate photographic portrait lenses. The set of four brass cylinders, as above described, and set of coloured glass, in leather case, packed in a fitted mahogany box, with lock and key and brass handle .. £15 15 0

322.—Double Combination Front Lenses.

The following gives about the average price for the ordinary makes Double combination achromatic, mounted in brass, with rack and pinion £0 12 0

With large size lenses to the back combination, and fitted with sliding shutter or flasher £0 15 0

The same, but with double pinions to the rack adjustment, and grooves to carry shapes of coloured glasses.. .. .. .. £1 4 0

323.—Brass Lengthening Tubes.

For use with long focus lenses to extend the range.

	s.	d.		s.	d.
2in.	4	6	6in.	7	0
4in.	5	6	8in.	9	0

324.—Single Achromatic Front Lenses.

In brass cells, for showing pictures at long distances (6in., 8in., 10in., and 12in. foci).

	s.	d.		s.	d.
1⅞in. diameter	5	6	2½in. diameter	15	0
2in. ,, ...	9	0	2¾in. ,, ...	18	0
2¼in. ,, ...	12	0	3in. ,, ...	24	0

325.—Compound Condensers.

Plano-convex, mounted in brass cells.

	s.	d.		s.	d
4in. diameter	10	6	8in. diameter	84	0
4½in. ,, ...	20	0	9in. ,, ...	115	0
5in. ,, ...	30	0	10in. ,, ...	160	0
6in. ,, ...	47	6	11in. ,, ...	210	0
7in. ,, ...	63	0	12in. ,, ...	270	0

PART III.

Gas Apparatus and Accessories.

OXYGEN gas is obtained by heating chlorate of potash chrystals ($KClO_3$), the heat driving off the oxygen and leaving the potassium chloride (KCl) behind. It is found that the decomposition is facilitated by the addition of black oxide of manganese (MnO_2), and a small quantity of this substance is usually mixed with the potassium chlorate for this purpose, although no chemical combination takes place with the manganese, which remains unaltered.

To obtain the oxygen in sufficiently large quantities for use with the lantern, it is necessary to use a considerable amount of the mixture, 2lbs. of potassium chlorate being required to produce ten feet of oxygen, which will, with the blow-through jet, be sufficient for a two hours' exhibition (considerably less—about one half—being requisite when the mixed jet is used). This has to be most carefully—as it is a highly explosive compound—pounded, and intimately mingled with the manganese dioxide. The mixture is then placed in a retort, care being taken to see that no small chips of wood or other extraneous substances of a like nature are included, the reason for this precaution being that, should any inflammable substance be present, the great heat to which the mixture is subjected would most probably cause it to ignite, and this, in the midst of a vessel full of oxygen, would cause an explosion, the consequences of which we should hardly care to speculate upon. As air and oxygen form together a highly explosive compound, it is also of the highest necessity that the retort should be entirely air-tight, and that the joints and tubes attached to it be accurate in their fit. The purity of the gas is another consideration with a view to a good light, and as a little chlorine comes off with the oxygen, the gas should be passed through a wash bottle, such as will be found illustrated and described in the following pages. The apparatus having been put together with the mixture in the retort, it—the retort—is placed over a fire or Bunsen burner, and in a few minutes the oxygen will come off and make the fact known by bubbling through the wash bottle. It should be allowed to escape for a minute or two into the air, in order that any air present in the retort and pipes may be expelled, and, when it is coming freely off, the tap should be joined up with that of the gas-bag.

The construction and shape of the gas-bag is clearly shown in the following pages, and before attaching it to the oxygen tube it should be rolled up and squeezed in all directions in order to expel any air it may contain. Should there not be sufficient mixture in the retort to fill the bag, the gas will cease coming off, gradually lessening in volume until only a few bubbles pass slowly out. When this is the case, or when the bag is full, the tap on the gas-bag should be turned off, and the connection between the wash bottle and the retort be broken before removing the latter from the fire. This is important, as, if the heat be removed first, a vacuum will be created in the retort which will cause the water in the wash bottle to be sucked back into it with dangerous results.

The gas-bag filled, it can be carried to wherever the entertainment is to be held, care, of course, being taken to properly secure the tap. To use it, the gas is required to be forced through the jet with some force, much more force being requisite with the mixed jet than with the blow-through, and in order to effect this the bag, which is wedge shaped, is placed between two pressure boards, the top one of which is provided with a shallow shelf on which weights of the required size may be placed. When using the blow-through jet the pressure on the gas main is usually found sufficient for the hydrogen, and it is only necessary to connect the hydrogen tube with the gas-pipe of the house, but for the oxy-hydrogen mixed jet another gas-bag should be filled from the main, and this, as well as the oxygen bag, placed under pressure. For this purpose double pressure boards are of the greatest utility, a strong point in their favour being the power of altering or adding to the weight on the bags without putting out the light, which cannot be done when two single pressure boards are used, except by two persons using very great care. The weight to be placed on the bags depends upon circumstances, 1 cwt. being about the average, although very much more is required at times. It is important that the pressure should be uniform, or the light will vary, and although weight can be added to the bag when requisite, should it be deemed advisable to reduce the weight the light should be turned off, as the great danger entailed by the use of the mixed jet is the chance of a weight dropping off, or a person standing on the bag and then getting off, which would have the same effect, viz., to cause a sudden expansion of the bag, which would produce a suck-back into the bag full of oxygen of the mixed—and therefore highly explosive—gases and flame, and in nineteen cases out of twenty a tremendous explosion, resulting, of course, in a sudden and disastrous finish to the entertainment.

A far cleaner, easier, and cheaper way of procuring oxygen is to purchase it compressed at very high pressure in strong steel cylinders—a system which is now rapidly making headway. The advantages of the gas cylinders are many, the chief objection to their use being the fact that the gas is forced into them at such high pressure that it escapes with much more force than is usually required, and as it escapes the pressure continually diminishes, becoming at last so low that some feet of gas in each cylinder are unusable, no extraneous pressure, of course, being possible with an inflexible steel casing. This latter objection, however, is obviated to a certain extent by the makers allowing for the unused gas when the cylinder is returned to them, and the difficulty of pressure is got over in a great measure by the use of pressure regulators, into which the gas passes from the cylinder, and from which it escapes at an uniform pressure. When the mixed jet is used, however, and an uniform pressure is required on both bags, it will be found best to simply fill a bag with the gas from the cylinder, when the pressure can be regulated in the usual way, and the whole of the gas—or nearly so—in the cylinder used. Hydrogen gas is also supplied compressed in cylinders, but as it is twice the price of oxygen, and twice the quantity is required, it is too expensive a luxury to be of much commercial use where coal gas can be readily obtained.

For the proper regulation of the gases for dissolving purposes, dissolvers, which have already been mentioned, are used. These are combination taps, by the turning of which the gas may be cut off one lantern and turned on in another simultaneously, and the different varieties and their construction and action will be found fully described and illustrated in the following pages.

Making Gas.

We have separated this part into two headings, of which the above comes first, though it only deals with retorts and purifiers.

326.—Gas Saucepan (Cubley & Preston).

The illustration shows clearly the idea of this, which is quite different to the ordinary retort. The lid has bayonet catches, and is screwed in tubes. For use the oxygen mixture is placed inside and the lid replaced. The lid is then made to slide as far as possible, a few taps with a hammer being a good help to make it fit more tightly Previous to covering the pan, a paste is made of white powdered clay, which is smeared round the edges of the pan to ensure an air-tight covering. We have thoroughly tried this

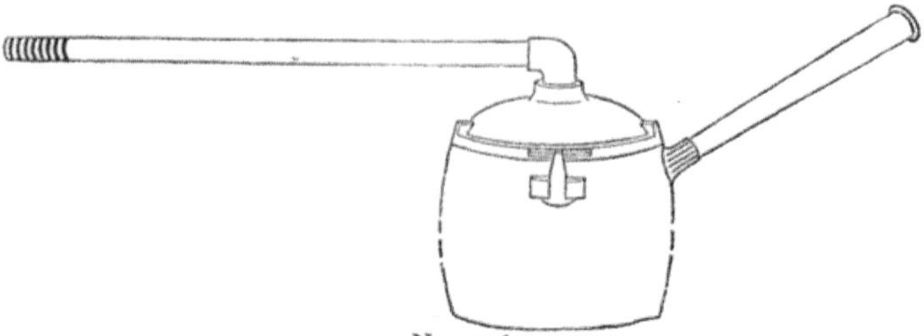

No. 326.

saucepan and can speak well of it. At first sight the clay business may seem a nuisance, but it is very simple, and takes no time worth mentioning. And even this is certainly very much counterbalanced by the ease with which the residue is removed. In the ordinary retorts, to do it by the "dry process" means much bumping of stick or poker to break the lumps to the necessary fineness to come out at the small hole at top. Of course the quickest way is to pour in hot water, but this necessitates careful drying afterwards—a thing not always easily managed away from home, and out of the question if a second immediate use be necessary. Out of this saucepan, however, the residue is removed almost instantly.

The prices are not excessive either.

	s.	*d.*
To hold 1 quart	5	0
" " 2 "	6	6
" " 4 "	7	6

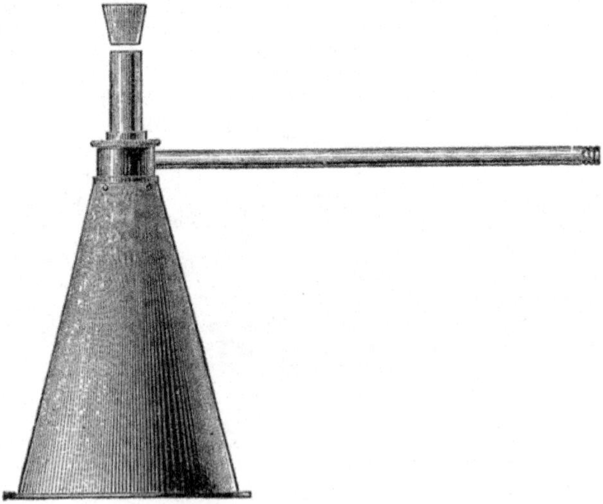

No. 327.

327.—Hughes's Iron Safety Retort (W. C. Hughes, Kingsland)

With brass screw head and blow-out, fitted with cork. Mr. Hughes, to prevent the chance of explosion, has constructed a brass screw head-piece to the retort, which being fitted with a cork would prevent an explosion , let the gas run off ever so violently, or the tubes choke up, or any foreign substance be mixed with the gas composition, no accident can possibly arise, for at a certain pressure or combustion of the materials in the retort the cork will blow out and the contents pass harmlessly up the chimney. This is strongly recommended for safety, and can be used by a child.

With double bottom, for durability, 12/-

Extra large size to hold enough to fill large gas bag at one filling, 16/6.

The large retorts are made with a ferrule so that a new bottom part can be replaced at any time for 8/6.

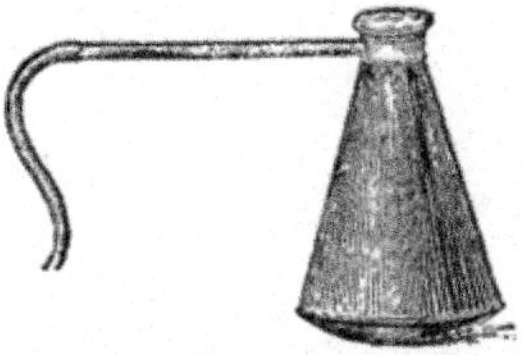

No. 328.

328.—The Optimus Retort (Perken, Son & Rayment).

	s.	d.
Made of iron, brazed	10	6
Made of copper, brazed	14	6

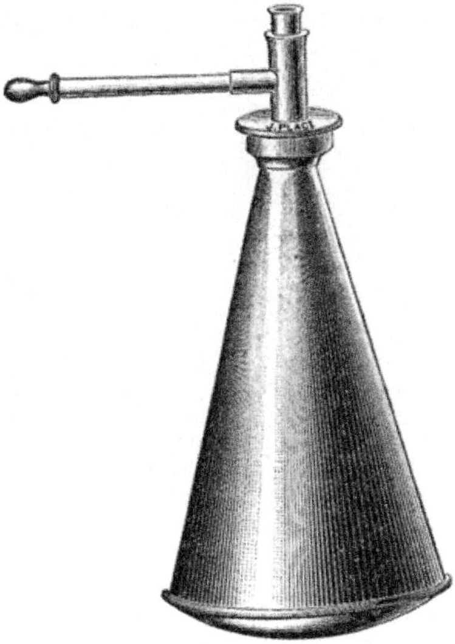

No. 329.

329.—Place's Safety Retort (J. Place).

The illustration is so clear, description is unneeded. We would point out the safety vent cork, however, and that the tube has good wide clear apertures £0 8 6

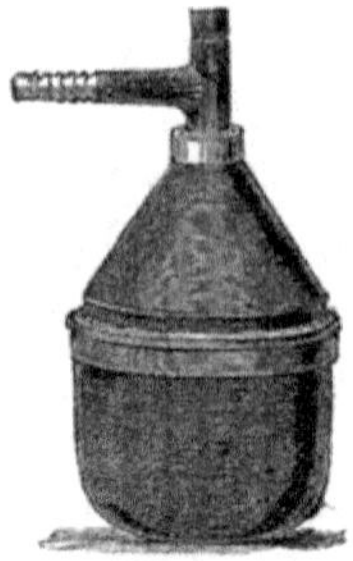

No. 330.

330.—Cast Iron Safety Retort.

This retort is made of cast iron, and is so constructed that it can be taken in half for cleaning purposes, and can then be screwed together again the reverse way, which makes it about half its size and, therefore, very portable. Being made of cast iron it will last for many years. The arm is made of brass, and has a large safety valve in case the gas should come off too quickly from the oxygen mixture £1 5 0

331.—Hughes's Zinc Purifier.

	s.	d.
Small size	4	6
Large size, with large inner cap and brass screw head ..	10	6

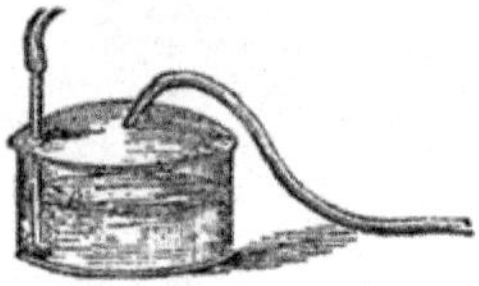

No. 332.

332.—The Optimus Purifier (Perken, Son and Rayment).

	s.	d.
Made of zinc	4	0

333.—Place's Glass Purifier (J. Place)

By means of this glass purifier the operator can always tell by a glance how the gas is coming off, whether slowly or quickly, and how much water is in the bottle. It is also more easily cleaned out than the ordinary tin or zinc patterns. Brass screw top. The price is extremely low .. £0 2 6

334.—Glass Purifier Tubes.

New shape brass tube for thoroughly purifying the gas, with indiarubber top, which fits into any wine or spirit bottle. A convenient method for quickly making a purifier out of any ordinary bottle £0 4 6

335.—Chadwick's Portable Oxygen Gas Generator and Holder (J. H. Steward).

This consists of an ingenious retort, made to contain prepared cakes of chlorate of potash and manganese, in a bell-shaped cap, supported by a cylindrical stand in which a Bunsen gas-burner of improved construction is fitted, placed on a portable gas-holder, which also forms the stand for the

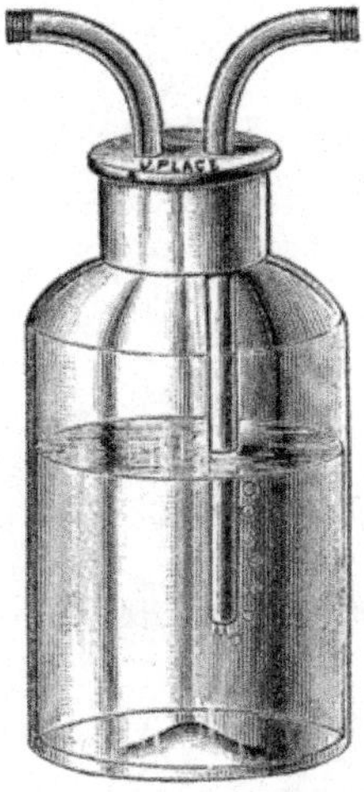

No. 333.

lantern. After being fitted up in position, and the water inserted in the reservoir, which is made with a displacement chamber, reducing the amount of water required for purifying to the minimum (about one bucketful being sufficient), an indiarubber pipe is connected from the house supply of gas to the retort, and the oxygen gas liberated by heat from the mixture, passing down by means of indiarubber and metal pipes to the bottom of the reservoir, thence bubbling up through the water into the dome of gas-holder, which is raised by it until full. The exhibition can now be commenced, and a fresh cake of oxygen mixture put into the retort, ready to go on generating when more gas is required to replace what is used. This is known by the height of the dome; and if the retort is kept warm by means of the gas burner attached, the quantity is driven off *very quickly and quite safely*. Each cake makes about two feet of gas, so that it is only required to recharge about three times. The apparatus when packed measures outside 25 x 15 x 12 inches, and the displacement chamber being always dry, is used for packing apparatus, or slides, screens, etc. Made of tinned iron, enamelled oak colour £5 10 0

336.—Portable Pure Hydrogen Generator (J. H. Steward).

This generator is made of copper, for obtaining hydrogen from sulphuric acid and zinc cuttings. The gas generated can be collected in a gas-bag, and used under pressure, giving the utmost light possible, or for certain experiments it can also be used with a separate or blow-through oxy-hydrogen jet, as it is generated, without passing into a bag or receiver at all. It is constructed on the principle of a gasometer; when closed it measures about 16 inches high and eight inches in diameter.. .. .. £2 2 0

Using Gas.

The advent of the new compressed gas cylinder, if it became universal, would sweep away many of the accessories and requisites now to follow. But the cylinders possess disadvantages, and this, coupled with prejudice in some cases, makes the inclusion of these things still a necessity. We shall deal with the better known articles first, and then treat of the gas cylinders with their special accessories.

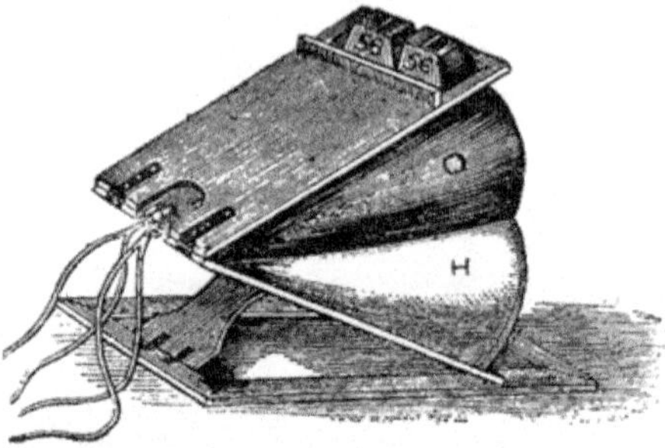

337.—The Excelsior Gas Bags.

These are the bags usually sold, of which two qualities are catalogued The best are made of black or drab twill, very strong—in fact, are the best made. The second quality are of lighter construction, and are not capable of so much pressure, nor stand the wear and tear of travel so well. The prices are as given below, and most dealers quote the same, with only very trifling variations

Length.	Width.	Wedge.	Cubic feet.	1st Quality. £	s	d.	2nd Quality. £	s.	d
30	by 20	by 17	3	1	18	0	1	10	0
30	,, 24	,, 20	4	2	2	0	1	14	0
36	,, 24	,, 20	5	2	4	0	1	18	0
36	,, 24	,, 24	6	2	6	0	2	2	0
36	,, 28	,, 24	7	2	10	0	2	6	0
36	,, 32	,, 24	8	2	18	0	2	8	0
40	,, 32	,, 24	9	3	6	0	2	16	0
40	,, 36	,, 24	10	3	18	0	3	8	0
40	,, 36	,, 26	11	4	4	0	3	14	0
40	,, 36	,, 28	12	4	8	0	3	18	0

338.—Dale's Bags (H. Dale, Limited).

Strongest black twill, guaranteed to stand 5 cwt. pressure, with improved lever cocks and screw union, having large gasway and arrangements for locking the cock when filled with gas. Two qualities.

Sizes.			Cubic feet.	1st Quality. £	s.	d.
30	by 24	by 18	4	1	18	6
36	,, 24	,, 24	6	2	10	0
36	,, 26	,, 21	7	2	12	6
40	,, 26	,, 22	8	2	17	6
				2nd Quality.		
24	,, 18	,, 18		1	7	6
30	,, 24	,, 18		1	10	0
36	,, 24	,, 24		2	0	0

339.—The Optimus Gas Bags (Perken, Son & Rayment).

Sizes.		Capacity, cubic ft.		Prices.
20in. x 15in. x 12in.	..	1ft.	..	21/-
24in. x 18in. x 12in.	..	1½ft.	..	24/-
24in. x 18in. x 16in.	..	2ft.	..	27/-
27in. x 20in. x 16in.	..	2½ft.	..	28/6
30in. x 20in. x 17in.	..	3ft.	..	31/-
30in. x 24in. x 17in.	..	3½ft.	..	34/-
30in. x 24in x 20in.	..	4ft.	..	36/-
36in. x 24in. x 20in.	..	5ft.	..	42/-
36in. x 24in. x 24in.	..	6ft.	..	44/-
36in. x 28in. x 24in.	..	7ft.	..	46/-
36in. x 32in. x 24in.	..	8ft.	..	50/-
36in. x 36in. x 24in.	..	9ft.	..	54/-
40in. x 32in. x 24in.	..	9ft.	..	55/-
40in. x 36in. x 24in.	..	10ft.	..	58/-
40in. x 36in. x 26in.	..	11ft.	..	59/-
40in. x 36in. x 28in.	..	12ft.	..	62/-

340.—Pumphrey's Gas Bags.

Made of thick black indiarubber cloth, threefold thickness with lock-tap (see 341) that may be padlocked so that it cannot be opened

Length.		Width.		Wedge.		Cubic feet.		Price.		
24	x	18	x	18	....	3	....	1	5	0
30	x	24	x	18	....	5	....	1	10	0
36	x	24	x	24	....	8	....	2	0	0
36	x	30	x	24	....	10	....	2	10	0

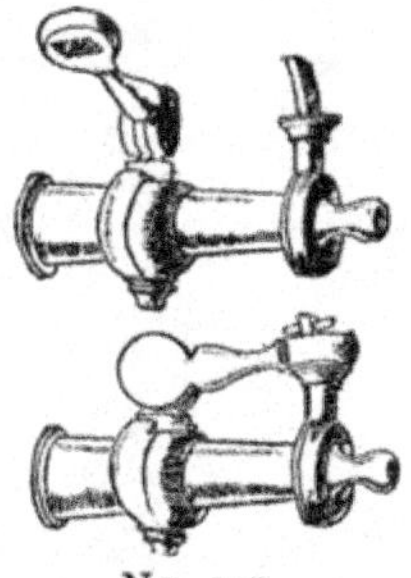

No. 341.

341.—Lock Taps to Bags.

These are a great relief to one's mind, as they save much anxiety as to tampering with the bag. Of course, the best thing with children, or even grown-up children, is to lay down the maxim—" Touch the bag and up you go sky-high." But even with this dreadful fate prognosticated, there is always the chance of somebody turning on the tap "just for fun." In these lock taps, a bar lever fits over a point near nozzle of tap, into which, when closed, slips a padlock. When the key is turned and safely stowed away, the operator cares for no one. In some catalogues the taps are an extra, in others are included in the price.

342.—Repairing Gas Bag Solution.

This is a solution of indiarubber, and is sold in tins about 1/- and 2/-

343.—Pressure Boards.

Pressure boards for bags are principally made in three varieties, viz.—the plain single, to hold the bag; the skeletòn single, the same, but much lighter and more portable; and the double skeleton for two bags. The average prices are—

	£	s.	d.
Plain single	0	12	6
Skeleton single	1	0	0
Skeleton double	2	0	0

The prices, however, vary much more than do gas-bags, owing to the different wood used and perfection of manufacture.

344.—Archer's Double Skeleton Boards.

These are made of selected American elm and canvas, with brass hinges and screws; two shelves, so that the weights cannot possibly fall off; metal quadrants, so that the bags can be adjusted and fixed any height from the floor. Large size £3 10 0

345.—Dale's Pressure Boards (H. Dale & Co., Limited).

An illustration of these, at all events of the double form, will be found at No. 252. The single form sizes are—

	s.	d.
26 x 20	12	0
37 x 25	15	0

The double pressure boards are £1 8 6

346.—Reversed End Pressure Boards (H. Dale & Co., Limited).

For a candid opinion we have pleasure in quoting Mr. Dale's own words:—"Reversed end pressure boards are very good, but they require more weight than the others, and also, by the motion of the gas-bags and absence of the leverage of the ordinary form of boards, are a little liable to cause the boards to cant and shift the weights. The advantage is that the top of the boards being level, various things can be utilised as weights, which on the ordinary form is out of the question." Size, 44 x 27 £1 10 0

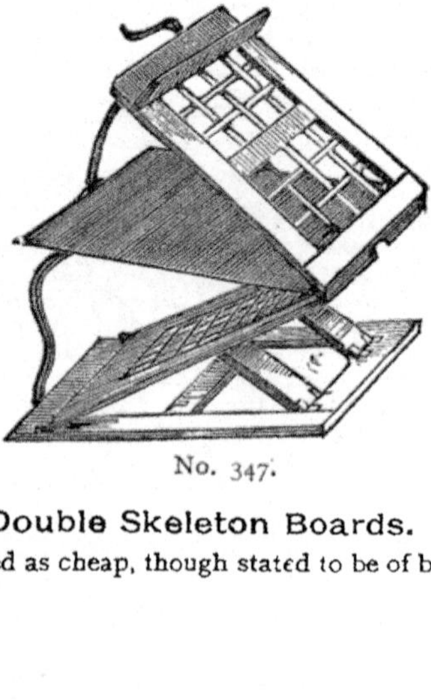

No. 347.

347.—Banks's Double Skeleton Boards.

This is quoted as cheap, though stated to be of best make .. £1 5 0

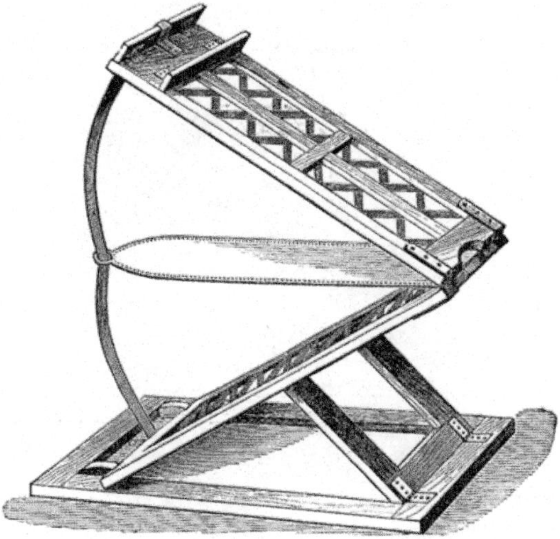

No. 348.

348.—Hughes's Double Skeleton Boards (W. C. Hughes, Kingland).

This apparatus was specially designed for lightness and strength by Mr. Malden, and is self-supporting, and will stand on the floor or on the tops of seats or pews, as may be desired. The gas-bags are retained in their positions by the strap which passes through a loop fixed to the centre sail cloth, and which also binds the whole for travelling. The pressure is equal on each bag, and a much steadier light can be kept up than by separate pressure boards for each gas Moreover, considerable weight is saved. The directions for use are . "Put the hydrogen bag at the bottom, while the lower board is flat on the frame; pull the centre sail cloth tight between them by the loop; then strap them up, and raise the boards into the position shown in the drawing by raising the hinged end, and the lower board will drop to its place of its own accord" Made in polished mahogany

46in. by 33in.		£6	6	0
40in by 28in.		5	5	0
Ditto in plain black	£4 10 0 &	4	4	0

349.—Pumphrey's Skeleton Pressure Boards (A. Pumphrey).

These boards act just as well as the more solidly constructed ones, those which are double sided throughout, and have the advantage of being only half the weight £0 12 0 £0 14 0 £0 16 0

350.—Limes.

	s.	d.
Best hard limes, per doz.	2	3
Oxy-calcium limes ,, ...	2	0
The Excelsior (hard in box) per doz. ..	3	0
Pumphrey's (2½ in. diameter) ,, ...	3	6

The Excelsior brand are the most popular. Pumphrey's are entirely of a special shape, see his jet No. 286.

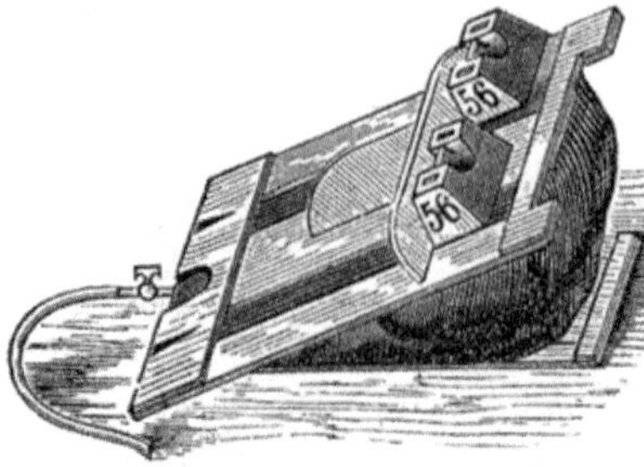

Fig. 1.

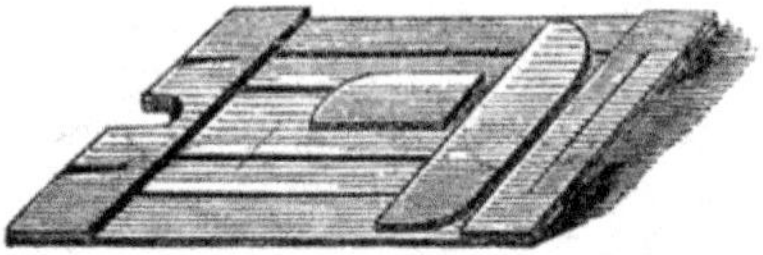

Fig. 2.

No. 349.

351.—Lime Cylinder Holders.

		s.	d.
Brass tube with sliding cap to hold six	..	2	6
Brass " " " twelve	..	5	0
Large size brass box with screw lid ..	..	6	0

352.—Chadwick's Safety Valve (J. H. Steward).

This consists of two parts made in brass, between which is placed a thin indiarubber or oiled silk diaphragm, through which four or five holes are made. The area of these orifices is in excess of the ingress tube, therefore the slightest pressure entering the valve from the gas-bag acts upon the diaphragm so as to open the valve. Any pressure the opposite way—the pressure to be feared—closes the valve, and the greater the pressure the tighter the valve is held shut. The diaphragm can be replaced or taken out for any purpose. £0 7 6

353.—Water Interceptor Safety Valve (J. H. Steward).

This valve is the result of a series of experiments conducted by Mr. R. M. C. James, of Watford. It was found, after spending some considerable time in practically testing the different known forms of safety valves, and trying those described in books, that this arrangement was the only one that would stop a return of *ignited mixed* gases, and yet keep up a *steady* pressure for ordinary use in a dissolving view apparatus without any objectionable diminution of force. This safety valve is in the form of a cylinder, and is partly filled with water, the chamber and pipes being so arranged that in a small space provision is made to stop the return of gas, either plain or mixed, at a point well removed from the bag, the contents of which, under every condition would be safe Price in brass .. .. each £1 1 0

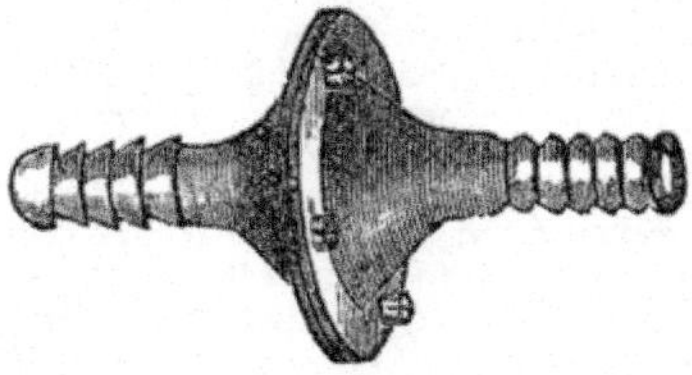

No. 354.

354.—Back Pressure Valve (W. Tyler).

	s.	d.
Small size to attach behind jet	4	0
Large size to place between pipes	7	0

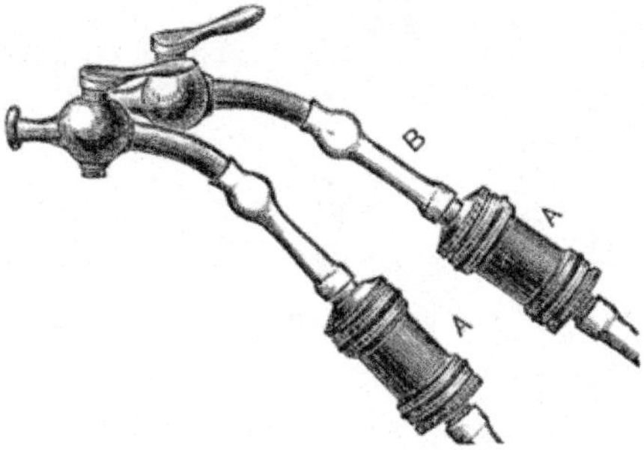

No. 355.

355.—Tested Flame Extinguishers (A. Pumphrey).

These consist of tubes filled with finely granulated pumicestone, which allows the gas to pass through one way, but through which no flame can possibly pass, even when it is an explosive mixture of gas, thus securing greater safety. Price per pair £0 10 0

The system of compressed gas cylinders (which take the place of the bags) is rapidly coming to the fore, and as fast as the conveniences for the hire and refilling of cylinders is extended, so we think their use will become the more common. The following details refer entirely to the oxygen gas and gas cylinders, and apparatus or fittings connected therewith.

356.—Brin's Oxygen Process (Brin's Oxygen Co.)

The usual method of producing the gas is by the decomposition of chlorate of potash in a retort, which passes into a bag. The objections to this are—the time it takes and the inconvenience, the difficulty of production on a large scale, and the contamination of the oxygen by the presence of free chlorine gas. This latter has also an injurious effect on the bags. These obstacles are overcome by Brin's Oxygen Company, who can supply, in unlimited quantities, oxygen absolutely pure, extracted from the atmosphere by mechanical and not chemical means. Their process is as follows:—One or more retorts are charged with an absolutely pure anhydrous oxide of Barium, these retorts being raised to a given temperature, atmospheric air, carefully purified, is then passed through the Barium oxide, which arrests the oxygen and permits the nitrogen to escape, thus effecting a complete separation of the two gases. When the Barium oxide has become sufficiently charged, or, in other words, peroxidised, the air pumps are reversed, and the oxygen that has been absorbed is yielded up by under vacuum, and passed into suitable receivers or holders." The process being, in reality, a mechanical one, and there being no chemical change in the substances operated upon, makes the purity of the product a certainty.

357.—Brin's Gas Cylinders (Brin's Oxygen Co.)

The following table gives very full information on the various matters connected with the purchase, hire, dimensions, &c., of the cylinders.

Cubic Contents in feet.	Diameter in inches.	Length.	Gross Weight.	Rent per week after 14 days.	Price of Gas per foot	Price of Gas per cylinder.
		ft. in.	lbs	s. d	d.	£ s. d.
3	3	0 7	4½	0 4	4½	0 1 1½
5	3½	0 11	6	0 4	4	0 1 8
10	4	1 4	15	0 5	3½	0 2 11
12	4	1 7	18	0 5	3⅜	0 3 4½
20	4	2 6	26	0 6	3	0 5 0
40	5⅜	2 6	28	0 8	2¾	0 9 2
80	5⅜	5 0	44	1 0	2⅜	0 15 10
100	5⅜	6 6	61	1 3	2¼	0 18 9
225	5⅜	8 0	76	1 4	2⅛	1 2 1½
225	5⅜	14 0	177	2 2	2	1 17 6

Rents at the above rates are charged on all cylinders detained more than 14 days out of the works. One key for the cylinder valve, and one set of nipple and union are sent with the first delivery, for which a charge of 5/- is made, and, in case of the order being repeated, should be retained. Keys will be charged 2/6 each, nipples and unions 2/6 the set.

Compressed coal gas is sold at the same price and on the same terms, whilst compressed hydrogen is just double in cost. The oxygen cylinders and all the connections thereto, such as gauge and valve, are all painted green, and the hydrogen or coal gas red. This makes distinction an easy matter. We have personally tried the cylinder and gas, and found it to give a beautiful light. It seemed to our eyes much better oxygen than we could make ourselves, and as the price of 10 feet, which is ample for a three hours' entertainment at least, is only 2s. 11d., it is really cheaper than buying the materials, to say nothing of the labour and trouble. An idea may perhaps arise in the reader's mind as to the difficulty of knowing how much-gas is left in a cylinder after partial use This can be readily ascertained by a glance at the table on next page, after testing the pressure with the gauge.

Size of Cylinder in Cube Feet.	Diameter in Inches.	Length.		Gross Weight.	PRESSURE IN LBS.																		
		ft.	in.	lbs.	1,800	1,700	1,600	1,500	1,400	1,300	1,200	1,100	1,000	900	800	700	600	500	400	300	200	100	50
3	3	0	7	4½	3	2¾	2½	2½	2¼	2	2	1¾	1½	1½	1¼	1	1	¾	½	½	¼	—	—
5	3½	0	11	6	5	4½	4¼	4	3¾	3½	3¼	3	2¾	2½	2	1¾	1½	1¼	1	¾	½	¼	—
10	4	1	4	15	10	9½	8¾	8¼	7¾	7¼	6¾	6	5½	5	4½	3¾	3¼	2¾	2	1½	1	½	¼
12	4	1	7	18	12	11¼	10½	10	9½	8½	8	7¼	6½	6	5¼	4½	4	3¼	2½	2	1¼	½	¼
20	4	2	6	26	20	18¾	17¾	16½	15½	14¼	13¼	12	11	10	8¾	7¾	6½	5½	4¼	3¼	2	1	½
40	5⅜	2	6	28	40	37¾	35½	33¼	31	28¾	26½	24¼	22	20	17¾	15½	13¼	11	8¾	6½	4½	2	1
80	5⅜	5	0	44	80	75½	71	66½	62	57¾	53	48¾	44½	40	35½	31	26½	22	17¾	13¼	8¾	4½	2
100	5⅜	6	6	61	100	94¼	88¼	83¼	77¾	72	66½	61	55½	50	44½	38¾	33¼	27¾	22	16½	11	5½	2½
125	5⅜	8	0	76	125	118	111	104	97	90¼	83¼	76¼	69¼	62½	55½	48½	41½	34½	27½	20¾	13¾	6¾	3¼
225	5⅜	14	0	177	225	212½	200	187½	175	162½	150	137½	125	112½	100	87½	75	62½	50	37½	25	12½	6¼

358.—Gas Cylinders (Howard, Lane & Co.)

This firm are the manufacturers and retailers of gas flasks or cylinders, into which they compress the gas by a powerful engine. They are thoroughly well made and sound, the metal being non-porous even with hydrogen gas. The cylinders and valves are severely tested by air pressure before leaving the works, and so far as we can judge are completely reliable. The details are—

Capacity in cubic feet.	Diameter. Inches.	Length. ft.	in.	Approximate Weight. lbs.	Price. £	s.	d.
3	3	0	8	4	1	3	0
5	$3\frac{1}{4}$	0	11	6	1	8	6
10	4	1	$5\frac{1}{2}$	15	1	19	6
12	4	1	8	18	2	2	6
20	4	2	8	26	2	19	6
40	$5\frac{3}{8}$	2	9	28	3	6	6
80	$5\frac{3}{8}$	5	2	44	5	15	6

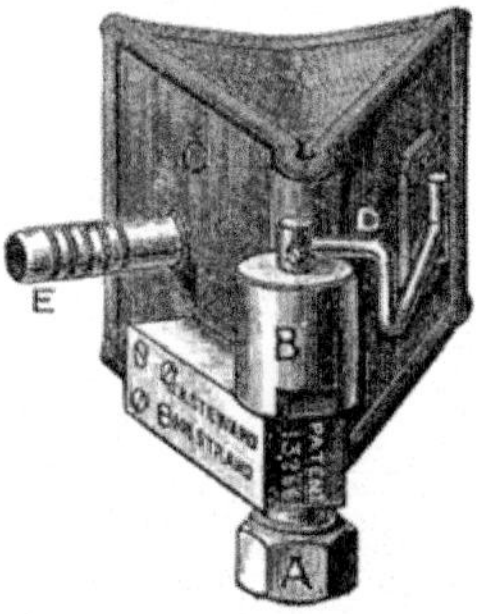

No. 359.

359.—Clarkson's Self-acting Valve.

In order that dissolving may be performed from the cylinder as easily as from a gas-bag, and to effect economy in size, a miniature gas-bag is placed on a self-closing valve, the gas from the cylinder supplying dissolver and jets at a useful pressure, and preventing the furious rush of gas that would probably take place on using only the valve of the bottle. As the gas enters the fitting A from the cylinder, it passes into the box or cock B, and from thence into the bag C, which, as soon as full, has, in the course of expanding, turned the valve in box B, and so stopped the opening. A strong spiral spring fixed in box B acts on an arm D behind the bag, and exerts a working pressure equal to from twelve to tweny-four inches of water, according to the number of lights it is supplying from pipe E, and so a constant pressure is kept up £1 5 0

360.—Beard and Oakley's Automatic Gas Regulator (J. H. Steward).

For screwing on to gas cylinder, limiting the rush of gas even when ordinary valve is full open. The gas enters the valve, and fills the india-rubber bellows. As this, in rising, causes the Archimedean screw to work down the point, directly the bellows is full the gas is cut off, and the pressure exerted by a spiral spring on the top of the bellows causes the gas to escape from tap at a workable pressure. If there is any undue pressure or stoppage, a little valve attached to the bellows will open so that it cannot burst. The bellows and screw work automatically—that is to say, as the gas is used in

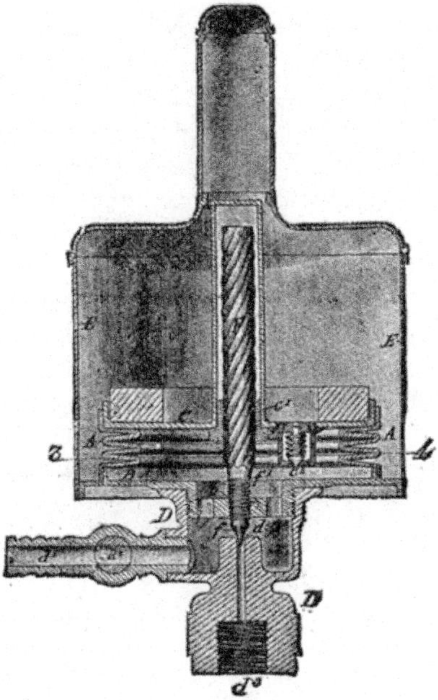

No. 360.

the lantern, it is supplied by the valve, and there is more than ever required to keep three lights going. In fact, it is like a miniature gas-bag, properly weighted, being placed between the gas supply and the lantern, and kept always filled, or nearly so £3 3 0

Dissolvers.

By the above heading we refer to those used with the gases only. They may be roughly divided into two sections—those used for biunials, that change the light from one lantern to the other; and the rather more complicated kind that deal, as in triples, with three lanterns. There is also a special one for use when the fountain spirit jets are employed, and one for use with the top lantern of a triple. Dealing with the more common form first—that used for a biunial or pair of lanterns—it is known as a "six-way" tap; that is to say, it consists of six tubes as follows:—Inlet for oxygen from bag, inlet for hydrogen supply, and two outlets to each lantern, one for each gas. Those used in triples have, of course, more tubes, and are necessarily more complicated. But they are not by any means always used, many operators preferring two dissolvers, one for the biunial portion, and a separate one for the top lantern. This brings the "four-way" into use. A third system for triples is to have three of these "four-way" taps, one to each lantern.

361.—Archer's Star (ARCHER & SONS).

Six-way. This consists of a single plug which cuts off the gases and turns them from one lantern to the other by means of a lever handle. When the lever is put to the right the right-hand lantern is working (or, if a biunial, the top lantern), and *vice versâ*, when in the centre both lanterns are full on, and when turned right round both are off. The two small bye-pass taps are to keep a little gas always burning, and so prevent the gases "popping" out £0 17 6

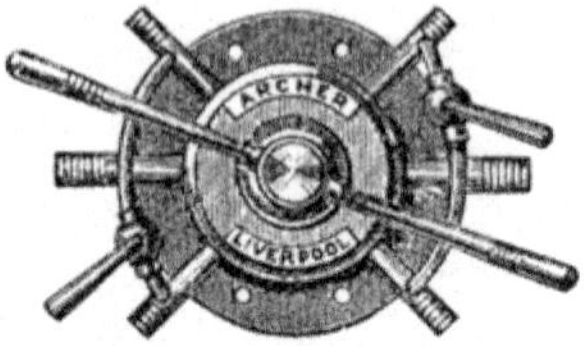

No. 362.

362.—Archer's Improved Star (ARCHER & SONS).

Six-way. This is similar in design to the other, but with the firm's latest improvement of so forming the plug that the rubber tubes conveying the gases to the jets do not cross each other. We can speak very highly of this dissolver; it is extremely well turned out in brass, and works perfectly £1 5 0

363.—The Malden (W. C. HUGHES, Kingsland).

Six-way. This dissolver was designed and invented by Mr. Malden, hence its title. It consists of a large plunger, so arranged as to put off and on the gas to the two jets of a pair of lanterns, and can be applied equally well to the oxy-hydrogen chamber jet, or the safety blow-through jet. In order to keep alight the hydrogen in each lantern a bye-pass cock is provided; this can be regulated at the commencement of the exhibition, so as to allow the desired quantity to be burning, as the case may be. With the handle properly placed both lanterns have their full quantum of gas, and are equally illuminated; but in the production of effects one lantern must be lowered to darken the view while lighting up the effect. Snow, aurora, or other effects can be introduced. When the handle is to the right, the right-hand lantern (or upper one, if a biunial lantern) is on, then, if the handle is slowly moved over to the left, the dissolving is effected in the most pleasing manner. From the simplicity of the parts, the plunger can be taken out and cleaned when necessary £1 2 6

364.—Hughes's Lever (W. C. HUGHES).

Six-way. The feature of this dissolver is that it takes a horizontal position on the lantern. It is provided with a horizontal lever handle; the bye-pass is so situated that it allows both gases to be turned off both lanterns, excepting sufficient to keep the limes hot, so that the operator can have both lights full on in each lantern immediately, or only in one £0 18 6

365.—The Noakes (D. NOAKES & SON).

For triple lanterns only. For those who prefer one dissolver only with which to work all three lanterns, the above is a very clever and useful piece of mechanism. It is worked by means of a coupler and handle. When in use, either the coupler or handle is moved to effect the dissolving. With this dissolver any two lanterns are the dissolving pair *at will*, as follows:—

Top and middle as the dissolving pair.
Top and bottom " " "
Bottom and middle " " "

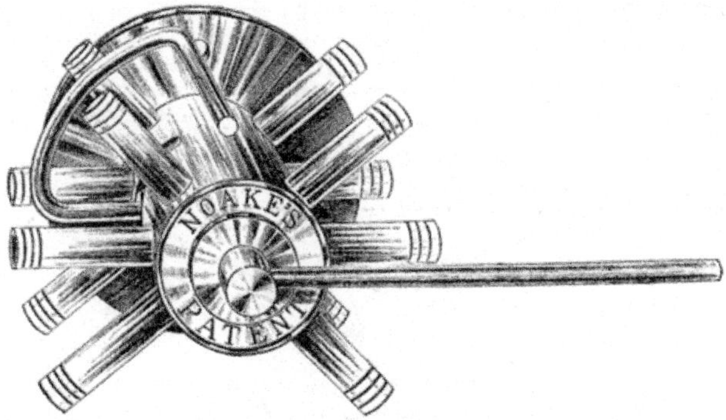

No. 365.

A single lantern can be illuminated, bottom and middle together, or all three simultaneously, in any order at will of operator. There is no dial plate to read or watch, and the invention is so simple that any person can work it after a few minutes' trial £3 3 0

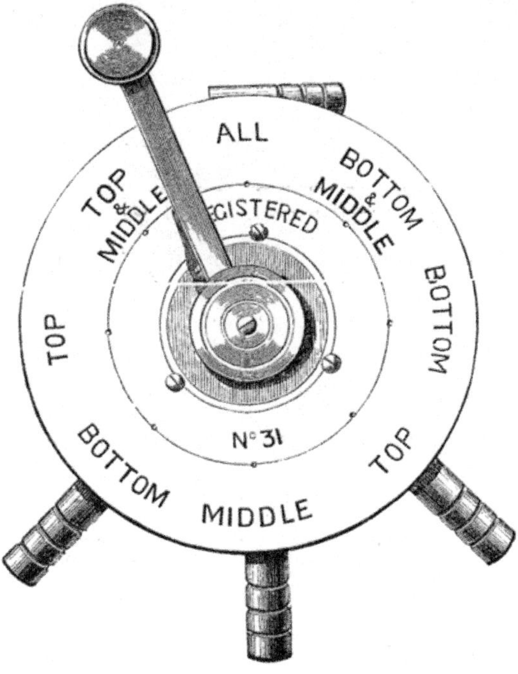

No. 366.

366.—The One Plug (OTTWAY & SON).

For triples only. This apparatus is so well known, and the illustration so clear, that a description is unnecessary. But an important improvement has been made which will enable the operator to turn on or shut off two lanterns at the same identical moment By this means a view and effect can be brought on simultaneously and taken off in the same way; also a further improvement which gives the power of shutting off all three of the lanterns without the jet taps being disturbed, after the lantern has been set up and adjusted, or when an interval occurs £3 3 0

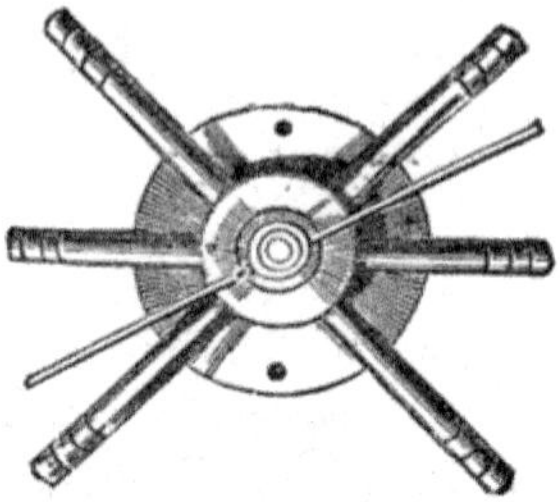

No. 367.

367.—The Optimus (PERKEN, SON & RAYMENT).

Six-way An ordinary pattern with regulating tap for bye-pass £0 18 0

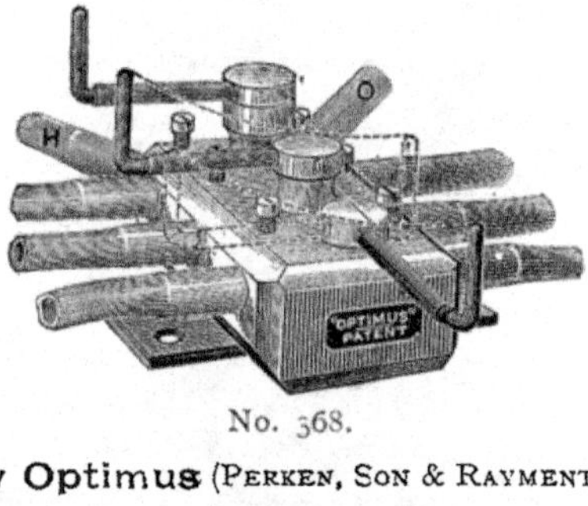

No. 368.

368.—The New Optimus (PERKEN, SON & RAYMENT).

Six-way. This dissolving tap has certain special advantages. It is small in size and thoroughly efficient. The screws (as shown on diagram) are for the regulation of the bye-passes, so that you always have a flame in each burner which can be turned up or down at will. In dissolving, the handle of one burner merely needs to be turned on as the other is turned off. (This is shown by dotted lines on diagram) When using only two lanterns the handle of the third burner is clamped down and the bye-pass screw tightened so as to stop the channel of supply. It is then impossible to turn the gas on to this jet by accident. By the same mode of stopping, one lantern only can be used, so that the one tap will do all that can be required. There are but three handles, and each controls the supply of both oxygen and hydrogen to one lantern. After a trial of this new pattern, we have no hesitation in asserting it to be convenience itself. We were especially pleased with the arrangement that enables the one dissolver to be used for a single, biunial, or triple at will. Two more commendable features we must mention—it can be readily taken to pieces for cleaning purposes, and in size it is extremely neat.

369.—The Chadwick-Steward System (J. H. Steward).

An arrangement for placing the dissolvers of triple lantern in combination, so that one movement will change from top to middle or bottom, and *vice versâ*, without lowering the light of the intermediate lantern. This is a great convenience, as it leaves the operator a hand free for the slides, either to turn a handle, as in the water-mill slide, or for bringing on a rising moon effect, etc. An illustration of this system will be seen at No. 239. Price, including the three dissolvers £4 15 0

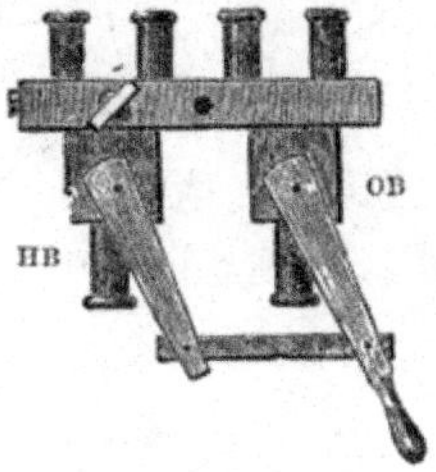

No. 370.

370.—Wood's Bye-pass (E. G. Wood).

Having ourselves frequently worked this dissolver, we can safely say that it is as simple as efficient, and is, moreover, convenient to the touch. The little tap shown in engraving is the bye-pass tap for hydrogen, which allows a small flame of hydrogen only to remain burning when the gases are turned off from that jet £0 17 6

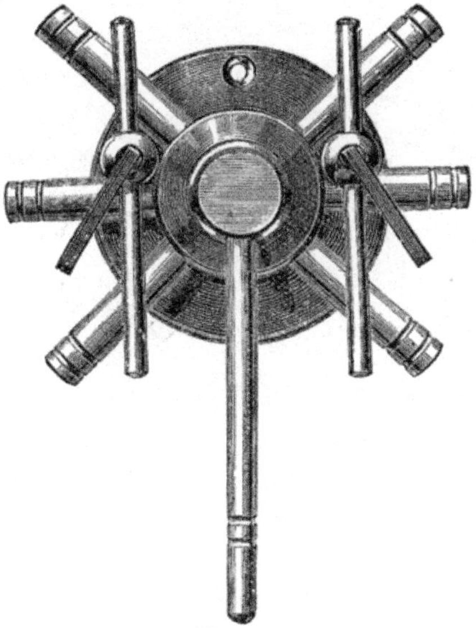

No. 371.

371.—The Star Six-way.

This is the ordinary trade dissolver, with double bye-pass, one for oxygen and the other for hydrogen. Brass £0 15 0

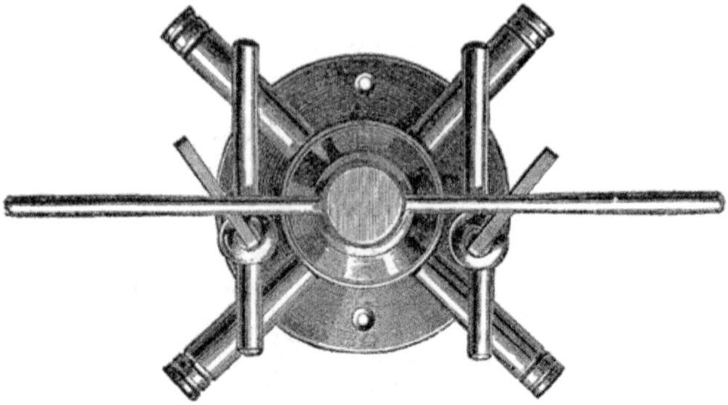

No. 372.

372.—The Star Four-way.

Same as foregoing in construction, but used only for the top lantern of triple, or with two of the same in conjunction. Brass £0 15 0

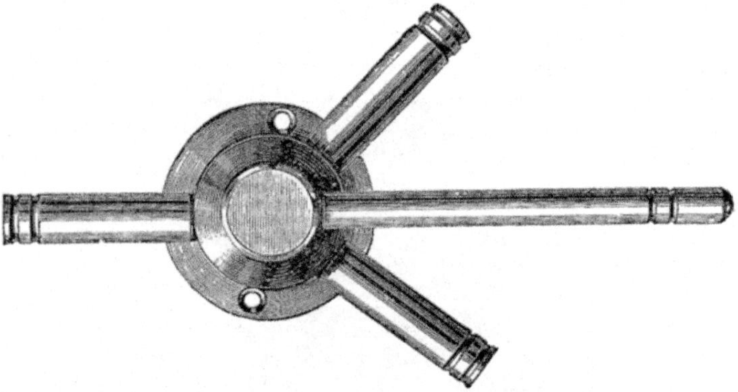

No. 373.

373.—The Three-way.

This is a special one, for use with a pair of the spirit jets only. Brass £0 12 0

PART IV.

Various Apparatus Used in Exhibiting.

A LANTERN picture, being something more than thin air, requires a resting place, and this is termed the screen or "sheet," and must be, of course, sufficiently large to take a picture of the desired size. In the earlier days of the optical lantern—when some amount of mystery was observed in order to keep up the character as well as the name of "magic" lantern—it was customary, and the custom still obtains to a small extent, to place the audience on one side the sheet and the lantern on the other, the sheet being rendered as transparent as possible by being wetted and kept wet. This, whilst it hid the operator and his apparatus from view of the spectators, and consequently added somewhat to the interest of the affair, necessitated the use of a long room, and the picture also lost somewhat in its transmission through the screen, whilst there was also the additional trouble of the wetting operations to be gone through; all of which have—now that the public generally have been let into the secret of how lantern pictures are produced—combined to cause a preference for pictures thrown on to the face of the screen in full view of the persons assembled to see them. The screen itself is usually of linen woven in very wide widths, for the smaller sizes made without a seam, and for the larger sizes having a large central square, and the rest of the sheet joined on round it so as not to have a seam down the centre or anywhere prominently within the picture. Better screens, so far as pure optical effect is concerned, are constructed with a facing of paper of very white colour, close texture, and smooth surface; but these opaque screens are impracticable when portability is a consideration. They do very well fitted on a roller (like a blind), and attached as a permanent fixture to the wall, and a screen of this kind in a room which is frequently used for lantern displays is of the greatest convenience. An even better result than this for constant use is to coat the wall itself with a smooth and even surface of white dull enamel. This, however, is expensive, and but rarely resorted to.

By far the majority of lanternists require their screens to be portable, and this necessitates the use of frames on which the screen may be stretched. Some amount of ingenuity has been shown in the construction of these, the chief points of desirability being that they may take apart or fold up readily and simply, and that, when erected and the screen upon them, they should keep it firmly and evenly stretched, presenting a smooth appearance at all parts of its surface. For the general uses of the exhibitor these frames are made with feet, so as to stand firmly upon the floor of the room; but at other times they are merely leant in as vertical a position as possible against the wall. To some connoisseurs of lantern work, however, this system—as well, indeed, as that of standing the screen upright on the floor—presents the objection that, as in most cases the lanterns have to be tilted somewhat upwards to get the pictures sufficiently high up on the sheet, the rays of light are not thrown evenly over the whole surface, and therefore, to get absolutely the best results, the screen should be tilted slightly forwards at the top, so as to place it at right-angles to the central ray of light from the lantern. This, however, although worth consideration when the finest results possible are sought to be obtained, is more apparent in theory than in actual practice, and may be, and is, ignored for all ordinary purposes.

For public entertainments where the exhibition of the pictures is accompanied by a descriptive lecture two persons are required to carry it through, one working the lantern at the back, whilst the other in front of the audience reads the lecture, and points out objects of interest in the pictures as they appear on the screen, and as all light, save that emanating from the lantern, is excluded from the room, it will be necessary for the lecturer to be provided with a reading lamp, several varieties being specially made for the purpose, the especial feature in their construction being the casting of their rays strongly on the book, whilst no light is allowed to emanate from them in any other direction, and the provision of some form of signal to secure the co-operation of the lantern operator. Nothing makes a lantern entertainment "hang" so much as any want of understanding between the lanternist and the lecturer, and consequent breaks in the continuity of the discourse, insertion of wrong slides, etc. At times, as when illustrating a story with the lantern, an arrangement may be made for certain slides to be thrown on the screen when certain words or sentences are read; but such a code is more frequently than otherwise impracticable, and resource has to be had to either a silent signal in the flash of a red light at the back of the lamp or an audible one in the shape of a gong. The latter system is the most certain, but the former is often considered preferable, as the

ORDER FORM.

....188

Messrs. ILIFFE & SON, 98, Fleet Street, London, E.C.

Please send me the new Penny Weekly Paper, "PHOTOGRAPHY," for

....*months.*

..............................*cop.* *Indispensable Handbook to Optical Lantern, 2s. 6d.* *Post free, 2s. 10½d.*

.... ,, *Science and Practice of Photography, 3s. 6d.* *Post free, 3s 10½d.* [*Paper covers, 2s. 6d. Post free, 2s. 10½d.*]

.... ,, *Photography for All, 1s.* *Post free, 1s. 2d.*

.... ,, *Photographer's Indispensable Handbook, 2s. 6d.* *Post free, 2s. 10½d.*

.... ,, *Collotype and Photolithography, 5s.* *Post free, 5s. 3d.*

for which I enclose..

Name..

Address..

....

attention of the audience is not called so forcibly to the fact of a change of scene being about to be made.

Besides many varieties of screens, frames and reading lamps, the following pages contain descriptions of a variety of supports and stands for the lantern and other items; but as we have fully dealt with the general principles and purposes of these in previous chapters we will only here refer our readers to the particulars which follow.

Screens and Screen Frames.

374.—Linen Screens.

A linen screen can be used in two ways—shown upon, or shown through The latter method is but very rarely adopted nowadays, for three principal reasons—1st, the result is not so good, owing to loss of light in transmission through the linen, 2nd, the trouble and bother of wetting the sheet, which is rendered necessary to get the best result; and 3rd, that in very few places is there sufficient room behind the screen to get back with the apparatus. The prices of linen screens do not vary very much, and therefore the quotations following may be given as average ones :—

	Best White Union Screens.	£	s.	d.	Best Linen Screens without Seams.	£	s.	d.
5ft	square .. .. .. each	0	4	0	.. .. each	0	7	6
6ft.	,, ... ,,	0	5	6	.. .. ,,	0	10	6
7ft.	,, ... ,,	0	7	6	.. .. ,,	0	16	0
8ft.	,, ... ,,	0	10	0	.. .. ,,	1	1	0
9ft.	,, ... ,,	0	16	6	.. .. ,,	1	7	0
10ft.	,, joined with wide width in centre ,,	0	18	0	.. .. ,,	1	13	0
12ft.	,, ... ,,	1	3	6	joined with wide width in centre ,,	2	5	0
14ft.	,, ... ,,	1	17	6	.. .. ,,	3	0	0
16ft.	,, ... ,,	2	8	0	.. .. ,,	3	15	0
18ft.	,, ... ,,	3	0	0	.. .. ,,	4	15	0
20ft.	,, ... ,,	3	12	0	.. .. ,,	6	6	0

375.—Opaque Screens.

These are made of calico, covered with white paper, having an ornamental border, and mounted on roller and lath as a map.

	£	s.	d.
6 feet square	0	15	0
8 ,, ...	1	0	0
10 ,, ...	1	10	0

376.—The Self-acting Rolling Screen.

These can be fixed to any wall, beam, window, roof, &c. The bottom edge is provided with strip of glazed material, so that, when it is rolled up, the white sheet can be kept clean and free from dust Each roller is fitted on a lath of wood ready for fixing. A lever (self-checking) is at the right hand side, so that it can be drawn up or let down any distance required by simply pulling or slackening the cord. A slight pull of the cord will raise the lever, and let the curtain or screen run down; by slackening the cord it is locked in any position required.

				£	s.	d.
Opaque screens	.. ..	6ft. x 6ft.	..	1	0	0
,,	.. ..	8ft. x 8ft.	..	1	5	0
,,	.. ..	10ft. x 10ft.	..	1	15	0
Transparent screens	..	6ft. x 6ft.	..	0	10	0
,, ,,	..	8ft. x 8ft.	..	0	16	6
,, ,,	..	10ft. x 10ft.	..	1	4	0
,, ,,	..	12ft. x 12ft.	..	1	10	0
,, ,,	..	14ft. x 14ft.	..	2	0	0

377.—Screen Frames.

The following are the usual trade frames. They are made of stout poles, with brass connecting tubes and strong brass corner pieces, with pulley wheels. Complete in box (space for screen), with lock, key, and handle.

		£	s.	d.
8ft. square, which will also make	4ft. and 6ft.	1	16	0
9ft. ,, ,, ,,	3ft. and 6ft.	1	16	0
10ft. ,, ,, ,,	4ft., 6ft., and 8ft. square	1	16	0
12ft. ,, ,, ,,	4ft., 6ft., 8ft., and 10ft.	2	5	0
14ft. ,, ,, ,,	4ft., 6ft., 8ft., 10ft., and 12ft. ..	2	5	0
16ft. ,, ,, ,,	4ft., 6ft., 8ft., 10ft., 12ft., and 14ft...	2	14	0
18ft. ,, ,, ,,	4ft., 6ft., 8ft., 10ft., 12ft., 14ft., and 16ft.	3	0	0
20ft. ,, ,, ,, and 18ft.	4ft., 6ft., 8ft., 10ft., 12ft., 14ft., 16ft., ...	3	3	0

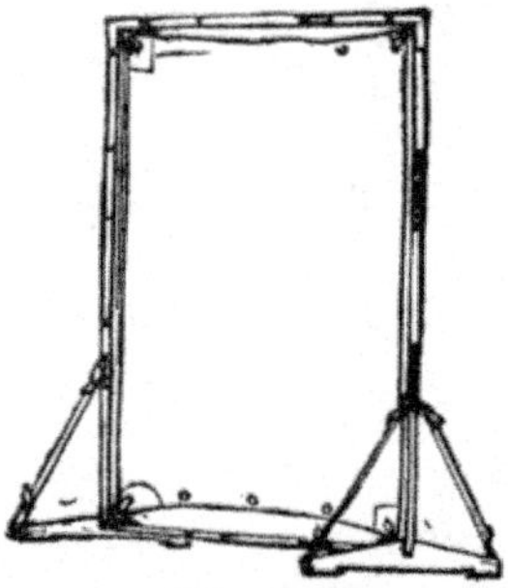

No. 378.

378.—Small Screen Frames.

These are of course very light, but are only suitable for the smaller sizes.

			£	s.	d.
5ft. screen stand in box,	2ft. 10in.	long with space for screen, each	1	1	0
6ft. ,, ,,	3ft. 4in.	,, ,, ,,	1	1	0
7ft. ,, ,,	3ft. 10in.	,, ,, ,,	1	1	0
8ft. ,, ,,	3ft.	,, ,, ,,	1	10	0
9ft. ,, ,,	3ft. 4in.	,, ,, ,,	1	10	0
10ft. ,, ,,	3ft. 8in.	,, ,, ,,	1	10	0

Boxes are 6in. wide and 6in. deep.

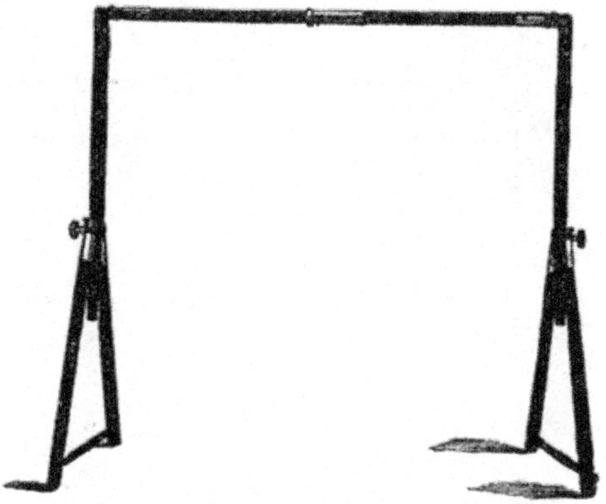

No. 379.—Fig. 1.

Fig. 2.

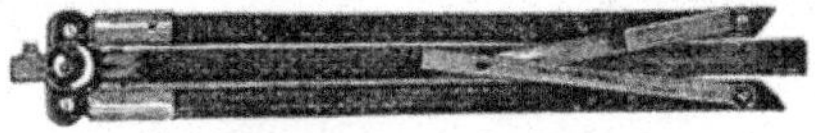

Fig. 3.

No. 379.

379.—Articulosus (W. C. Hughes, Kingsland).

The frame will stand alone, solid, and firm, as shown by the drawing. There are also peculiarly constructed joints which hold the top, sides, &c., together, that the whole is only divided into two separate pieces, which join in the centre of the top piece by aid of a special bayonet joint. The legs, which support the uprights, cannot recede farther, neither can the corner top pieces—they are held in check by a powerful and ingenious contrivance, so that the whole screen frame can be adjusted ready for use in less than two minutes. Everything is firm and steady, and being held together by stationary and immovable joints, no portion can give way, neither can any of the pieces be lost. The upright rods are made to slide up and down, and kept firm by aid of a powerful screw, so that an additional height may be obtained.

Fig. 1 shows the screen-frame erected, and figs. 2 and 3 when closed and packed for travelling.

	£	s.	d.
9 and 10 feet	2	2	0
12 feet	2	10	0
14 and 16 feet	4	4	0

380.—Screen Elevators (A. Pumphrey).

Exhibitors are often called upon to show in a room where there is no convenience for putting up a screen; without a ladder or assistance the work of erection is oftentimes uncertain or laborious. He recommends opaque screens as decidedly the most satisfactory. A sheet of calico nailed to a flat wall, and then pasted over with white paper, allowed to dry, and then rolled up like a map, with a lath at the top and a roller at the bottom. Such an

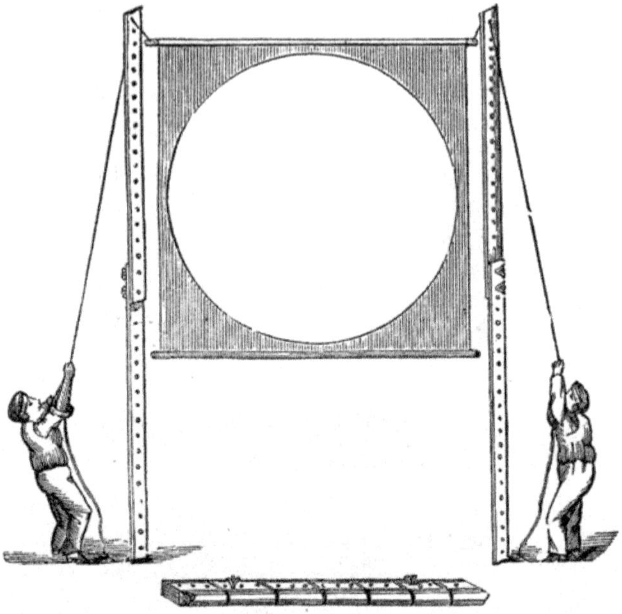

No. 380.

one, rolled up, is represented in the illustration. Along with it are four pieces of wood, about 3in. wide and 1in. thick, pierced at intervals of six inches with holes. Through these holes a pair of screws with fly nuts are placed; supposing the screen to be 10 feet square, two of the pieces of wood placed together will reach a height of 19 feet. These are placed on either side of the screen, and a string is run through the top holes; by means of this the screen can be hoisted up to any required height. The position of a picture in a room is often of more importance than the size. If hoisted up so that the bottom is 8 feet from the ground, in a large room with flat floor, all will see equally well; if the subject is made too large, so that the bottom of the picture is low, those at the further end of the room will not see. If there is no wall to rest the supports or elevators against they may be tied to a form and leaned backwards, and the strings may be brought forward so as to support the screen. The illustration represents the supports used as above, and the screen being hoisted. £0 8 6

381.—The Universal (J H. Steward).

This is adjustable to any size from half its length to full length, and is entirely self-supporting, has no loose pieces, but folds and slides into position. Can be inclined to or from the lantern, so that it shall be parallel to the front, and so correct distortion occasioned by the tilting of lantern. It is made of superior pine, stained and varnished. The mounts and hinges are of brass and gun-metal. Sizes 6ft. to 20ft, price per square foot £0 5 6

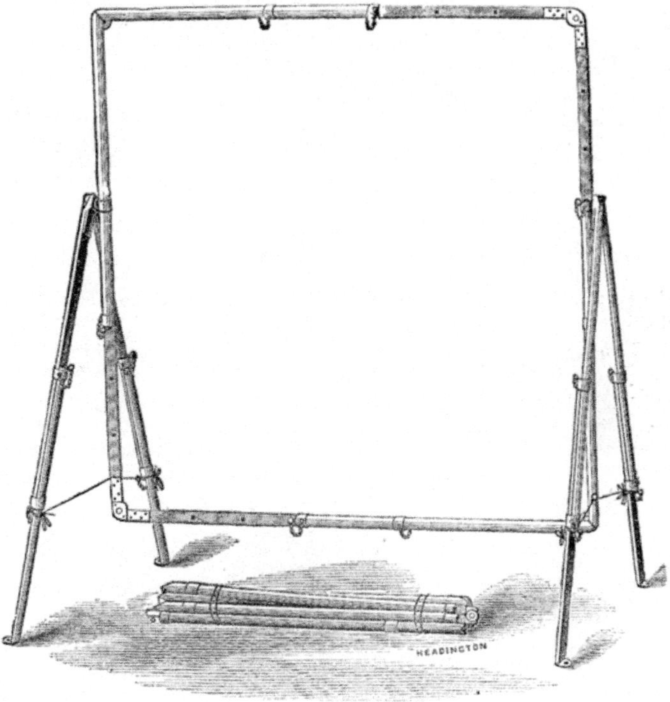

No. 381.

If in plain brass, with bolts for joints and brass mounts, price per foot £0 4 0

8ft. Plain	..	£1 12 0	..	Best	..	£2 4 0
9ft. ,,	..	1 16 0	..	,,	..	2 9 6
10ft. ,,	..	2 0 0	..	,,	..	2 15 0
12ft. ,,	..	2 8 0	..	,,	..	3 3 0
16ft ,,	..	3 0 0	..	,,	..	3 17 7

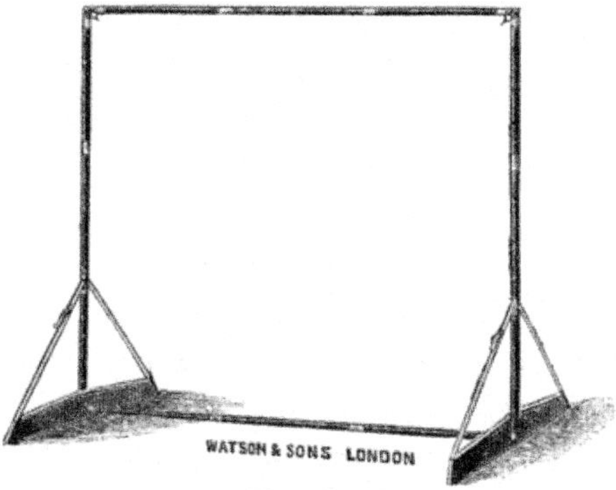

No. 382.

382.—Watson's Screen Stand (W. Watson & Sons).

This is a simple pattern to erect, and rigid too.

For Screen.	Length	Black Poles.	White Polished Poles.
8ft.	3ft.	£1 15 0	£2 10 0
9ft.	3ft. 4in.	2 2 0	2 15 0
10ft.	3ft. 8in.	2 2 0	2 15 0
12ft.	..	2 2 0	2 15 0

Reading Lamps, Desks, &c.

These may be termed the personal comforts of the lecturer, for they certainly tend to render his task a pleasure instead of a misery. The object aimed at is to provide a good powerful light or easy desk, which, whilst allowing the lecturer to see easily his notes or manuscript, will not deteriorate the picture on the screen. Some of the lamps have also a bell, red signal, or both, to indicate the required change of slide to the operator.

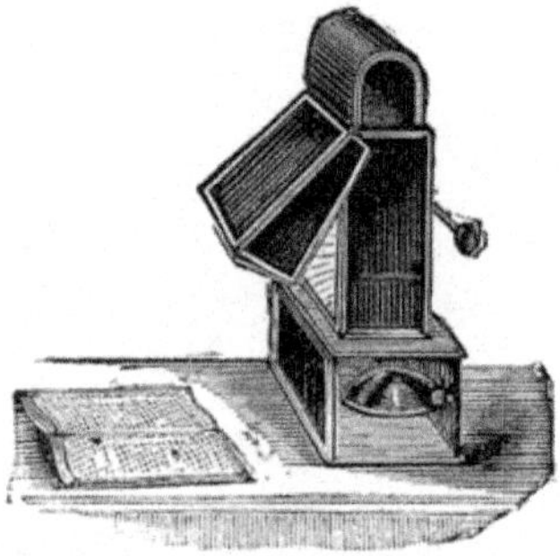

No. 383.

383.—Reading Lamp.

With oil light, bell lever and red signal to operator £0 12 6

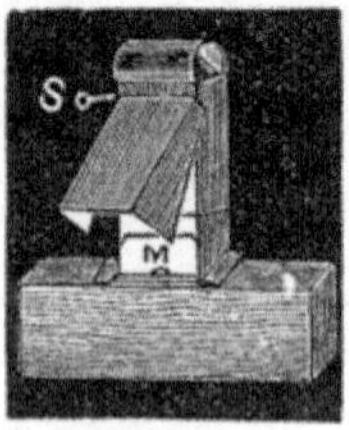

No. 384.

384.—Archer's (Archer & Sons).

The features of this lamp are the light is cast only upon the book. (S) Silent signal (red light) may be used. (M) Match-box always at hand. A screw cap fits over burner and acts as extinguisher; when screwed down, oil cannot run out even if turned upside down. Will burn for twelve hours without refilling. Can be used upon table, desk, stand, piano, organ, or harmonium, and only illuminates the part the light is directed upon. Being open, does not impede the sound of lecturer's voice. The whole packs in box 3in. by 3in by 8in., forming stand, and so made that a night-light could be used instead of oil. Price 7s. 6d., or with bell £0 9 6

385.—Archer's Candle (Archer & Sons).

To burn night-lights or candle only, with new reflector and adjustable shade. In box complete £0 5 0

386.—Armstrong's (Armstrong & Brother).

This is a cheap lamp, being of same design as No. 383, but the price only £0 9 0

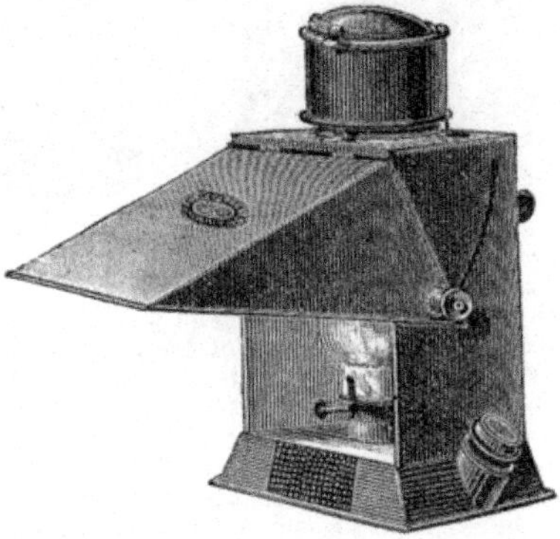

No. 387.

387.—The C.P.S. Perfection (Cubley & Preston).

A lamp to burn mineral oil. The following are its principal features:—

(1.) It possesses a regulating reflecting screen, which throws a perfect light over a large or small surface of the book or manuscript as required.

(2.) By means of an outside supply-tube, it can be readily filled, or as readily emptied, without soiling the hands, avoiding, as it does, the oily and sooty condition so common to reading lamps.

(3) A candle adapter is provided, so that it may be used as a candle lamp.

(4.) It has a light-excluding and ventilating chimney.

(5) It has a ruby signal for communicating with the operator, or, if desired, a signal bell in addition.

(6) The inside is bright metal, and free from the offensive odour of scorched paint.

We tested this lamp recently, and find it to give a capital light, with little or no attention, after once lighting. The compiler has used it for hours together when compiling this Handbook. Without bell £0 10 6

388.—The Refulgent Benzo Lamp (W. C Hughes, Kingsland).

A new form of reading lamp for burning benzoline or other mineral oil. It has the following features:—It will not fail nor snuff out, the flame retaining its brilliancy as long as the cistern contains any spirit; the spirit cistern, placed outside the lamp and connected with the interior by a tube for feeding the flame, supports a light for four hours. Risk or danger from overheating the spirit is minimised, and the lamp can be easily filled or

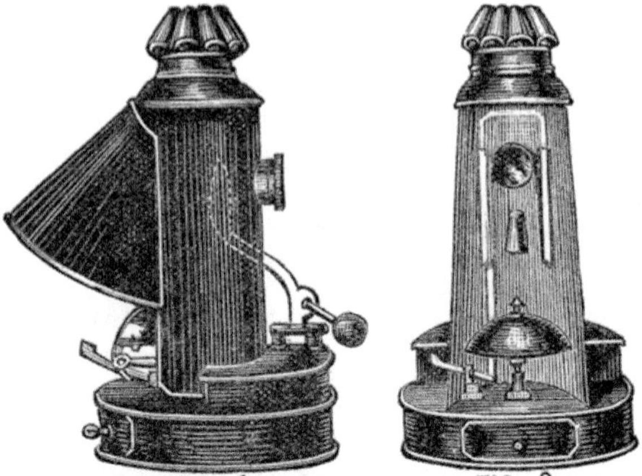

FIG 1 FIG 2

No. 388.

emptied; the reflecting screen can be regulated to cover any portion of a book or manuscript; on the back part of body is a coloured signal to communicate with the operator, the shutter of the signal being worked from inside the lamp by a lever at the reader's hand, and close by is a signal bell, and underneath a drawer and striker for matches, all light is excluded, except on the book, by a ventilating shaft; the wick can be raised or lowered by rack and pinion to regulate the light; the japanned inside prevents an offensive smell. In use, the cap is unscrewed, and the absorbing material in the cistern saturated, the residue to be poured out (away from any artificial light). Benzoline gives the best results, other oils do not need the absorbing material in the cistern £0 12 6

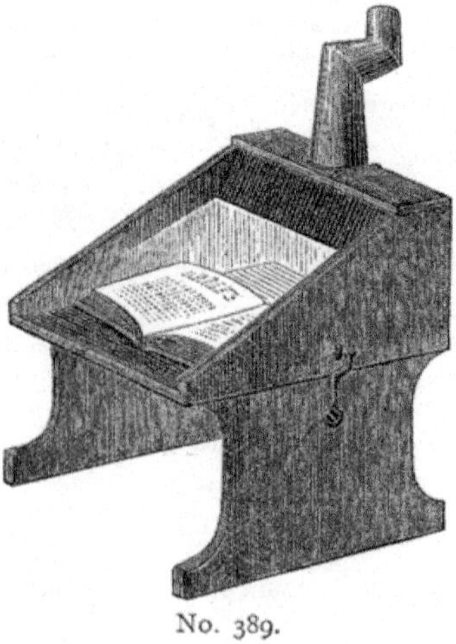

No. 389.

389.—Lamp and Signalling Desk (E. G. Wood).

This consists of a light wooden case, stained and varnished, large enough to take paper 13 × 11, screened at the sides, and with a hinged hood, covering about one-fifth of the length, fixed to the back. Underneath this hood (which is provided with a small chimney) is situated the lamp. By pressing a small stud on the right of the chimney a red light is displayed to the operator, and a silent signal given. When not in use the sides and top fold down flat for portability £0 12 6

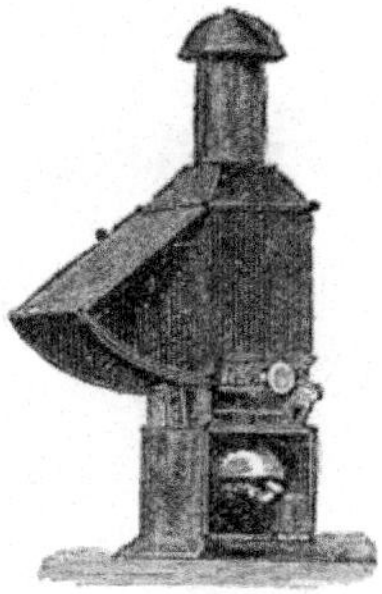

No. 390.

390.—New Shape Mineral Oil Lamp (Wholesale Firm).

This lamp is so constructed that it will burn paraffin oil instead of sperm oil, as in ordinary reading lamps, thus giving a far more brilliant light and being much cleaner to use, and at the same time requiring no lamp glass, which is a very great improvement on ordinary paraffin lamps, the glasses of which are so often breaking. Silent signal, bell, and match box In box, each £0 10 6

391.—Portable Reading Stands (W. C. Hughes, Kingsland).

The illustration (fig. 1) gives a clear idea of this reading stand. The lamp itself is made entirely of brass, which, with the desk, revolves round to any side, at the operator's discretion. The lamp burns ordinary colza or paraffin oil, an advantage over candles. The lamp shutter reflects the light only on the book. The base socket has a screw passing through the foot into the ground, which gives stability It will divide and take to pieces, and pack portably. The upright rod is polished ebony ; the reading desk is japanned. The lamp apart is useful for other purposes than those mentioned ; for instance, it can be taken off the stand and used for illuminating music on a pianoforte during an exhibition £0 17 6

Fig. 2 represents a stand much the same in general detail, but is slightly cheaper, being only £0 15 0

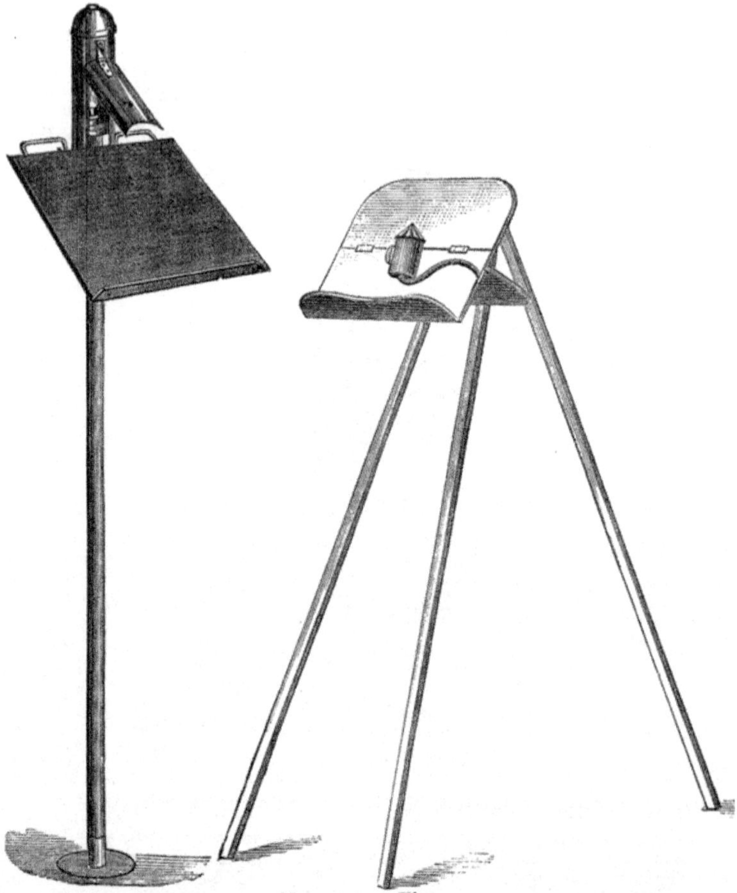

Fig. 1. No. 391. Fig. 2.

Tripods, Lantern Supports, Tables, Cases, &c.

Under this heading will be found the various supports for the lantern when in use. Tripods are convenient and portable, but in the majority of cases, as the lantern must be carried about from place to place in some sort of box or case, that case is made to serve the double purpose of a stand. The extended legs of a stand, too, are awkward to deal with in the dark, and a further disadvantage is the absence of any place to put slides and other things down upon. In some designs a sort of compromise is effected by a combination of tripod and case. The supports for a lantern may thus be divided into

Tripod stands,
Cabinets or cases to stand on floor,
Cases provided with supports on legs.

As will be seen in reading those described, some cabinets are most elaborately fitted up for convenience.

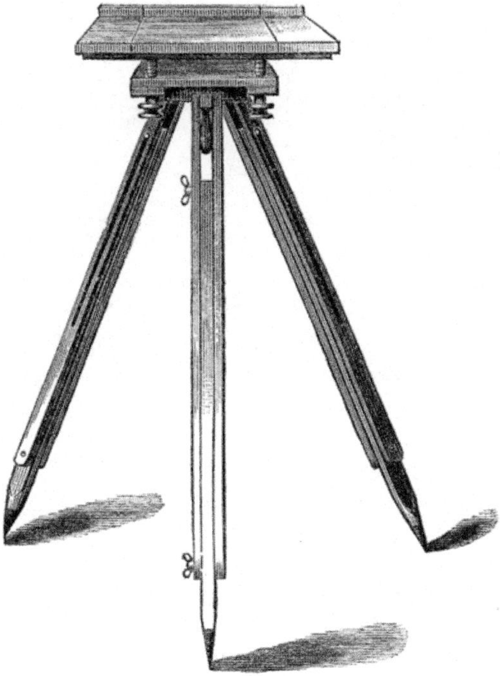

No. 392.

392.—Hughes's No. 1 Tripod (W. C. Hughes, Kingsland).

A tripod with sliding legs, and screws to elevate to any angle. The top works on a centre, and can be removed at pleasure by turning the screw underneath. The top boards are hinged, and will fold down if required. The front can also be raised or depressed by the two front screws. It is one of the most substantial and firmly constructed stands for using lanterns in pairs, biunial or triple £3 10 0

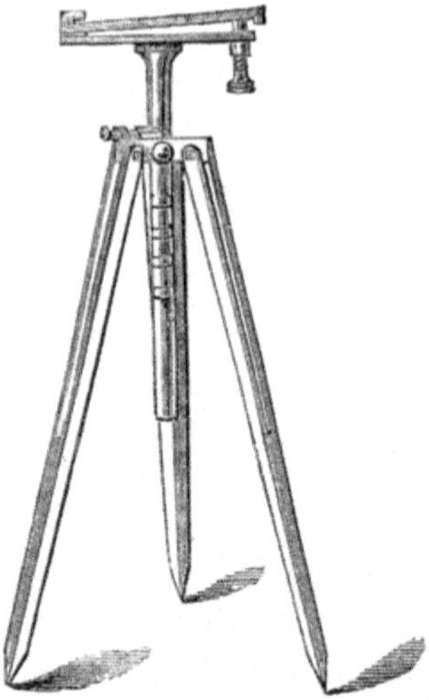

No. 393.

393.—Hughes's No. 2 Tripod (W. C. Hughes, Kingsland).

A cheaper but reliable tripod. The platform top is raised by screws to throw the lantern to any angle The centre support can be adjusted to different heights, and kept stationary by two screws. Stained oak £1 15 0

394.—Mason's Cabinet (Mason & Co)

Illustration of this will be seen at No. 163. The case is supplied with legs (which unscrew) and a shelf. A small handle screw, turned from inside the case, tilts the lantern as required. This is a very steady cabinet, and there is plenty of room for the slides. These can be arranged previous to the exhibition, and the lid of the box put on and locked, thus preventing subsequent disarrangement.

395.—The Noakes Triple Case (D. Noakes & Son).

This really consists of two cases When travelling, the lantern is packed in one and the various accessories in the other. For use, one box is placed flat on the floor (or benches, pews, &c , as required) and the other is firmly stayed to it by rods in an upright position. These rods are used to fasten the boxes for travelling as well. The upright box is conveniently arranged with shelves. Pads are so arranged that the lantern is firmly held when the stays are screwed on. Altogether an excellent arrangement.

396.—Ottway's Cabinet.

An illustration of this will be found at No. 230. It needs very little further description. In addition to the shelves, it has a convenient drawer for lenses and odd apparatus. The tilting screw is in front, and works from the outside of case.

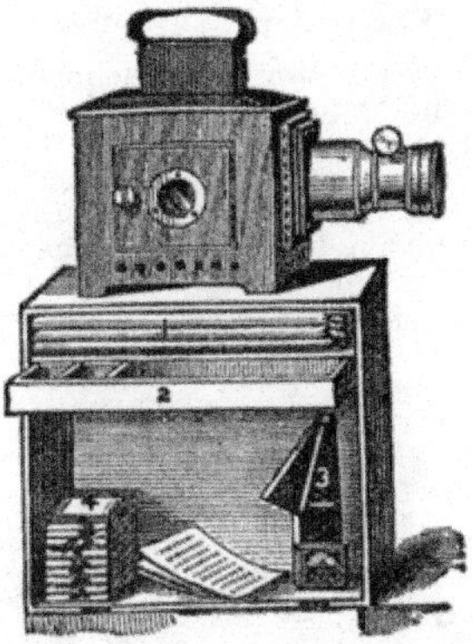

No. 397.

397.—Pexton's Exhibitor's Cabinet (Chatham Pexton).

In the illustration, 1 is the sheet, 2 the tool drawer, 3 the reading lamp placed in box when the lecturer is also the operator, and 4 the slides arranged. A strong partitioned case, with spaces for lantern, sheet, tools, extra lamp glasses, paraffin oil, &c., and forming stand to which the lantern may be screwed when in use. The price includes a 9ft sheet with rings and cord for suspending, a' can for paraffin oil, hammer, bradawl, supply of screw-eyes and holdfasts, lamp scissors, three lamp glasses, and six lamp wicks

Single lantern. Case	£1 15 0
Double lantern. Case	2 0 0.

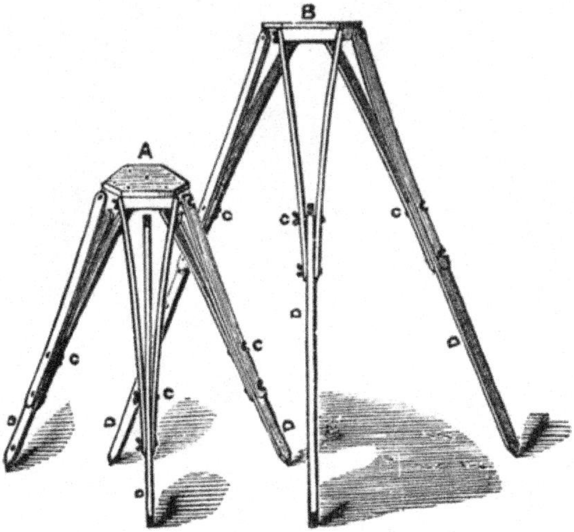

No. 398.

398.—Pumphrey's Tripod (A. Pumphrey).

This tripod has the power of adjustment to work at two different heights—either at standing height, or at five to nine feet high when the exhibitor is on the same side of the screen as the audience and wishes to be above them—as at A the stand can be adjusted to a height of from four to five feet, and at B it will stand nearly double that height. In order to alter the stand from single to double height, the screws at C are taken out, and the last joint of each leg D is turned down and the screws C replaced. Each leg is pointed at both ends, and made to stand four and seven feet, or five and nine feet, as desired.

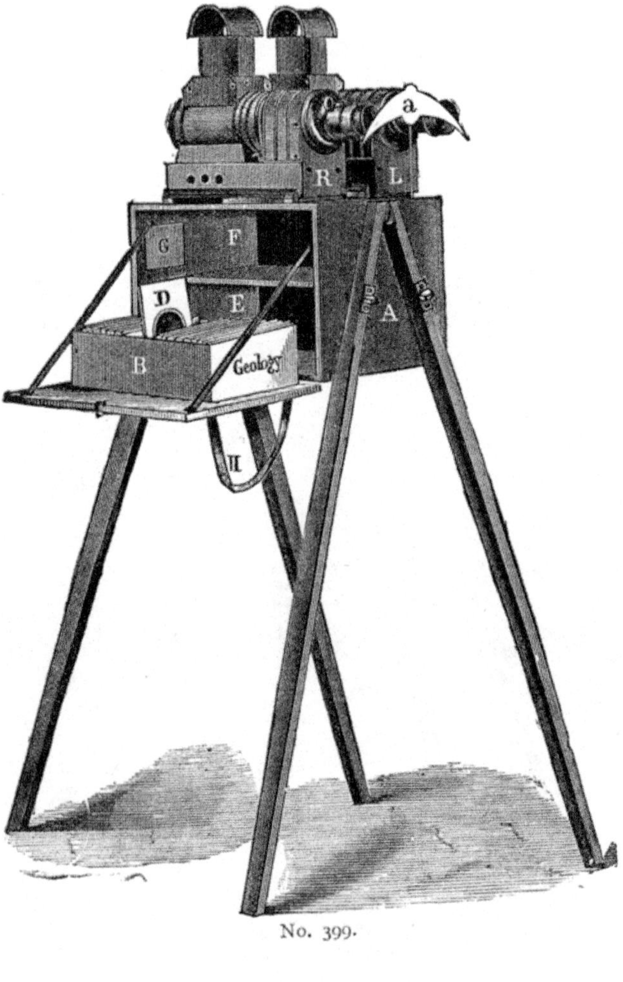

No. 399.

399.—The Sciopticon Dissolving Stand (Sciopticon Co)

The above stand, though represented as for Sciopticons only, ought to be suitable for any oil lanterns that have not large dimensions. It consists of a well-made box, A, mounted on two pairs of legs, attached by thumbscrews and nuts. When the apparatus is taken down, the legs swing together on their hinges, and are tied in a bundle; the open side of the box becomes the top; the instruments occupy the divisions E and F; the dissolver is drawn apart and placed alongside; the tray for slides B occupies the space in front; the swing shelf C becomes the lid, and is locked down. But as a stand, as seen in the diagram, the front of the box becomes the baseboard, and, like any other 15 by 17 board, affords suitable standing room for the apparatus; it is more likely to keep level than a separate board, as it is firmly fastened in place. In place of the legs represented, a tripod stand, with sliding legs, may be used, fastened by means of a thumbscrew to the box. This affords the facility of raising the lanterns to any desired height, and also of tilting them when necessary. Although they will bear tilting to a considerable angle without danger, it should be avoided, unless the screen can also be placed parallel with the slide, otherwise the definition will suffer. Still greater firmness is obtained by fitting the tripod legs direct to the box. Or the box can be simply placed on a table which will raise the lanterns to a convenient height. The swing shelf C is made removable, the divisions E F thus forming convenient receptacles in which to arrange the slides for exhibition, while the tray B will be a handy place for dusters, loose apparatus, &c. The fronts of the lanterns R and L hold firmly by claws to two screw-heads; the flanges in the rear slide under two similar screw-heads holding the instruments down, but allowing them to spread till their discs coincide on the screen.

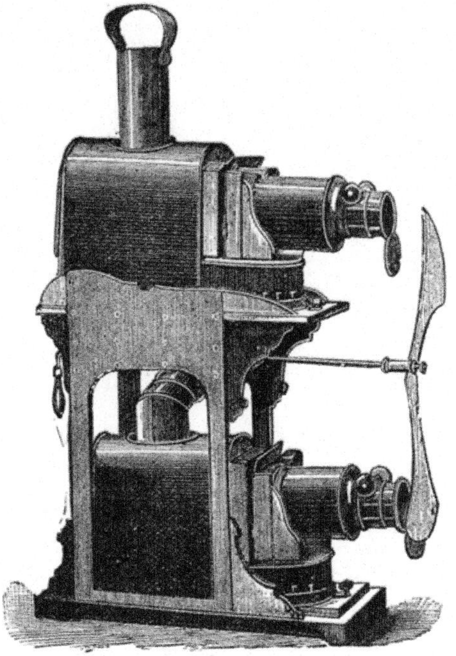

No. 400.

400.—Bi-dissolving Stand for Oil Lanterns (W. C. Hughes, Kingsland).

An arrangement for using oil lanterns in the form of a biunial. The top and bottom lanterns are placed in a mechanical platform frame, made of polished mahogany , a non-conductor of heat. The chimney of the lower lantern is curved and brought quite outside, thus preventing any connection with the upper lantern The platform and frame have every accommodation for fitting the lanterns to cause coincidence of discs Between the spaces allowed currents of cool air pass, and non-conductors are placed in requisite positions to prevent any radiation of heat. Everything is so devised that after several hours' burning the upper and lower lantern lamps remain cool. The lantern can be removed and used singly or side by side, including mechanical dissolver. Any two or three-wick lanterns can be worked in this stand. The price for No. 1 bi-dissolving stand, including dissolving arrangement, with handle for working at the back, for either two triplexicons or duplexicon lanterns £2 2 0

Second quality £1 10 0

The price of the stand without the dissolver, best £1 5 0

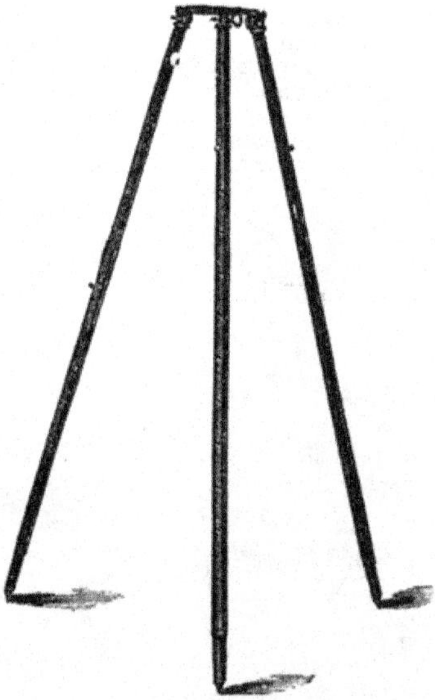

No. 401

401.—Portable Brass Tripod (Wholesale Firm).

This tripod stand is made entirely of brass, being a brass tube with a telescope sliding tube, adjustable at every 3in. from 3ft. 9in. to 6ft. 6in., and can be fitted to any lantern box or board. It is most portable and solid, and can be packed in the screen stand box when not being used .. £2 12 6

402.—Revolving and Canting Table.

These are sold pretty generally. They will fit any lantern box or stand, and can be moved round with great ease or inclined up or down as wished by

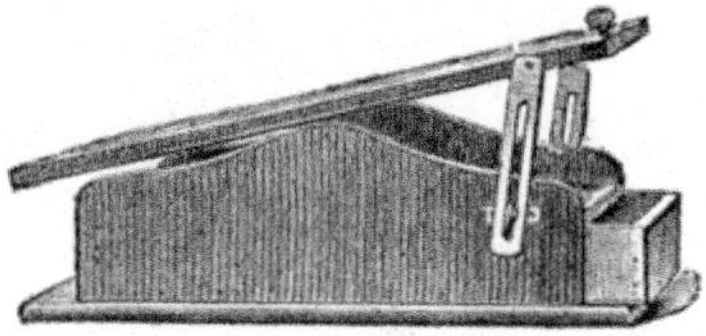

No. 402.

simply loosening the screw and moving to position required, and the screw again tightened. Made in polished mahogany, with drawer between the tables for keeping tools, etc., size 18in. by 11in., with two brass quadrants £1 2 6

Various Accessories to Lanterns.

403.—Tinter Openings in Lanterns (OTTWAY & SONS)

These are placed so that a graduated or other tint slide can be passed through the lantern between the condenser and the luminant. This produces a pretty effect upon the screen, as, being *behind* the condensers, it simply colours the rays of light before they are acted upon by the lenses, and gives a superior effect without the consequent loss of light in tinting in front of the nose of the lantern A graduated slide, 18in long, painted to represent the different tints of sunset down to the deep blue of night, produces a fine effect

404.—Rack and Pinion Curtain Diaphragm (OTTWAY & SONS).

This is an improvement on the usual drawn diaphragm. It is made with rack and pinion movement, actuated by a neat brass and ivory winch handle, which ensures a smooth and even motion, not always attained by the sliding form.

405.—Curtain Diaphragm for pair of Lanterns (W. C HUGHES, Kingsland).

This is a novelty, but of which we have no details. It is made for the purpose of obtaining the same effect in side-by-side lanterns of a curtain or picture rolling up, as the ordinary form in a biunial. It is made in brass, but requires some slight alteration in the feet and body of each lantern £2 2 0

406.—Brass Fronts (T. S. & W. TAYLOR).

Brass fronts for biunial optical lanterns, with adjustable slide stages, opening at the top and sides , hinged to one plate, and with fittings for 4 or 4½in. condensers.

Single tubes	£1	18	0
Double draw tubes	2	8	0
Triple draw tubes	2	18	0
Rolling curtain diaphragm, slide movement extra	0	5	0

PART V.

Scientific Accessories to Lanterns.

As has been explained in previous chapters, it is not necessary that lantern slides should be merely pictures on glass, the optical lantern being adaptable to a variety of interesting and scientific purposes by the use of specially constructed prisms, microscopes, etc., etc., and not only can slides which are not transparent be exhibited, but small machinery in motion, and even living insects and organisms illuminated and enlarged upon the screen, together with a large number of scientific experiments illustrative of the phenomena met with in the study of chemistry and light. In the following pages will be found illustrations, descriptions, and prices of the various scientific attachments to the lantern which have been devised and placed on the market to obtain these effects.

407.—Electrical Decomposing Trough (Cubler & Preston).

This is a very well arranged trough for showing electrical decomposition —such as that of water, for instance. Mahogany stand, glass sides and solid indiarubber walls, mounted in brass frames with binding screws, upright brass electrodes, with universal joints and electrical screws complete. This trough can be very readily taken to pieces and cleaned, having many advantages over the ordinary kinds, and being more solid and durable £1 1 0

408.—The Aphengescope.

This is an apparatus for showing opaque objects, and is sold universally. The prices following are a fair average. It is adaptable to any of the larger lanterns (having 3in. condensers and upwards) by removing the ordinary front lens, and fixing it in its place. It consists of a japanned tin body, having an aperture for the introduction of a carte-de-visite, watch, ferns, flowers, coins, medals, etc., just where the rays of light can illuminate them,

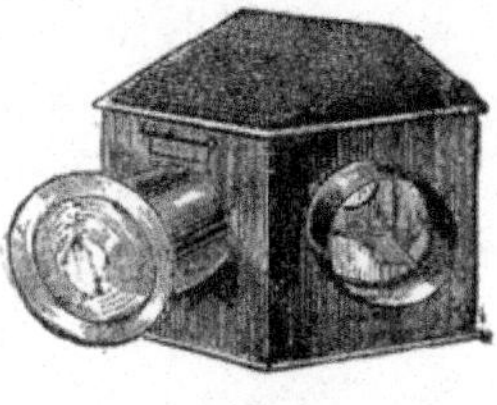

SINGLE.

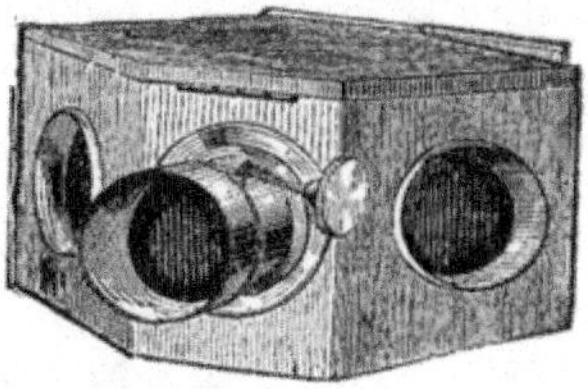

DOUBLE.

No. 408.

and at such an angle that a combination of lenses in a brass adjusting mount

magnifies and projects them on the screen	£0	17	6
If for lantern with rack and pinion front	1	1	0
Ditto, ditto, superior, in mahogany, with brass fittings ..	1	12	6
The Aphengescope only, to take ordinary front lens of lantern, tin body	0	12	6
Ditto, ditto, mahogany body..	1	4	0

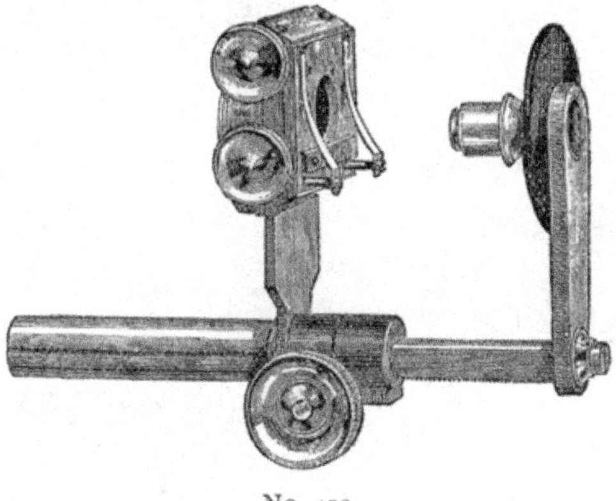

No. 409.

409.—Microscopic Attachment.

This microscope screws into the nozzle of the lantern, in place of the rack front lens, and is supplied with a double achromatic objective in brass case, and fitted into a mahogany case, with lock and key. The microscope is made entirely of brass, and is fitted with a revolving diaphragm and a mechanical stage, working vertically and horizontally, and with a substage condenser, fitted in a telescopic sliding brass tube for adjustment of focus, according to the length of the front O G. tube of lantern, thus allowing all the light to be concentrated on the object. It can be applied to any lantern. Microscope, with revolving diaphragm, mechanical stage, substage condenser, and one best quality double achromatic objective, in brass case, fitted in mahogany case, with lock and key, as fig. 1.. .. .. .. £5 5 0

Extras.—Fine adjustment to the above microscope, 7/6. Prism for showing the objects on the ceiling or floor, 16/- The following extra objectives can be supplied —No. 1, 2in , 2, 1½in. ; 3, ½in., at 45/- each. Alum trough, for use with microscope, to prevent the heat injuring the slides, 3/6. Glass trough, for showing animalculæ in water, 4/6.

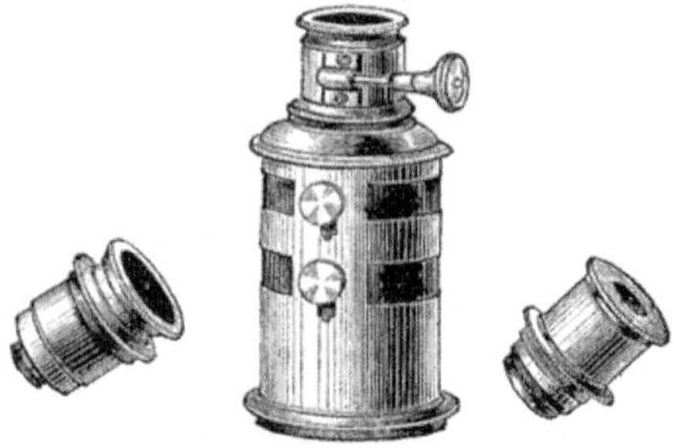

No. 410.

410.—Lantern Microscope.

This is the simplest form sold by all dealers. Brass body microscope, with one shallow power and one deep power, to fit any of the lanterns with 4in. compound condensers £1 2 6

Extras.—Mahogany box with lock and key, 6/- Very deep achromatic power, with rack adjustment for above microscope, for use with limelight only, 22/6.

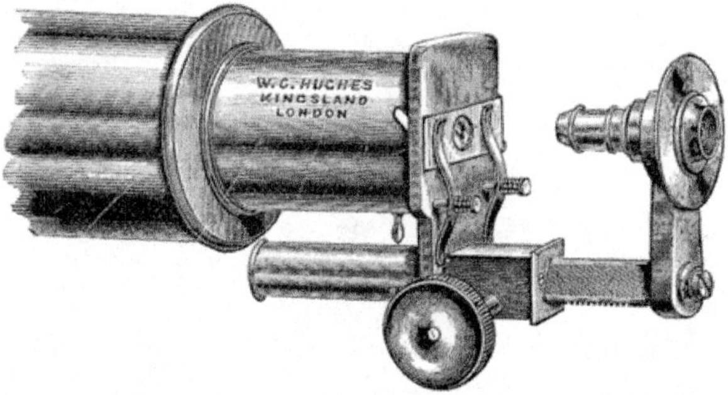

No. 11.

411.—Hughes's Cheap Microscopic Attachment (W. C. Hughes, Kingsland).

A form of microscope taking the place of the front lens to show ordinary microscopical slides. It is fitted with one best double achromatic objective, but the extra powers mentioned below can be used equally as well. Mr Hughes says that with an $\frac{1}{4}$in. the spiral formation of a blow-fly's tongue can be shown, and a sheep-tick is shown 6ft. long, exceedingly sharp and well-defined. Sections of wood, spiders, flies, scorpions, and each hair on a flea or other small insect are brought out with great distinctness, and that pond life is easily demonstrated, *volvox globator* showing young inside, and hydra 6ft. to 7ft. long. It has a movable substage condenser that enables it to be used with different object glasses, and a new form of spring on the stage by means of which the thinnest objects can be held as firmly as the glass zoophite troughs. The bar, with rack motion, is constructed on the best principle, by which wear and tear can be compensated for by simple adjustment of the screws, thus incurring absolute absence of all shake. If desired, the image, by a special contrivance perhaps superior to the usual right-angle reflecting prism, can be thrown directly on to the paper for drawing. It has a new form of diaphragm arrangement, by which the aperture can be changed with great facility The instrument is made entirely of brass and gun-metal. This microscope can be adapted, say, to the centre lantern of a triple, while the other two can be utilised for showing ordinary photographs and micro photos to consecutively illustrate a given object under different phases without leaving the screen blank £6 10 0

Extras—Alum trough, to prevent the heat injuring the objects, 6/6. Glass trough for pond objects, 2/6. Reflecting apparatus to throw the image directly on to the paper for drawing, 15/- polished.

412.—Lantern Microscope (Newton & Co.)

For illustrations see No. 260, the front attachment only. Complete with alum trough and parallelising lens, revolving diaphragm stage, rack adjustment to substage condenser, coarse rack and fine screw adjustments for focussing, adapter, with R M.S. standard screw for objectives, standard 1½in substage condenser and amplifying lens, but without objective. The whole packed in polished case £15 15 0

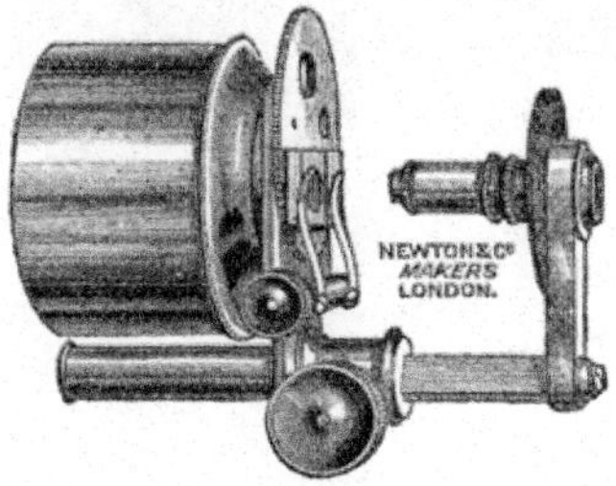

No. 413.

413.—Newton's Microscopic Attachment (Newton & Co.)

This new instrument screws on to the nozel of a lantern in place of the front lens, and with the limelight will show an ordinary microscopic slide on the screen 8ft in diameter, better defined than the old form of lantern microscopes Any good microscope objectives can be used with the instrument, which will show wood sections, scorpion flies, spiders, &c., 8ft. in diameter, sheep-tick 5ft. long, exceedingly sharp and clear, every hair on a flea or other small insect shows with perfect distinctness on the screen. Pond life is easily shown, hydra 6ft. long, *volvox globator*, showing young inside, &c. One double achromatic objective is supplied with the instrument, and additional achromatic objectives of different powers can also be had as below. With these powers the transverse striations of trachea of dytiscus and the spiral formation of blow-fly's tongue, can be shown. It is suitable for any lantern, and has a large rotating diaphragm, forming entirely open stage, which facilitates the manipulation, and gives a sharp edge to the disc. It is advisable to use an alum trough, as the most delicately mounted balsam slides may then be used without fear of injury from the heat, and pond life can be shown to much greater advantage, the water being kept perfectly cool. If desired, the image can, by means of a specially made prism, be thrown down direct on to paper for drawing, &c. The instrument is made entirely of polished brass and gun-metal £4 14 6

414.—Watson's Lantern Microscope (W. Watson & Sons).

The illustration shows a lantern fitted with the microscopic attachment and prism, which throws the enlarged picture down on a sheet of paper

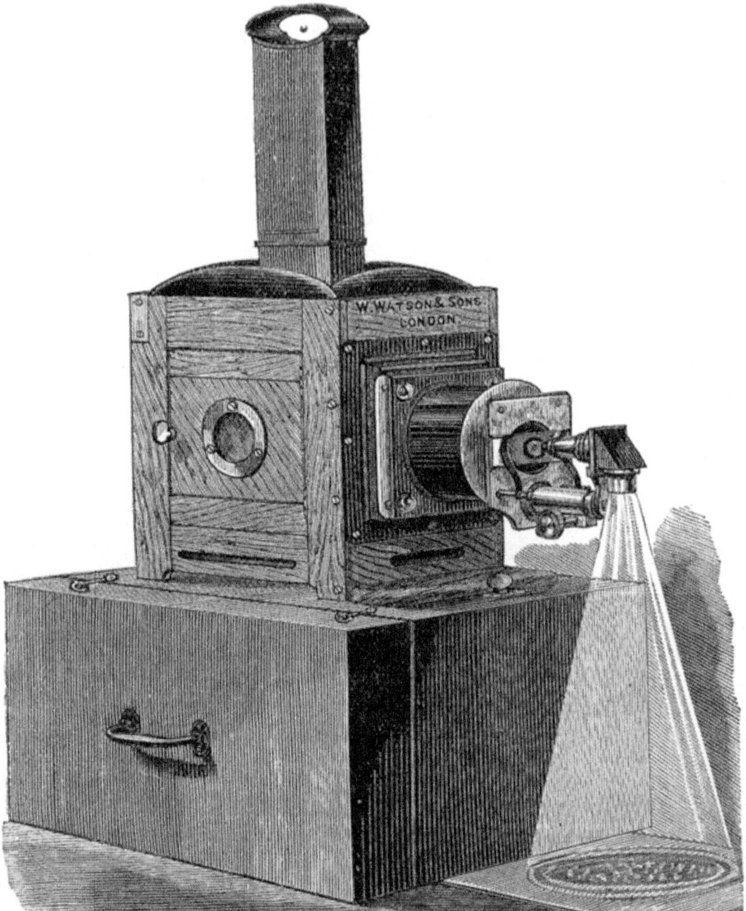

No. 414.

spread below to receive it. By this means an object can be conveniently examined by several people at once, and can be drawn or traced very easily. It can be fitted to any lantern. No object glass is included with this apparatus, but the powers from any microscope may be used .. £2 10 0

Extras.—1in. object glass, £1 15/- Additional achromatic objectives, constructed specially for microscopic projection—3in. achromatic objective (in case), £2; 2in., £2 2/-; 1½in., £2 5/- $\frac{8}{10}$in achromatic objective (in case), £2 7/6, $\frac{6}{10}$in., £1 10/-; $\frac{4}{10}$in., £1 10/- Alum trough, for use with the new microscopic attachment, to prevent the heat injuring the slides, 8/6. Glass troughs, for pond life, 1/6 to 2/6. Reflecting prism, in revolving brass mount to fit the above, to throw the image direct down on the paper for drawing, 17/6. Polished case, to contain instrument and alum trough, if required, 5/-

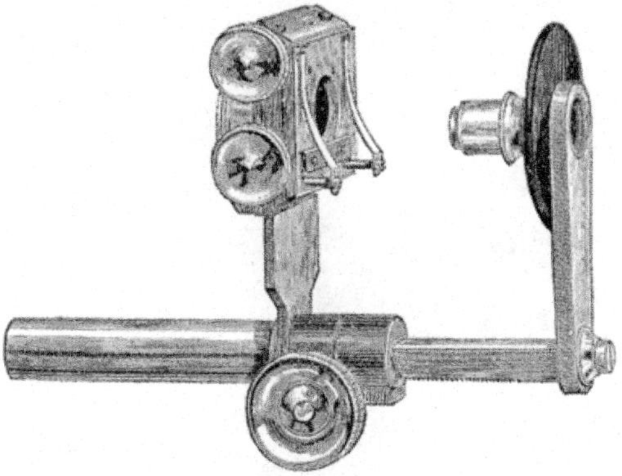

No. 415.

415.—Lantern Microscope (Mason & Co.)

With mechanical stage, rack adjustment for focussing, substage condenser with adjustments, revolving diaphragm and one achromatic objective. The whole in polished mahogany case £5 0 0

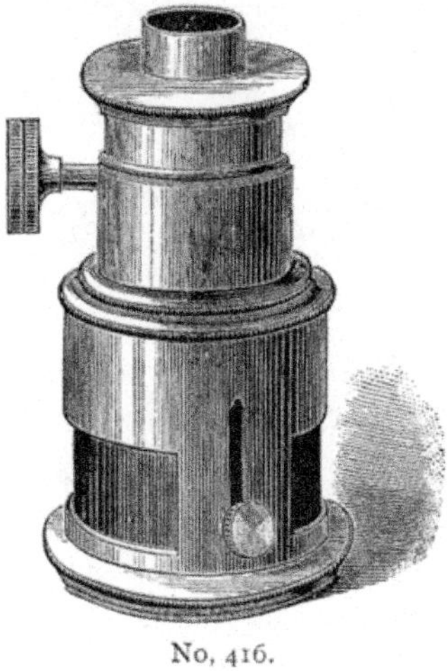

No, 416.

416.—The Lewis Wright Optical Front (J. H. Steward).

This front has spring clip tube for slides, troughs, &c., rackwork focussing arrangements, and lenses for projection £2 2 0

417.—The Kaleidoscope.

Another instrument sold by all. It is an adaptation of the well-known hand instrument, by which the beautiful and ever-changing chromatic designs can be projected on the screen. It is so arranged that it will fit any 3½in. or larger size lantern by simply unscrewing the front lens and screwing that in its place. It is made of brass, and has focussing adjustments. In case £2 2 0

Extras.—Slide containing coloured glass, beads, net, and other transparent objects, moved by rackwork, 7/6.

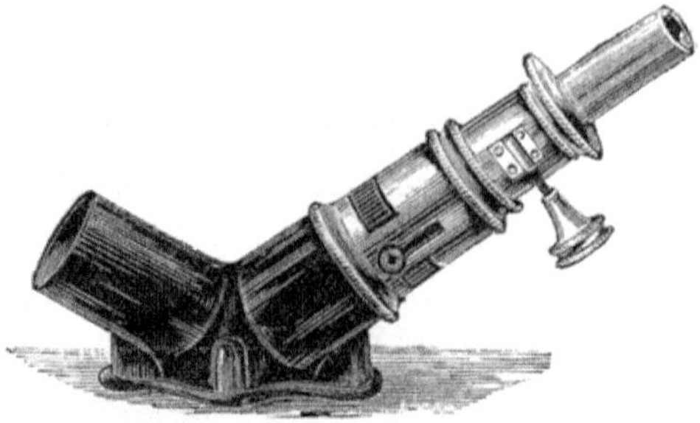

No. 418.

418.—Elbow Polariscope (Newton & Co)

Of best quality and construction, for illustrating the various phenomena of polarised light, with polarised glass plates, prism and lenses, mounted in brass, with rack adjustment to focus tube. This instrument is suitable for use with any good optical or magic lantern, and forms an excellent table polariscope In case, complete £6 16 6

Extras.—Brass adapter for using the front portion of the Elbow Polariscope as Wright's optical front, 7/6. Crystal stage for the above, for exhibiting crystals, 12/6. Case to contain microscope and troughs, 7/6.

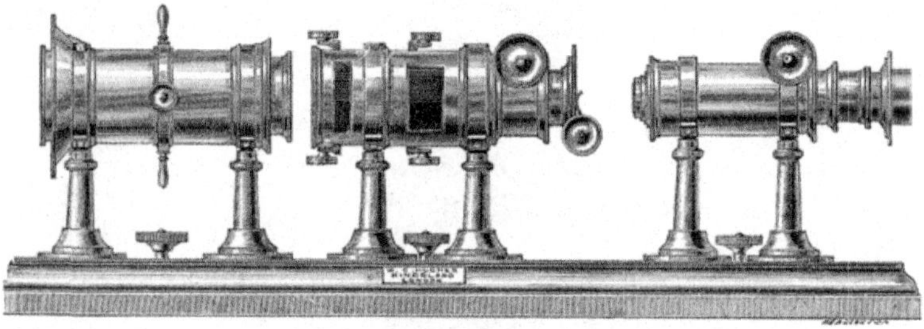

No. 419.

419.—Polariscope Direct Nicol Prism (W. C. Hughes, Kingsland).

This is fitted with large Nicol prism as polariser, a large aperture, special flat-end prism as analyser, rack adjustment for focussing the lenses, rotating motions to polariser and analyser, circles divided on German silver, crystal stage, double system of convergent lenses for exhibiting rings in wide angle bi-axial crystals, adjustment by rackwork when focussing, selenite stage, &c, on mahogany base-board with adjustments. In case complete £45 0 0

420.—Erecting Prisms (W. C. Hughes, Kingsland).

These are mounted on the brass stand as represented on the Gilchrist lantern (No. 257). They can be used with astronomical diagrams, mechanical diagrams, science slides, etc.

No. 1	£1 10 0
,, 2	2 2 0
,, 3	3 3 0
,, 4	4 10 0
,, 5	5 15 0

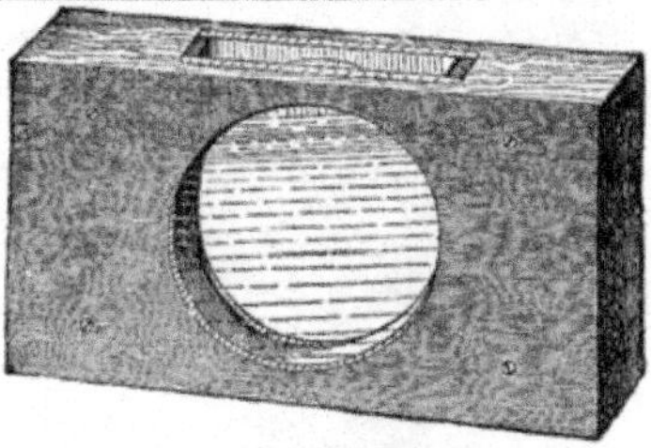

No. 421.

421.—Animalculæ or Chemical Tank.

These being made of glass and indiarubber are not affected by any chemicals.

422.—Chemical Tank (Sciopticon Co.)

A very good form, plate glass being used. This form takes to pieces very easily for cleaning purposes The price includes two pipettes. .. £0 5 0

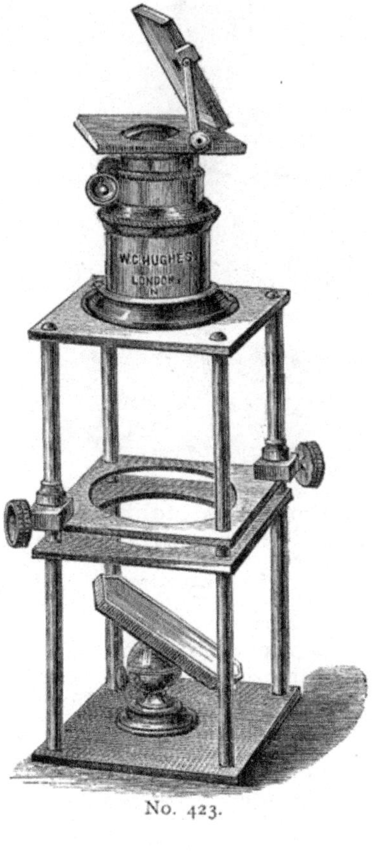

No. 423.

423.—Hughes's Vertical Apparatus (W. C. Hughes, Kingsland).

This is for showing horizontal projections, and consists of a new form of stand with sliding arrangement, reflecting mirror, large adjusting brass stage with glass plate for magnetic and fluid experiments, and is fitted with the new adjusting mirror, with special appliances for placing it at any angle to obtain entire or vertical reflections The lens is so arranged as to take on and off easily, and allows the mirror to be shifted round to any desired point. £8 8 0

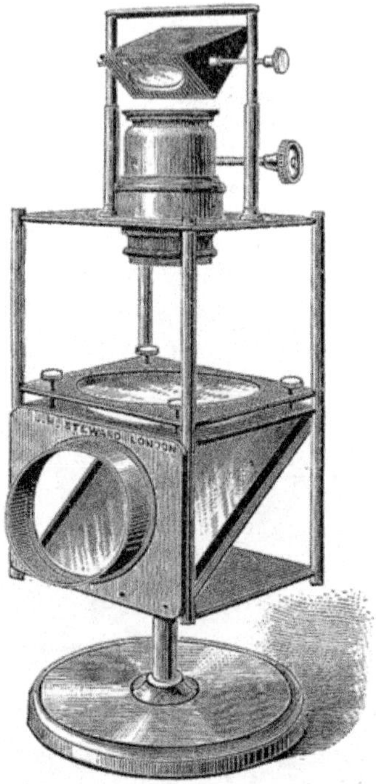

No. 424.

424.—Steward's Vertical Attachment (J. H. Steward)

This apparatus is of improved form, with skeleton brass frame and brass telescopic stand, reflecting mirror, condenser, large adjusting glass stage-plate for magnetic and fluid experiments, improved adjusting prism and telescopic pillar attachment, and has an achromatic objective with rackwork. Complete in case £13 13 0

425.—Polariscope Accessories (Newton & Co.)

Direct Nicol prism polariscope, designed by Lewis Wright, Esq , for use with the lantern, with large size Nicol or compound square-end prism as polariser, special flat-end prism of large angular aperture as analyser, with rack adjustment to focussing lenses, rotating motions to polariser and analyser, circle divided on German silver, crystal stage, double system of convergent lenses for exhibiting rings in wide-angle bi-axial crystals, with rackwork focussing adjustment, selenite stage, etc , on adjusting mahogany base-board, in case complete £45 0 0

	£	s	d.
Apparatus for showing on the screen the number of degrees of rotation caused by various percentages of sugar solution. The index-hand points to the degrees on the screen as the prism revolves. With biquartz, divided glass, pointer, three tubes, etc., fitted to polariscope	5	15	6
Double-image prism of Iceland spar, mounted in brass tube, with aperture plate for polariscope	0	16	0
Huyghen's apparatus fitted to nozzle of polariscope, with two rotating double image prisms, selenites, etc.	2	2	0
Biquartz plate for use with the above tubes	0	15	0
Crystallising frame, with plates and forceps	0	7	6
Metal slide for heating crystal (Mitscherlich's experiment) ..	0	5	6
Apparatus for heating glass while in the polariscope	0	10	6
Apparatue for compressing glass in the polariscope	0	10	6
Improved apparatus for compressing, bending and breaking squares of plate glass, showing most beautifully the lines of strain, with two keys for obtaining increased power	1	1	0
Glass plates for ditto .. .. per doz.	0	6	0
Crystals mounted in cork, for use with tourmaline forceps or polariscope, arragonite, calc. spar, nitre, quartz, borax, bi-chrom potash, etc .. .. each	0	3	0
Crystals, topaz, quartz, showing straight lines, etc.	0	5	6
Unannealed glass of various shapes, showing the permanent polarising structure of glass that has been uniformly heated and suddenly cooled .. .. each	0	7	6
Rotating frames for rotating objects in the polariscope .. each	0	4	6
MICA DESIGNS, WEDGES, ETC —			
Fox's wedge of 24 ⅛ wave films, showing Newton's first three orders in ⅛ wave differences .. .. £2 2s. to	3	0	0
1½ wave film to superpose on ditto	0	6	0
Circular wedge, circle divided into 24 sections of different colours .. .. £1 10s to	2	2	0
With A. superposed this gives apparent revolution of the colours, with G superposed, beautiful contrary revolutions, with a concave selenite superposed, beautiful spiral figures.			
Wedges, eight divisions each, for crossing and other superpositions .. .. per pair	1	16	0
Double wedge, giving curious floorcloth patterns and beautiful phenomena when A is superposed .. .. per pair	2	5	0
Circularly-polarised geometrical designs .. .. 9s and	0	15	0
These are far superior to the selenite stars or cube usually employed. They may be used to show either the complementary colours only, as with the latter, or in the reverse position give in rotation all the colours of the spectrum, with magnificent effect			
Double combination circularly-polarised designs. The extraordinary kaleidoscopic changes in pattern and colour of these preparations baffle description. They present the most gorgeous and brilliant phenomena yet produced in polarised light. The components of each pair can also be used singly. Per pair .. .. £1 10s and	2	2	0
Mica and selenite preparations for convergent light. These are mounted on 3 by 1 glass slips, as being more convenient for use. They can be kept in the ordinary rack microscopic slide boxes			
Norremberg's series of crossed micas, showing how by crossing bi-axial films the uni-axial rings appear. Six preparations of 1, 2, 4, 6, 8 and 24 films	1	13	0
A single preparation only, showing uni-axial rings, with 24 crossed films	0	13	6
Various combinations of crossed micas, giving complicated coloured fringes .. .. 5s. to	0	10	6

	£	s.	d
Mica and selenite combinations after Norremberg, giving fringes of the most varying patterns and colours .. .. .. 5s to	0	10	6

OBJECTS FOR THE POLARISCOPE MOUNTED IN MAHOGANY BLOCKS.

	£	s.	d
Selenite, even	0	2	6
,, uneven	0	2	6
,, showing mackle	0	2	6
,, cube	0	4	6
,, six-pointed star	0	10	0
,, flowers and fruit .. .. 10/- to	1	15	0
,, tulip	0	12	0
,, chameleon	0	15	0
,, butterfly	0	9	6
,, parrot	1	10	0
,, wedge, showing that colours are due to thickness 7/6 and	0	10	6
,, pair of wedges to counteract each other, one in revolving slide	1	10	0
,, concave plate, showing Newton's rings	0	10	6
,, Sweep turned Baker	1	10	0
Crystals cut to show the rings surrounding the optic axis .. 7/6 to	0	15	0
Quartz, showing cross crystallisation .. .. 2/6 to	0	3	6
,, ,, Airy's spirals	0	8	6
,, ,, Savart's bands	0	8	6
,, ,, Hyperbola	0	8	6
Benzonic acid	0	3	0
Salicine	0	3	0
Granite	0	3	0
Horn, Organic	0	3	0
Rotating selenite plate in frame, for use with the mica preparations	0	10	0

MICA OBJECTS FOR THE POLARISCOPE—QUARTER-WAVE PREPARATIONS.

	£	s.	d
A Is for ordinary experiments in circular polarisation	0	6	0
Whole or half-circles of half-wave thickness at the same price.			
B Gives contrary rotations	0	7	6
C Illustrates the dislocation of rings in crystals	0	9	0
D, E Superposed upon a concave selenite give magnificent phenomena, which may be further varied by adding a rotating selenite film	0	12	0
F Gives beautiful contrary rotations with the circular wedge, or when superposed upon C, D, or E, each segment will appear alternately white and black in succession, on rotating the analyser, showing the effect of half-wave retardations.. ..	0	7	6
G Circular ring of quarter-wave mica	0	7	6

426.—Accessories for Experimental Lanterns.

These are various pieces of apparatus, &c., for use with such Lanterns as Nos. 261 to 264

	£	s.	d
Brass slit front with adjusting tube and adjusting slit	0	15	0
Brass diaphragm front with adjusting tube and rotating diaphragm	0	13	0
Bottle prism for bisulphide of carbon on adjusting stand	0	18	6
Flint glass prism, large, fitted with universal joint, on stand for lantern work	0	17	6
Focussing lens, 4in. diameter, on adjusting stand	0	17	6
Glass tanks, to contain liquids for showing absorption bands, &c., in spectra of blood and various solutions	0	3	6
Colour disc, with multiplying wheel, arranged for showing composition of white light with the lantern .. .. } Oscillating reflector fitted to same multiplying wheel for re-composing the spectrum into white light on the screen .. .. }	2	10	0
Wright's optical front fitted to any lantern	2	10	0
Brass adjusting slit for use with Wright's optical front	0	15	0
Brass diaphragm plate ,, ,, ,, ...	0	5	0

	£	s.	d.
Rotating stand, with adjusting platform, to support lantern and experimental apparatus	1	5	0
Erector, or vertical apparatus for use with the optical lantern, for showing experiments, etc, on the screen when the object is in a horizontal position, with 5in stage, 5in. plano-convex condenser, and achromatic lens with rackwork adjustment for focussing	4	4	0
This is also very useful for making diagrams with pencil on ground glass, as they can thus be shown on the screen while being drawn.			
Metal screen on stand. Black one side, white the other, for receiving images or stopping off light	0	10	6
Ground glass screen, for receiving images on adjusting stand ..	0	10	6
Wright's refraction trough, for experiments in refraction and total reflection	0	12	6
Luminous cascade for illustrating total reflection	0	17	6
White light slide for showing the composition of white light on the screen	0	12	
Universal rod stand with adjusting fittings for holding prisms, mirrors, lenses, etc, vertically and horizontally..	0	10	0
Dense flint prism, face 2in. by 1in, mounted to fit rod stands ..	0	10	0
Plane mirror, 5in. by 4in, mounted to fit rod stands	0	6	6
Focussing lens, 4in, mounted to fit rod stands	0	10	6
Wave slide, for showing the motion of light and sound waves, and retardation of wave vibrations, on the screen	0	15	0
Lantern phoneidoscope, for showing the vibrations produced by sound waves in a soap film	1	1	0
Interference prism, for showing interference of light rays	0	5	0
Diffraction gratings, photographic .. ..21/- each, per pair	1	17	6
Newton's rings apparatus, for showing the phenomena of Newton's rings on the screen, mounted on adjusting stand	0	17	6
Lantern kaleidophone, on stand	0	10	6

427—Accessories to Lantern Microscope.

Objectives —

	£	s.	d.
Objective for large objects	0	12	6
3in. achromatic objective	2	0	0
2in. ,, ,, ...	2	2	0
1½in ,, ,, ...	2	5	0
$\frac{8}{10}$ths ,, ,, ...	2	7	6
$\frac{6}{10}$ths ,, ,, ...	1	10	0
$\frac{4}{10}$ths ,, ,, ...	1	10	0

Substage condensers These are necessary, whether the specially-made objectives be used or ordinary microscope powers.

	£	s.	d.
No. 1 Substage condenser in cell, for double plano power, and 3in achromatic objectives	0	8	6
No. 2. Substage condenser in cell, for all powers from 2in. to $\frac{5}{10}$ths in	0	8	6
No. 3. Substage condenser for $\frac{8}{10}$ths, ½in, and $\frac{4}{10}$ths objectives ..	0	17	6
,, 4. ,, ,, for ¼in. objectives	1	7	6
,, 5. Same as No 4, with extra front for immersion objectives..	1	12	6

Amplifier —

	£	s.	d.
No. 1, in cell	0	9	0
,, 2 ,, ...	0	9	0
,, 3 ,, ...	0	9	0
Extra carriers for objectives (useful for rapid change of powers) ..	0	3	0
Live cage, large size, specially made with flush foundation plate for exhibiting live insects, animaculæ in water, etc.	0	12	6
Glass troughs for pond life, etc... .. .. 1/6 to	0	3	0
Frog plate for showing circulation of blood in living frog, special pattern for use with lantern microscope	0	17	6
Spot lens for use with $\frac{6}{10}$ths objective in cell	0	17	6
Lieberkuhn for opaque objects	1	1	0

Section III.

SLIDES.

PART I.

Apparatus Connected with Slides.

A GREAT deal of the effect and success of a lantern exhibition is in a measure dependent upon the manner in which the change from one picture to another is made. Nothing looks so bad, or causes the audience to "think things," and occasionally "say things" as to the operator's abilities and the character of the show, as to have frequent breaks in the continuity of the pictures, during which nothing but the white disc appears on the screen, or to have the change of slides effected visibly in a jerky and erratic manner. When one slide is withdrawn and, after a few seconds' interval, a piece of another appears for a moment, only to be hastily withdrawn as soon as it is seen—by the audience as well as the operator—to be upside down, the spectators are wont to laugh, but a frequent occurrence of this performance is apt to pall upon the appetite, and matters are not improved when the slide, having been inserted right way up at a second attempt, is pushed into its place with a series of jerks, and then, being found to be too low down or too high up, is jumped about again to get it right in that direction. Yet at some amateur entertainments—and even at *some* professionally conducted ones, too—this sort of thing is more or less frequently a feature of the show. To prevent this, and to effect the change of scene as smoothly, rapidly and pleasantly as possible, is the object of most of the inventions set forth in the present chapter. One of the most popular systems of changing slides is by dissolving, as described in previous pages, but, even with a biunial lantern and the most perfect dissolving system, the beautiful results of "effect" slides will be marred if the slides do not register exactly one with another, for, although the lantern fronts may be adjusted so that the discs coincide, it is also necessary that the slides themselves exactly superimpose on one another, and to ensure this, various forms of registering apparatus, which secure the desired results with the expenditure of the minimum of time and trouble, have been invented and placed on the market.

With single lanterns, of course, the difficulty is greater, as the operator has not the dissolver to fall back upon, and each slide has to be inserted after the one on view has been removed. A difficulty, too, exists in the varying sizes of slides, for whereas hand-painted slides are always framed, it is the rule with photographic slides not to frame them; and, again, whilst the frames of the former class of slides may be of a multitude of sizes, photographic slides, too, differ in dimensions. The rule in this country is to make them 3¼in. square, whilst on the Continent they are usually made on 4¼in. × 3¼in. measurements. This variation in sizes necessitates some means of readily ensuring each picture being brought at once to the centre of the disc of light, and we have described several devices to that end. Photographic unframed slides are usually inserted in what is termed a carrier, which takes the place of the frame, and enables them to be easily handled, and in carriers, too, there is considerable difference, the aim being to render the insertion of the slide and its accurate centring on the screen void of the slightest difficulty or uncertainty, whilst some varieties of this most useful instrument, by very ingenious means, cover up the light automatically whilst the change is being made.

We have in the present chapter dealt with a few varieties of slide tinters, but as in a previous part we have already dealt with the whys and wherefores of the tinter, we need not here dwell upon the subject, but pass on at once to an examination of the varieties offered for our selection.

Registering Apparatus.

428.—Hughes's Register Clip (W. C. Hughes, Kingsland).

This is a small piece of apparatus for placing slides when in a frame in a central position in the slide holder of lantern. The bottom projecting pieces slide a little, and have screws to hold the ground piece securely, while the screw in the side clamp ensures the slide being kept in its proper place—that is, central £0 2 0

429.—Hughes's Registering Apparatus (W. C. Hughes, Kingsland).

This can be adapted to any upright or other lantern, and worked in conjunction with the special register carriers, will bring every picture a dead centre on the screen, and once set will always remain so. This form of register is perfectly rigid and firm, and will, by an ingenious contrivance, fall over in order to allow of panoramic frame effects being centred without in any way interfering with the register of carriers, &c. Each carrier can be raised or depressed by means of the side screws, and special platform arrangement, which is attached to the registration apparatus. This

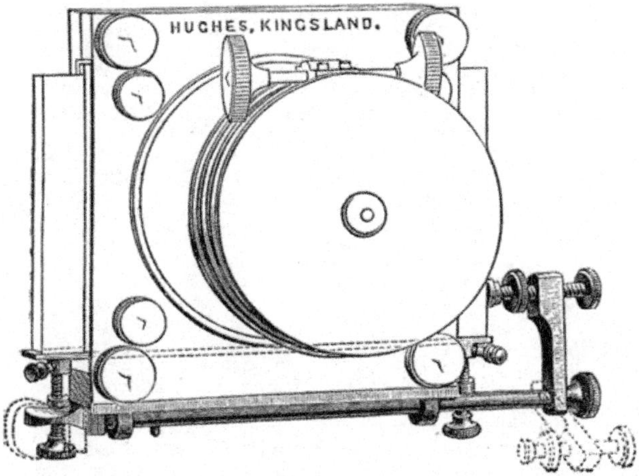

No. 429.

produces true coincidence of discs, and compensates for the possible inequality of the stages of the lantern, and thus enables the operator in any emergency to register his slides and effects in any particular slide-holder of his biunial or triple. Special private directions will be given with each set of registering apparatus, the carriers of which are most carefully centred in sets of twos or threes, as may be required, one set for square pictures and the other for circular—the point gained being this, that after the carriers are once centred all other framed slides and effects can be registered likewise to them with absolute precision and ease without fear of failure or disappointment.

A set of three registering apparatus for triple lanterns, including the six special carriers, three squares and three circles £4 10 0

A set of two, for biunial or other lanterns, with four special carriers £3 0 0

430.—Steward's Centring Stages (J. H. Steward).

An arrangement of vertical and horizontal adjustments and locking nuts, so that each lantern stage is set perfectly true at any angle or distance from the screen. We have carefully experimented with this excellent apparatus, and can assure our readers of our complete satisfaction with the results. It is very little trouble in the first place, and once done is always done:

For slides may come, and slides may go,
But the stage is true for ever.

The following description of the apparatus is from the *British Journal of Photography*, and needs no fresh writing up on our part:—" For existing lanterns, these registering adjustments are fitted to a metal back plate that slides into the ordinary stage, and can be rigidly fixed in position by clamps. It also holds the metal masks for cutting down pictures to a definite size. So that without sending the apparatus to the optician, the owner can fit the patent improvements himself; and if he marks the sliding tubes so that they can always be placed back in the original position if moved, he can preserve the same relative optical angles. For it is found that a partial rotation of the draw tube disarranges the centring. In order to secure the preliminary adjustment of the lanterns prior to an exhibition, or before registering new lantern slides, special slides are employed. They can be of various designs, but for convenience those employed with the apparatus described are plane cross lines and letters in the open spaces. These are photographed by contact so as to be absolutely alike, and are then (by carefully calipering) fixed with putty into a frame very accurately made to a given size, say 7 × 4½ (or, if the smaller frames are preferred, 7 × 4). With these frames each absolutely alike in the lantern stages, the clamps (milled head) are released, and the adjusting screws raised or lowered, or moved backwards and forwards, until *all* the crosses and circles agree. The clamps are then screwed home, and the registration is perfectly fixed. For perfect dissolving of views and effects, it is a *sine quâ non* that all slides must be framed. A simple method of getting photographs and pictures central in their frames, is to have the grooves made full large, so that if the glass is not very exactly cut true, or the mat is not exactly central, the glass can be raised, or lowered, or moved right and left, and the difference between it and the wood filled up with cardboard. If the view and the effect be held up to the light a very close approximation to superposition can be obtained, and then, when the slide is put in the lantern (previously adjusted as explained), the frame can be planed off or added to, until the view and the effect, or the two similar views, exactly agree. For this purpose a shooting board and plane are required, as well as narrow strips of mahogany and glue. From experience it has been found best for new, unframed views and effects to use mahogany frames having about one-sixteenth more wood one edge than the other; this, then, in nearly every case, prevents any addition being necessary, and the process of planing off this superfluous wood is simple until just the amount is left. Pieces can be added to the end of the slides that are short, and those that are long can have the part cut away just where the registering stop comes, or with photos down to a uniform size. There is a circular mask also included. The advantage of this carrier is, that all slides are inserted and withdrawn from the same side, and each is automatically centred.

For biunial £1 15 0
For triple 2 12 6

Extras.—Metal plate with two registering masks and four setting screws, 10/6; photographic registering crosses and standard frames for setting lanterns prior to use, per pair 7/6, and for triple, 10/6.

431.—The Noakes System (D Noakes & Son).

This firm make a specialty of correcting the lantern at the outset. They do not believe in any arrangements for attaining the end of perfect registration, except by correcting each biunial and triple individually This they do by a peculiar method which they keep secret. Whatever the value of their method, or the correctness of their views, may be, there can be no doubt as to the result. They attain registration as near as one can expect perfection in this world.

Carriers and Frames.

432.—Single Carriers.

To hold one slide, which is slipped in at time of exhibition, and readily emptied and refilled £0 1 0

433.—Slide Frames.

These are for permanently framing a slide, as distinct from the single carrier. Prices average about £0 0 6

434.—Double Carriers.

Designed to hold two slides, one being exhibited whilst the other space is being re-filled .. .. £0 1 6.

435.—Theobald's Double Carrier (THEOBALD & Co)

A very cheap double frame, well made.. .. .. .. £0 1 6

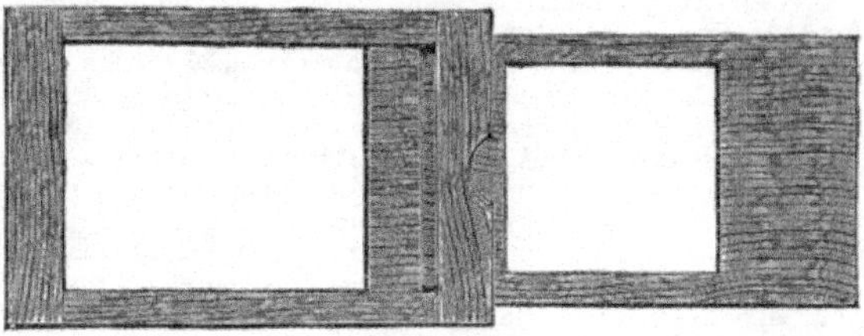

No. 436.

436.—Single Registering Carrier.

These are useful for biunials or triples when using a number of unframed slides, such as landscapes or views £0 2 6

437.—Skeleton Panoramic.

A plain grooved slide, through which the slides are pushed one after the other £0 1 3

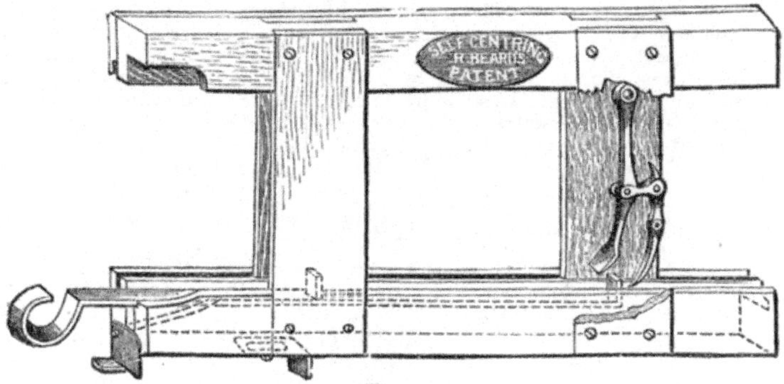

FIG. 1.

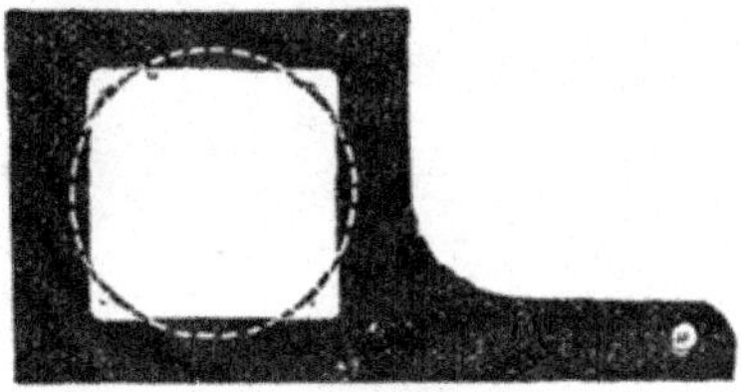

FIG. 2.

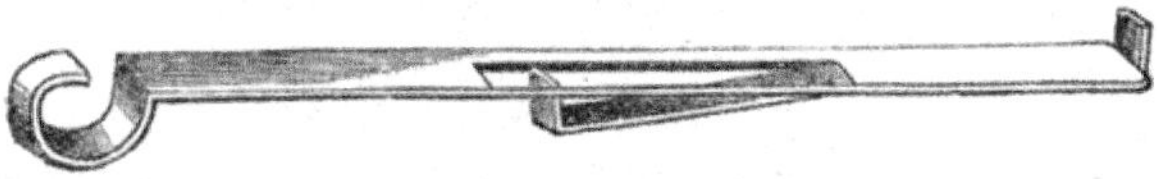

FIG. 3.

No. 438.

438.—Beard's Self-centring.

A well-known slide carrier, sold by most people. The advantage of this form is that, no matter what the length of the lantern photo or slide is, it becomes automatically centred on being placed on the brass runner and pushed home. By this means any of the commercial photographs (English, Scotch, French and American) can be used alternately or successively in the dark with perfect safety, and with infallible results. Fig. 1 gives a general view of carrier, and fig. 3 the runner with the spring catch that raises and holds the slide when pushed into its place. Fig. 2 is a little addition, consisting of two metal masks, sliding in a grove immediately behind the photo, so as to show the exact amount on the screen with each slide, and thus compensate for any little difference in the position of the mats on the slides. One cushion shape for Scotch and French photos, and one circular for English, &c.

The plain form without the masks £0 5 0
With the metal masks 0 7 6

439.—Beard's Eclipse.

A carrier for single lanterns, self-changing and accurately centring. A metal runner is drawn out, the slide placed in the first groove, and the runner pushed home, when it will be centred, irrespective of the usual differences in lengths. The runner is then pulled back, and, by means of automatic levers, the slide is pushed into the second groove, and is in focus. When this has been exhibited, another slide is placed in the first groove and the runner again pushed home. In so doing, the fresh photograph passes in front of the other, and, when centred by the levers, the runner is withdrawn, bringing with it the photo exhibited. A great advantage of this carrier is that all slides are inserted and removed on the one side only £1 1 0

440.—Archer's Quick Action (Archer & Sons).

A double carrier, an outer frame being fixed in the lantern stage. An interior sliding part travels to and fro to be refilled from alternate ends. With brass stop. Very well made in mahogany. £0 2 6

Extra.—For adjustable stops to fix firmly in centre, 1/3.

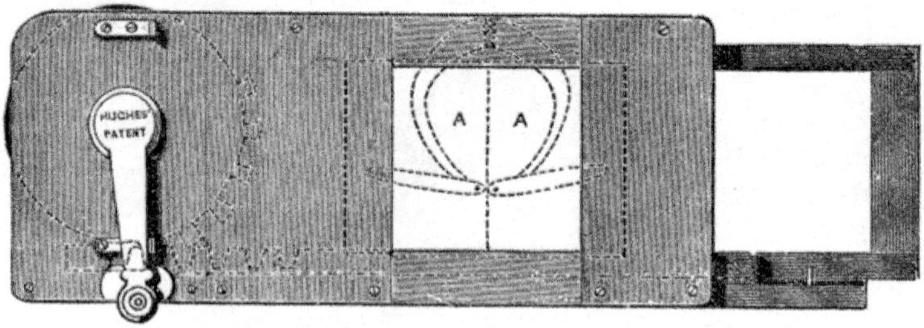

No. 441.

441.—The Presto (W. C. Hughes, Kingsland).

An instantaneous slide carrier for use with a single lantern, by means of which the pictures are changed from one to the other without any movement being shown upon the screen.

This new carrier is an improvement on the original "Presto." Instead of the slide or shutter working outside the lantern in front of the nozzle and lenses, it is now manipulated from the inside, and covers and uncovers the condensers while the double carrier frame is moving, and consequently by this arrangement is much more portable for carrying about. The manner of performing the change is novel and ingenious The vulcanite shutters are closed and opened by aid of toggle levers, acted upon by the large wheel spindle, and the crank moves the rackwork frame, into which gear the teeth of the wheel rotate, which enables the operator to change the picture instantaneously with ease. As the frame moves along the shutters close, but when the picture is home the shutters open. The action is rapid and precise, and the audience are much surprised and astonished at the marvellous change effected without their having seen the picture move £1 12 6

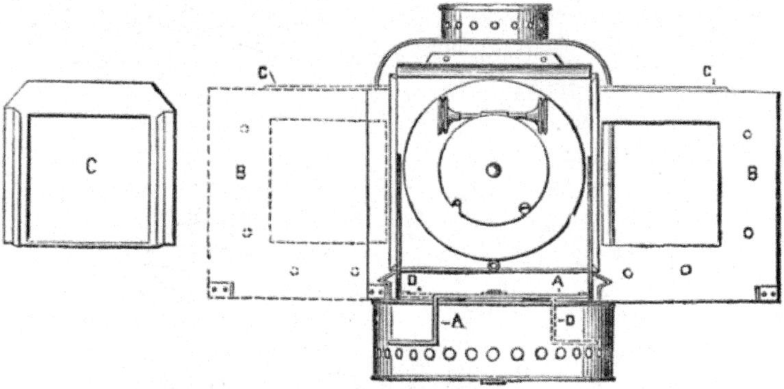

No. 442.

442.—The Simplex Lever (W. C. Hughes, Kingsland).

This is a frame made and fitted to the Pamphengos and triplexicon lanterns only. It is so arranged that the picture is changed instantaneously, although no shutter passes in front of the lantern nozzle. B is the frame, which is double, made of solid and polished brass, and moves on a brass runner. By sharply turning the brass lever handle A, the next picture appears, and so on. The large frame itself can be removed from the lantern at will, so that mechanical and other slides can be introduced. One advantage of this frame is that both French and English photographs can be used at pleasure, as two extra frames, C, are provided for that purpose, which easily adapt themselves in the larger slides £1 5 0

443.—Hughes's Automatic.

This ingenious carrier will take all the standard sized photographs, French, English or Scotch. The action regulating this accomplishment is simple, effective and complete, each photograph inserted in the runner when drawn out will immediately, on its being pushed back again, fall into its place exactly central, and be held securely by the spring arrangement at the top of the inner side of the frame. The frame is very portable, being no larger than an ordinary single carrier. In polished mahogany and boxwood, £0 5 0

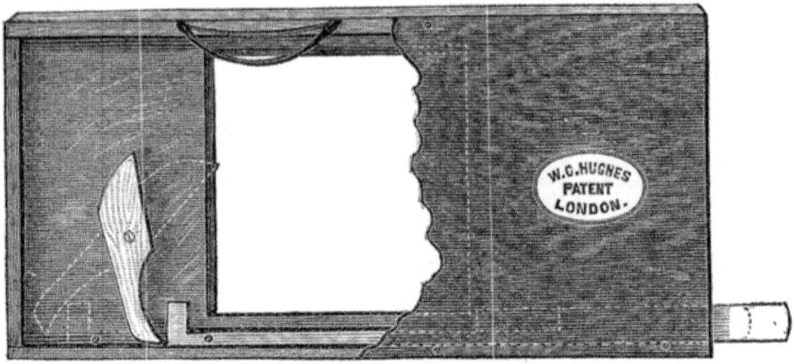

No. 443.

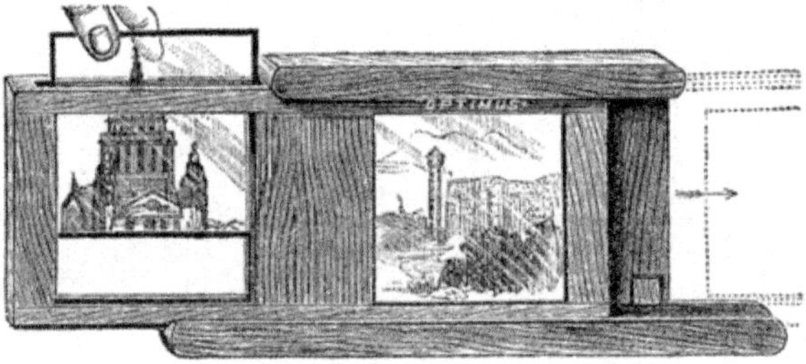

No. 444.

444.—Sliding Carrier (Perken, Son & Rayment).

Double carrier, sliding in grooved frame. The illustration shows method of use £0 2 6

445.—The Facile (A. Pumphrey).

A carrier that has many friends. A non-elastic band lies in the bottom of the carrier, and is kept tight by a steel spring, on this band the pictures rest; it can be moved in either direction by turning the handle, and the tension given to the band by the steel spring causes the pictures to move with great precision. The clip which holds the pictures steady is represented at B B B; this is pressed down while the picture is centred, and the moment it is released it holds the picture in its place top and bottom at C C. So long as the picture is held by this clip it will not move, even if the handle A be turned and the traversing band of rubber be moved—thus other pictures can be moved up to the one which is being exhibited, or taken away, without moving the picture which is seen on the screen. The best way to use this carrier is to introduce the slide at D; it then falls on to the traversing band, and is drawn towards the exhibitor, and held when in the centre; the next picture is then introduced, and drawn close up to the one being exhibited,

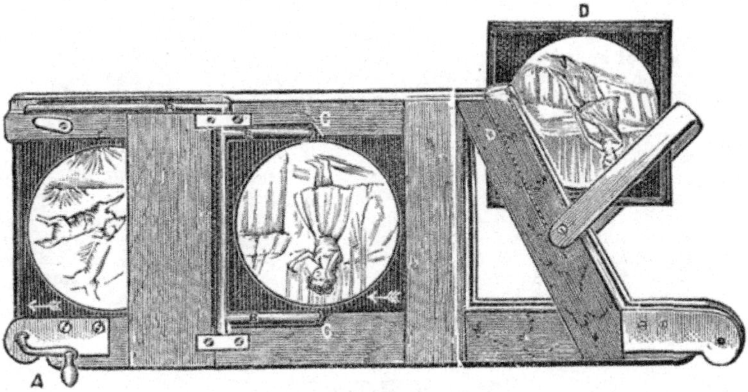

No. 445.

ready to follow it closely. The slide which has been shown is delivered into the hand of the exhibitor. The carrier will exhibit any pictures which are 3¼in wide, no matter whether 3¼ × 3¾in., or 3¼ × 4in., or 3¼ × 3¼in.

£0 5 0

446.– Place's Quick Changing (J. Place).

A convenient carrier, as we know from experience. It registers any size slide on its being placed in the holder and pushed along. It has two openings at the top, and the sliding part moves backwards and forwards, the slides being inserted at each end in turn £0 10 0

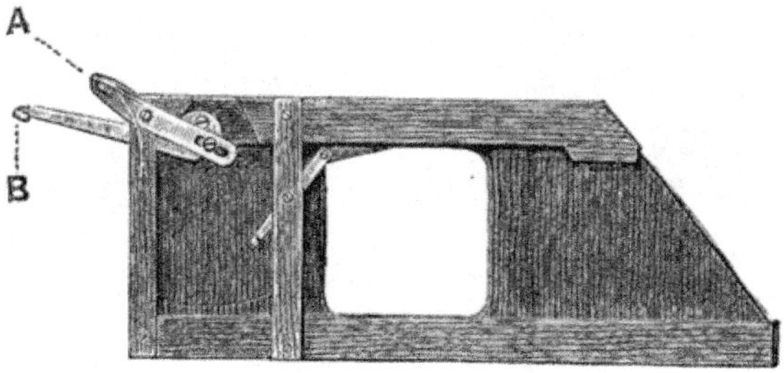

No. 447.

447.—New Panoramic Frame (Wholesale Firm).

With this carrier frame, photographs of any size can be passed through one after another, each one being properly centred as it is passed in. The frame is so constructed that when one slide has been exhibited, and a second slide is being shown, it is not possible to put in a third slide until the first has been removed, thus preventing the first slide from being thrown out of the carrier and breaking, as is often the case with ordinary panoramic carriers. Also, when one slide is being exhibited, it is not possible to pass it out till another slide is ready in the frame to take its place, unless, of course, it is desired by the operator to do so. This quite prevents the possibility of seeing the white disc of light on the screen. The slide to be exhibited is placed a short distance into the carrier frame, and by pressing the spring A, the lever arm B forces the photograph into its place, and centres it. The spring is then released, and a second slide is inserted in the same way as the first. The spring is again pressed, and the lever arm forces the second slide into the place of the first, and centres it. The two slides being now in, a third slide cannot be inserted until the first one (which has been shown) is taken out. Should the operator make the mistake in thinking that he has already placed a second slide into the frame ready for exhibiting when he has not, and presses the spring to make the change, the slide then showing will not move, which will at once show him his mistake, and will not allow the white disc of light to be seen on the screen, which is so unsightly in an exhibition, and which so often occurs by one slide being taken out before another is ready in its place.

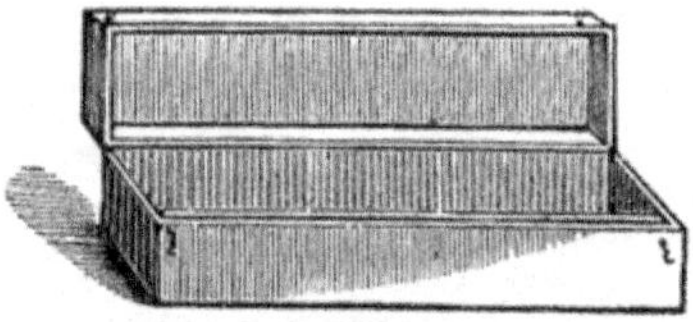

FIG. 1.

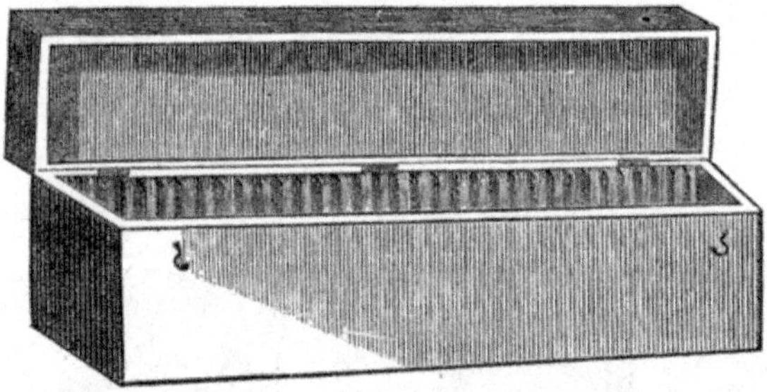

FIG. 2.

No. 448.

448.—Grooved Boxes.

These are for holding unmounted slides in grooves, ordinary photographic negative boxes being unsuitable for this purpose, owing to their being the wrong size and the groove too narrow. They are made to hold different quantities, the price being thereby guided.

To hold	12	24	36	50	60
Plain deal (Fig. 1)	1/3	1/6	2/3	2/6	3/-
Mahogany (,,)	2/-	2/3	3/3	3/6	4/6
Cloth covered (Fig. 2)	1/6	2/-	2/6	3/6	4/6

449.—Numerical Grooved Boxes (Chatham Pexton).

Each groove in this pattern is numbered, and consequently the box is much more convenient in use. In placing the slides back into the box after use, for instance, it is necessary to find the first and then follow on with the remainder in strict order. In Mr. Pexton's box any slide can be at once slipped in, and thus a sorting-out process is obviated. Another point worthy of notice is that two extra grooves are always provided, as a glance below will show, to allow for introductory, good night, or extra slides, in addition to the set on lecture.

To hold	14	26	38	52	104
Ordinary finish	1/-	1/6	2/-	3/-	5/-
Stained and varnished	1/9	2/9	3/-	4/6	8/6

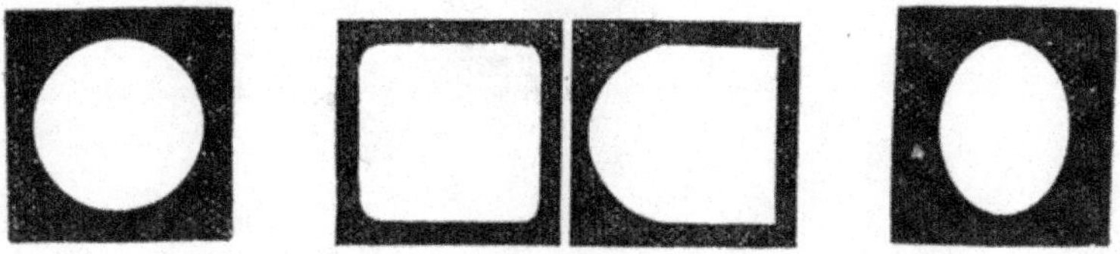

No. 450.

450.—Mounts or Masks for Slides.

Those who make, by photography, their own slides, or paint others, require these masks to insert between the actual photo and the covering glass. They are usually made in four shapes—circles, ovals, domes and cushions, in varying diameters. Price per gross, about £0 1 0

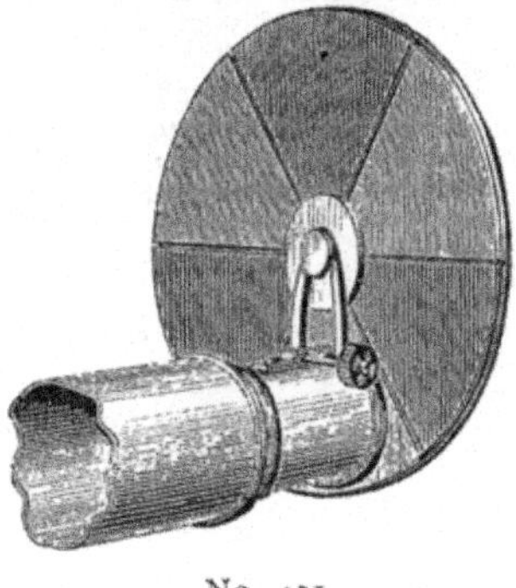

No. 451.

451.—Tinting Diaphragms (W. C. Hughes, Kingsland).

An arrangement by which a plain photograph, such as one of statuary, can be tinted, blue, green, orange, or violet. The diaphragm is mounted on a brass plate, which revolves steadily. There is a circular tube, which will fit to front of lantern nozzle. The colours are well selected, and when fixed to two lanterns produce some lovely effects. No. 1 is plain glass, 2 blue, 3 ruby, 4 orange, 5 green, 6 violet, as the coloured glasses are brought round. The introduction of a continuous change of colour is produced on one picture without having to take each glass away, and so leaving the picture white £0 15 6

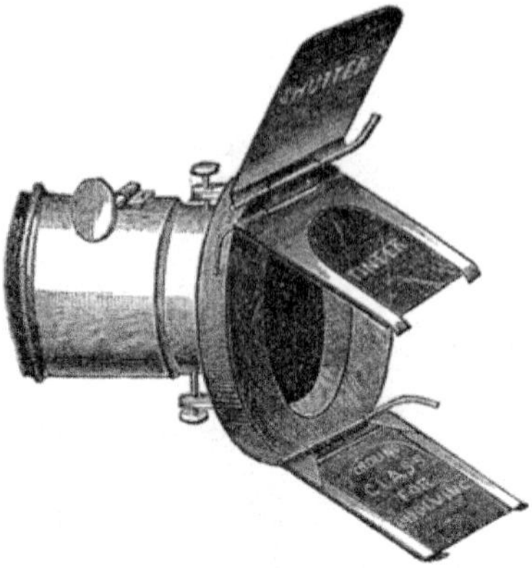

No. 452.

452.—Tinter and Dissolver.

A rim or circle of wood to which is hinged a shutter to cover the lens, and two frames which carry tinted glasses or gelatines. It is used like No. 451 to tint statuary, moonlight effects, &c., but it is also adapted for partial tinting. If the top frame be partially turned down with a blue tint, it will give a blue sky. The bottom frame likewise with green or brown tints the foreground. The frame not being close to the slide, indeed, really furthest away, no sharp line of division is seen, but a vignette or fading away effect is produced. By placing in one frame a piece of ground glass, a very good method of dissolving is arrived at; the lens being rapidly covered and uncovered, whilst the slide is changed by means of a double sliding carrier. This method avoids plunging the room into pitchy darkness at each change, as is the case when an opaque shutter is used £0 7 6

453.—Tinter Openings in Lantern.

For description of this system of tinting slides see No. 403.

PART II.

Lecture Slides and other Sets.

SLIDES and slide making form the subject of a later chapter, but, although to anyone acquainted with photography the manufacture of lantern slides is an easy, pleasant and interesting occupation, and many thousands are made by amateur photographers annually, by far the greater number are turned out by firms who make slide making a principal part of their business. As will be seen from the following pages, the selection of subjects offered to the critical public by these houses is large and varied, almost every object under the sun being dealt with. The majority of slides sold are photographically reproduced, as the once universal and expensive hand-painted ones have had to give way before the onslaught of photography, and slide painting bids fair soon to be reckoned amongst the lost arts. Owing to the vast number of slides now commercially on the market, and the frequent incongruity of their nomenclature, we have in the following pages endeavoured to show, in as plain and clear a manner as possible, exactly what each "set" comprises, so that persons desirous of purchasing sets of slides for any particular purpose may be assisted in their selection, and be enabled to readily light upon just what they want without having to actually see each set before purchasing. We may add that with the majority of the "lecture' sets a brief descriptive reading accompanies the set, which will be found very useful.

454.—Arrangement of Sets and Prices of Slides.

We have made a gigantic effort to arrange the almost unlimited number of lecture sets, stories, poems, &c., into something like discoverable order. Whether the divisions will meet the views of our readers we do not know, but if some do not we must crave their indulgence, for it has been an extremely difficult task. At the very outset we found an alphabetical arrangement unworkable, even were that a good system—a thing we doubt. Some sort of classified division was imperative, but it was not until we were in the thick of the fray that we found the real difficulty of the task. To start with, titles are of no value whatever to rely upon. To parody a well-known line—

"Sets are not what they seem."

But with the question of title overcome, there are still some sets which partake of two natures, or even three. To give one example, let us take, say, "The War in Egypt." One set under that title will be so constructed that it can well be placed into the history division, but another contains perhaps nearly all views of places, and really only one or two views which could be directly connected with the war. This would make the set more suitable for our first division—that of scenery, places, etc. Troubles like these have also cropped up in each division Still we have tried our best, and we'll leave it at that.

With respect to the prices of sets or individual slides, any attempt to quote them would be beyond our space. For the most part uniformity reigns supreme, but when, as in special sets or odd slides in a set, they vary, there is no mistake about the variance. The general price of the great majority of the slides in Part I., which are mostly photographic, is 12s. to 18s. per doz. plain, 36s per doz. coloured. There are, however, differences more or less marked, the greatest probably being in Mr Pumphrey's, which are 6s. per doz. plain; 12s. per doz coloured. However, for particulars as to cost, we must refer our readers to the firms who offer them.

Some further information on these and other points may be found in an article at the commencement of Part III., entitled "Slide Makers, Painters, and Sellers."

We must say a few words about the name which will be found following the title. Our aim has been to give, as far as lay in our power, the actual makers or proprietors of each set, and the name given in the majority of cases carries out that idea. At the same time the vast number of slides made by York, Pumphrey, Valentine, Wilson, and others are sold by every dealer and optician—Messrs. York's slides especially. List after list, catalogue after catalogue, perused in the course of the compilation of this part, have turned out to be simply York's productions. The above firms may therefore be termed trade producers. In fact, as a general rule, and with but very few exceptions, the slides noticed in our pages may be obtained from any dealer.

Scenery, Travels, Places, &c.

Some of the collections that follow are hardly lecture sets, but merely a series of slides from which a selection may be chosen. In order, however, to keep them under their proper heading, they are herein inserted.

GREAT BRITAIN.

455.—Some Interesting Places in English History (VALENTINE, 36 slides).

We see here Stonehenge, Salisbury Cathedral, Carisbrooke Castle, Ely and Canterbury Cathedrals, Durham Castle, Alnwick Castle, Castle Howard, Ulleswater, Fountains Abbey, Kenilworth, Tintern Abbey, Anne Hathaway's Cottage, etc.

456.—Sketches of English Life and Scenery (York, 40 slides).

The Quantock Hills, Cockermouth, giving a scene illustrative of "Fountain heads and pathless groves"; another scene on the same hills, as "There is a pleasure in the pathless woods"; and others, such as a Somersetshire cot, "Oh! for my lowly thatched cottage"; a cottage interior, "The whitewashed walls and nicely sanded floor, The varnished clock that click'd behind the door."

457.—English Scenery (Newton, 418 slides).

This is not a lecture set, but only a series from which sets may be made up, or single slides purchased. The list is, of course, too comprehensive to give an account of.

458.—English Scenery (Pumphrey).

This also is a long series of slides from which lecture sets may be made up.

459.—John-o'-Groat's to Land's End (Riley, 57 slides).

It will be sufficient to say of this set that it gives many views between the two places, but not always, we are afraid, strictly on the route. However, it eventually reaches Land's End, and therefore we cannot grumble. We shall by that time have seen many of the beauties of our "sea-girt isle."

460.—The Mansions of England in the Olden Time (50 slides).

The mansions are represented by Holland House, Kensington; Haddon Hall (3); Knole House, Kent (3); Hatfield Hall, Aston Hall, Birmingham (2); Kenilworth, etc.

461.—London, Ancient and Modern (Newton, 178 slides).

A large series from which sets can be made up.

462.—Tower of London (Sciopticon Co., 12 slides).

A tour round the Tower, showing most of the usual pictures with warders, the Horse Armoury, etc.

463.—Here and There about London (Valentine, 40 slides).

Proceeding from St. Paul's Cathedral (2), we pass the Tower of London (4), Westminster Abbey (5), Windsor Castle (6), "Star and Garter" at Richmond, Kew Gardens, Horse Guards, etc., etc.

464.—Rambles Round London (Wood, 60 slides).

The outlying villages of London, such as Enfield, Elstree, Edgware, Stanmore, Shoreham, Waltham Abbey, Cookham, etc.

465.—The Sights of London (Wood, 46 slides).

We start with a Map of London, followed by Buckingham Palace, St. James's Palace, House of Peers, Commons, Somerset House, General Post Office, Lambeth Palace, Westminster Abbey, St. Paul's, Guildhall, Crown Jewels, Tower, British Museum, etc

466.—Old London (Wood, 97 slides).

This is a series of views of various places in and about London connected more or less with the past history of the great city. Some of the slides, however, represent items more directly of historical interest, such as Latimer Preaching, the Great Fire, Last Heraldic Court, Maria Wood, Plan of Roman London, and Wren's Plan for the Rebuilding of London.

467.—Westminster Abbey (YORK, 36 slides).

In the Abbey we have Henry VII.'s Chapel, views of the Choir and Transepts, Tombs of Queen Elizabeth, of Mary Queen of Scots, the Grave of Charles Dickens, Poets' Corner (3 views), Livingstone's Grave, Tombs of Edward III. and Henry III., etc., and outside, views from the West and North.

468.—The Tower of London (YORK, 30 slides).

These slides illustrate the various towers of the Tower, such as the Byward and Beauchamp, also the Traitor's Gate, the Crown Jewels, and so on.

469.—A Day in London (YORK, 60 slides).

Here we have views of Westminster Abbey (5), the Tower of London (5), St. Paul's Cathedral (2), The Guildhall (2), Buckingham Palace, Houses of Parliament, interior of the House of Lords, interior of the House of Commons, Buckingham Palace, Thames Embankment, etc. We have also twelve views of animals in the Zoo, seven of British Museum scenes, winding up in Regent Street.

470.—General Description and Statistics of London (YORK, 48 slides).

This set includes four views of the Tower of London, five of the Albert Memorial, one of Buckingham Palace, St. James's Palace, Houses of Lords and Commons, Westminster Abbey (2), St. Paul's (2), Guildhall (2), Mansion House, Bank of England, Royal Exchange, Custom House, National Gallery, Temple Bar.

471.—London Street Traffic (YORK, 79 slides).

The streets include Holborn, Oxford (7), Regent Street, Piccadilly, Pall Mall (2), Trafalgar Square, Strand (7), Fleet Street (6), Ludgate Hill (2), Westminster, Euston Road, King's Cross, London Bridge, Waterloo Bridge, etc.

472.—London and Neighbourhood (YORK, 888 slides).

The slides illustrate a large number of scenes from which lecture sets may be made up.

473.—London to Liverpool (WOOD, 55 slides).

A coasting voyage, commencing of course with views on the Thames and on its banks, and passing round by Margate, Dover, Brighton, Portsmouth, Southampton, Torquay, Plymouth, Falmouth, and the Welsh Coast to Liverpool, in which town two views are given.

474.—Windsor Castle (BROOKS, 70 views).

Taken by special permission during the Jubilee, 1887, by Mr Brooks. The set shows a general view of the Castle, St. George's Chapel, the Queen's Entrance, the Long Walk, St. George's Hall and the Throne, the Presence Chamber, Audience Chamber, State Bedrooms, Cenotaphs of Prince Albert and the Duke of Albany, etc. In one of the State Bedrooms here shown the King and Queen of the Belgians slept during the Jubilee.

475.—A Day's Holiday at Windsor (YORK, 30 slides).

The portrait of Her Majesty is followed by Henry VIII.'s Gateway, St. George's Chapel (2), the Round Tower, Frogmore House and Grounds, the Queen's Breakfast Cottage, three views of Virginia Water, views of Windsor Castle, etc.

476.—The Thames, below Bridge (NEWTON, 13 slides).

London Bridge by Billingsgate, The Tower, Greenwich, Gravesend, and Southend to the Nore Lightship.

477.—The Thames, from London Bridge to its Source (NEWTON, 90 slides)

A series of views of the bridges, principal buildings on banks, locks, churches, abbeys, ferries, etc , up to Thames Head Bridge and the source of the Thames, showing iron pump over the spring.

478.—Oxford to London Bridge (YORK, 50 slides)

Starting with Oxford University, down the Thames, we pass Abingdon, Wallingford, Pangbourne Weir, Reading Abbey, Henley, Maidenhead, Bray Church, Eton, Windsor, Hampton Court, Twickenham (Pope's villa), Kew, Mortlake (finish of race), Putney (starting point of race), Houses of Parliament, Blackfriars Bridge, London Bridge, etc.

479.—Manxland and its Beauties (VALENTINE, 57 slides).

Furness Abbey (3) and Barrow-in-Furness (3) (all once in England) commence this set, and eleven views of Douglas, the famous Tynwald Hill (where the I.O M laws are read), Glen Helen (3), Peel (5), Sulby Glen, five of Ramsey, Laxey Wheel (1), Castletown, s.s "Mona's Isle," are among the pictures dealing with the Isle of Man proper.

480.—Ten Days in the Lake District (RILEY, 50 slides).

On the first day the tourist is at Ambleside (6), and then he proceeds by several charming spots, such as Colwith Force and Rydal Water, to Grasmere (3); then Keswick (9), Ulleswater, etc., to Windermere, and thence to Coniston (4), finishing up with Furness Abbey. No 49 is a view of Camping-out.

481.—The English Lake District (YORK, 50 slides)

Firstly, we have a Map of the Route. We see Windermere from different points, Ambleside, Grasmere, Ulleswater, Keswick, Derwentwater, Buttermere, Elterwater, Thirlmere, Bassenthwaite, Rydal Lake, Rydal Mount (the late residence of Wordsworth), etc , etc , and finally we have views of Coniston (2), Furness Abbey (2), and Grange-over-Sands

482.—Views in Bradford (RILEY)

Views of the buildings and parks of this town.

483.—Rambles about Yorkshire (RILEY, 50 slides).

A series from entirely new negatives. York (4), Scarborough (4), Whitby (2), the Three Falls at Aysgarth, Ingleton (4), Malham, Bradford and District (4), Bolton (3), Harrogate (2), Knaresborough (4), Ripon (2), and Leeds (2).

484.—Various Views in Yorkshire (RILEY, 50 slides)

Here we have views of Aysgarth (4), Barden (2), Bingley (3), Bolton (2), Ilkley (4), Ingleton (3), Kirkstall Abbey (interior and exterior), Leeds (2), Masham (5), Shipley (2), Scarborough, Studley Royal Church, five views of York, etc.

485.—Holiday Tour in Yorkshire (VALENTINE, 56 slides).

In this set we begin with Whitby Abbey, then we come to Scarborough (4 views), Bridlington Quay, Rievaulx Abbey, York Minster and York (9), Beverley (3), Leeds (5), Ilkley (3), Harrogate (5), Knaresborough (3), Ripon (3), Fountains Abbey (2), Richmond (3), Easby Abbey, etc.

486.—Liverpool, Ships and Shipping (HUGHES, 21 slides).

The set comprises six named ships, five views of docks, etc.

487.—Instantaneous Pictures of Life in Liverpool (PUMPHREY, 12 slides).

These include James Street (crowded with Liverpool men), Church Street (crowded), Mr. Gladstone's birthplace, and steamers (named).

488.—Liverpool (YORK, 42 slides).

St. George's Hall, the Free Library and Museum, the Custom House, a view of the Mersey, Sefton Park, Birkenhead Landing-stage, three views of New Brighton, etc., etc. These give one an idea of England's second largest city and neighbourhood.

489.—Manchester to Alderley, Buxton, and Matlock (RILEY 54 slides).

This arrangement of scenes, commencing at Manchester, passes through Alderley and Beechwood to Buxton (12), then amongst the dales, Castletown, Peak Cavern, to Eyam (3), Chatsworth (2), Haddon Hall (2), Matlock (2), Alton Towers, Dovedale, and ending up with Tissington village.

490.—Haddon Hall (YORK, 18 slides).

Entrance, Chapel, Banqueting-hall, State Bedroom, Dorothy Vernon's Walk, etc.

491.—Haddon Hall (BROOKS, 36 slides).

General Views, Entrance, Courtyard, Views of Banqueting-hall, Dorothy Vernon's Door, etc.

492.—Rambles Round Warwick (VALENTINE, 50 slides).

Of Warwick and its castle we have fourteen views, of Leamington seven, Guy's Cliff seven, Kenilworth seven, Charlecote two, and Stratford-on-Avon twelve.

493.—Norfolk Broads (WOOD, 20 slides).

The general scenery of the well-known Norfolk Broads.

494.—A Visit to Cambridge (WOOD, 20 slides).

We have here two views of Queen's College, one each of St. John's, Pembroke, Corpus Christi, Christ's and Jesus Colleges, four of Trinity, two of Caius, two of King's, and one of King's and Clare College, etc.

495.—Oxford (YORK, 78 slides).

Here are six views of Balliol College, four of Oriel, six of Magdalen, four of Worcester, three each of Merton College and Christ Church, two each of New College, All Souls', Brasenose, Corpus Christi, Trinity, Jesus, nine views of the Botanic Gardens, and one each of the Bodleian Library, University Museum, Chemical Laboratory, and the River Isis with University barges.

496.—A Trip down the River Wye (YORK, 36 slides).

We start at Hereford (3 views), and pass Ross (2), Goodrich Castle (2), Whitchurch, Monmouth, Raglan Castle, Tintern Village and Abbey (7), reaching Chepstow, of which we have three views.

497.—A Trip to Brighton (YORK, 40 slides).

The Railway Station, King's Road, Aquarium, Royal Pavilion, and other views at Brighton.

498.—Brighton Aquarium (YORK, 14 slides).

Gives the Entrance, Concert Hall, Corridors, &c

499.—A Trip to South Coast (NEWTON, 54 slides).

Starting from Victoria Station, we pass on to Hastings (6), and then with glimpses at St. Leonard's, Eastbourne, Beachy Head and others to Brighton (7), then *via* Arundel to Bognor, Chichester, Portsmouth and the Isle of Wight, ending up with Pulborough and Box Hill.

500.—Isle of Wight (CATLIN, 50 slides).

Dr. Croft writes a lecture under the title "The Garden of England; or, A Summer Ramble in the Isle of Wight. The views include Osborne House, Whippingham Church, Carisbrooke Castle, with the apartments of Charles I. (3), Shanklin Chine, Sandown Bay, Ryde, Freshwater (2), Ventnor (4), Yarmouth, Tennyson's House at Freshwater, etc.

501.—Devonshire—(THE SCIOPTICON Co., 24 slides).

The set includes a Devonshire Cottage, two views of Ilfracombe, eighteen of Lynmouth, one of Clovelly, and one of Saltash Bridge.

502.—Devonshire (CATLIN, 50 slides).

The set includes Barnstaple, Lynmouth, Lynton, Ilfracombe (3), Bideford, Westward Ho, Clovelly (4), Tavistock (2), Plymouth (3), Ivy Bridge, Totnes, Dartmouth (3), Brixham (2), Torquay (4), Teignmouth, Exeter Cathedral (2), etc., etc., giving a very good idea of the county.

503.—Along the Devonshire Coast (VALENTINE, 36 views).

The Exeter Cathedral starts this list with three views, Dawlish continues with two, Teignmouth five, Torquay eight, Brixham three, Dartmouth five, and Babbicombe, Anstey's Cove, Berry Pomeroy Castle, and others vary the scenes.

504.—A Tour through Cornwall (YORK, 50 slides).

Views of Saltash, Launceston, Boscastle, Tintagel, Truro, Falmouth, Penzance, St. Ives, and Land's End. We also get some views of the Scilly Isles, including the Tresco Gardens, and a wrecked steamer on St Agnes Island.

505.—A Month at the Lizard and Land's End, with a Trip to the Scilly Isles (VALENTINE, 50 slides).

Falmouth has here six views, the Lizard fourteen, Penzance four, the Scilly Isles thirteen (including the Long Walk at Tresco), Land's End three, with Helston, St. Ives, etc.

506.—Guernsey (STEWARD, 11 slides).

The views are of Petit Port (4), Vale, St. Martin's, and Forest Churches, etc.

507.—The Channel Islands (YORK, 60 slides).

Jersey, Guernsey, Sark, and Alderney. Harbours, Castles, Streets, Churches, Caves, Cows, Cottages, Bays, Ports, etc. Commences with steamer entering Guernsey Harbour, thence to Jersey, St. Aubin's, Mount Orgueil Castle, Gorey, Moulin Huet Bay, Creux Harbour, Sark, Isle of Marchands, etc.

508.—Scotland (NEWTON, 177 slides)

This series embraces a large number of well-known places of interest in Scotland from Abbotsford northwards to Wick, and from Inverary eastwards to Aberdeen ; in fact, we have here a general view of the country, and from the slides the ordinary number for lecture set can be made up.

509.—The Stately Homes of Scotland (VALENTINE, 36 slides)

The "Homes" include Balmoral Castle, Abergeldie, Hamilton Palace, Inverary Castle, Abbotsford House, and Holyrood Palace.

510.—From the Forth to the Clyde (RILEY, 50 slides)

A summer tour in Scotland Edinburgh (11), Granton, Leith, Linlithgow (3), Stirling (6), Callander, Brig o' Turk, Loch Katrine, Loch Lomond, Oban (4), Greenock (2), and Glasgow (3)

511.—The Lowlands of Scotland (YORK, 50 slides)

Edinburgh is here represented by five views, Roslin and the Falls of Clyde by three each, Ayr by four, including Burns's Cottage and "Alloway's Auld Haunted Kirk." We have also Dryburgh Abbey and Sir Walter Scott's Study at Abbotsford, etc

512.—Landscape Illustrations of "Lady of the Lake" and "Lord of the Isles" (WOOD, 50 slides).

Sir Walter Scott here shows us Ben Ledi, Loch Achray, Loch Katrine (4 views), Stirling Castle (5), Loch Lomond (4), Oban, Skye (10), Staffa, Iona, Arran, etc.

513.—The Land of Scott and Burns (VALENTINE, 51 slides).

Four views of Edinburgh are here followed by Linlithgow Palace, Roslyn Chapel, Abbotsford Study, Abbotsford from the Tweed, Melrose Abbey (2), Dryburgh Abbey (2), with Tomb of Sir Walter Scott, Kelso Abbey, Burns's Mausoleum, Dumfries, Auld Brig o' Ayr, Burns's Birthplace, Alloway Kirk, etc , etc.

514.—Through the Trossachs with Sir Walter Scott (24 slides).

The scenes include Ben Ledi. Loch Vennachar, Loch Achray, Ben Venu, Loch Katrine (5 views) and Loch Lomond.

515.—The Rivers, Glens, and Waterfalls of Scotland (VALENTINE, 31 slides).

Here we have the Grey Mare's Tail at Moffatt, Falls of Clyde (3 views), Bothwell Bridge, Auld Brig o' Doon, Pass of Glencoe, Pass of Killiecrankie, on the Tay above Dunkeld, etc.

516.—The Highlands of Scotland (VALENTINE, 51 slides).

Following Glasgow (3), we see Rothesay, Kyles of Bute, Arran, Glencoe (celebrated for its Pass and massacre), Falls of Foyers, Gairloch, Loch Maree, Aberdeen, Balmoral, Dunkeld Cathedral, Trossachs, etc.

517.—The Highlands of Scotland (YORK, 52 slides)

Eight views are here given of Glasgow and places, etc , on the Clyde, five of Aberdeen, one of Inverary Castle (seat of Duke of Argyle), a view of the Pass of Glencoe, of Inverness, Balmoral, Pass of the Trossachs, Loch Katrine, Loch Lomond, Stirling Castle, etc

518.—Orkney and Shetland (WILSON, 40 slides).

Here Kirkwall starts the list, followed by its ancient Cathedral of St Magnus (built 1137), Bishop's Palace, "Old Man of Hoy," Lerwick (the most northern town in the British Isles), etc , the last two of the set showing Shetland Ponies and a Shetland Knitter.

519.—Irish Scenery (PUMPHREY, 56 slides)

Kingstown, Dublin (7), Killarney (2), Ross Castle (2), Muckross Abbey (exterior and interior), views on the Shannon (2), Sir Walter Raleigh's house, Youghal, Natural Bridge at Ross, Blarney Castle, Kilkenny Castle, etc., etc.

520.—Ireland (NEWTON, 73 slides).

We encounter, after the first views, a characteristic group of beggars, possibly preparing for the next scene, Blarney Leaving the other beggars, we encounter some Connemara peasants in holiday costume. We see six views of Dublin, including the Phœnix Park. We pass on to Galway, the Giant's Causeway, Holy Cross Abbey, and at a cottage witness an Irish wake. Passing Kilkenny, we may imagine ourselves in a jaunting car till we reach Killarney, where we fall on seven views, including Innisfallen. A marriage party, with bride on pony, Powerscourt Waterfall, the deserted village of Lissoy, Belfast, Carrickfergus Castle, four views of Galway, Limerick, Londonderry, Sligo, Waterford, Youghal Abbey, etc , etc , make up the set.

521.—Ireland. Lecture No. 1 (YORK, 60 slides).

Dublin, Wicklow, Killarney, etc , with Kingstown Harbour, the Vice regal Lodge, Phœnix Park, Dublin, Kilkenny (celebrated for its cats), fifteen views of the neighbourhood of Killarney, etc , etc.

522.—Ireland. Lecture No. 2 (YORK, 50 slides).

The North and West is here shown, as illustrated by Belfast, Antrim, Londonderry, Garron Tower, the seat of the Marchioness of Londonderry, the Horn Head at Donegal, Lakes at Connemara, etc , etc

523.—Liverpool to Dublin and the Lakes of Killarney.—(RILEY, 60 slides).

Liverpool Landing-stage, Dublin (12), Kingstown (2), Bray (3), Powerscourt (2), and nine at the Lakes. Then to Ross Castle, Muckross Abbey, Cork (9), and Queenstown (5).

524.—North Wales (PUMPHREY, 51 slides).

Commencing on the Great Orme's Head, Llandudno, then on to Conway, Penrhyn, Menai Bridge, and Carnarvon. We have a rest to master a Welsh name, Llanfair-Pillgwllgwngwell. After this struggle we resume through Llanberis, Snowdon, Aberglaslyn, Harlech, and Barmouth, Bettws-y-Coed, finishing up at Binglog Falls.

525.—A Tour through North Wales (YORK, 50 slides).

The tourist is supposed to start from Llandudno, and proceed to Conway, Bettws-y-Coed, Snowdon, Bangor, Carnarvon, Llanberis, Aberglasyn, Barmouth, Dolgelly, and Llangollen, finishing up with Ruabon.

526.—British Isles (SCIOPTICON Co., 104 slides).

The thirteen slides referring to Scotland include Sir Walter Scott's Tomb, Robert Burns's Cottage, a Room in the Prince of Wales's Castle of Abergeldie, and Balmoral Castle. There are two views in the Isle of Man, one each in Guernsey and Jersey, and the remaining 87 views in England, include eight of the Thames, four at Richmond, and one at Windsor Castle, six of Bramshill House, five of Bath, three of Clifton Suspension Bridge, two each of Cheddar Cliffs, Chepstow Castle, Scarborough, one each of Kenilworth Castle, and the Head of Lake Windermere, etc.

527.—The Land of the Rose, Shamrock, and Thistle (VALENTINE, 50 slides).

Scenes in England, Scotland, and Ireland. The nineteen of *England* include Windsor Castle, Haddon Hall, Warwick Castle, a View in Liverpool, one in Manchester, and Lakes Derwentwater and Ulleswater; the fifteen of *Scotland* include Glencoe, Loch Lomond, Loch Katrine, Abbotsford, and Edinburgh; and the twelve of *Ireland*, the Viceregal Lodge, the Giant's Causeway, and Lakes of Killarney. The set winds up with the British Army, Navy, Parliament Houses, and Queen.

528.—Round Britain in a Yacht (50 slides).

After showing the Yacht, we view the Needles, Isle of Wight, Dartmouth, Penzance, Land's End, Ilfracombe, and thence, noting a Lighthouse and two points in the Isle of Man, we touch the Clyde, onward then till we reach Skye, and rounding the North *via* Thurso, we come to Aberdeen Dundee, Edinburgh (which is not on the coast, however), Tynemouth to Norfolk Broads (four views), and end at the mouth of the Thames.

Cathedrals, Castles & Abbeys.

529.—The Cathedrals of England (VALENTINE, 36 slides).

This set of Cathedral Views starts in the far North at Carlisle, and Durham, and takes them almost in geographical order down to Exeter. In some cases interiors are given, but the majority are exterior views.

530.—English Cathedrals (WILSON, 44 slides).

One exterior and one interior view of each of the following cathedrals: Canterbury, York, Durham, St. Paul's, Westminster Abbey, Winchester, Lincoln, Lichfield, Hereford, Worcester, Salisbury, Ely, Gloucester, Wells, Bath, Bristol, Exeter, Rochester, Peterborough, Beverley, Carlisle, and Ripon.

531.—Cathedrals (THE SCIOPTICON Co, 39 slides).

From negatives by F. M Good and others; seventeen of Canterbury, nine of Winchester, four of Worcester, etc.

532.—English Cathedrals (PUMPHREY, 38 slides).

The first slide of this set is a Plan of a Cathedral, and then the usual views of different cathedrals follow.

533.—Our English Cathedrals (YORK, 50 slides).

The set contains two each of Canterbury, Winchester, Exeter, Wells, Bristol, Gloucester, Worcester, Hereford, Llandaff, Manchester, Durham, Ripon, York, Lincoln, Lichfield, Peterborough, Norwich, Ely, and St. Paul's, and one each of Rochester, Salisbury, Truro, Bath, St. David's, St. Asaph's, Chester, Bangor, Carlisle, St. Alban's Abbey, and Christ Church, Oxford.

534.—Ruined Abbeys and Cathedrals of Scotland (VALENTINE, 24 slides).

The set includes Roslin Chapel, Paisley Abbey, Dryburgh Abbey, and Sweetheart Abbey.

535.—Abbeys (SCIOPTICON CO., 41 slides).

The series contains five of Tintern Abbey, five of Melrose, two of Roslyn, three of Jedburgh, four of Calder, four of Rievaulx, ten of Fountains Abbey, four of Kirkstall, and one each of Dryburgh, Bylands, St. Mary's at York, and St. Patrick's in the Isle of Man.

536.—The Abbeys, Monasteries, and Priories of Ancient Britain (WILSON, 36 slides).

This set includes in Scotland, Iona Cathedral, St. Andrew's Cathedral, Cambuskenneth Abbey, Melrose Abbey, Holyrood Chapel; and in England, Lindisfarne Abbey, Furness Abbey, Fountains Abbey, Tintern Abbey, and Glastonbury Abbey.

537.—Among the Ruined Castles and Abbeys of England (VALENTINE, 30 slides).

This set gives, *inter alia*, Kenilworth Castle, Scarborough Castle, Tynemouth Priory, and Haddon Hall.

538.—Abbeys and Castles of England (YORK, 50 slides).

This set includes three slides of Netley Abbey, one of Carisbrooke Castle (Isle of Wight)—interesting for its association with Charles I.—three of Tintern Abbey, three of the far-famed Kenilworth Castle, three of Byland Abbey, five of Fountains Abbey (near Leeds), and two of Furness Abbey.

539.—Canterbury Cathedral (BROOKS, 27 slides).

From negatives specially taken by Mr. Brooks we have the Entrance Gate, a General View, views of the Nave, Aisle, Choir, Crypt; Tombs of Edward the Black Prince and King Henry IV., etc.

540.—Exeter Cathedral (BROOKS, 16 slides).

The set includes the West Front, Nave, Pulpit and Choir, Lady Chapel, etc. Negatives specially taken by Mr. Brooks.

541.—The Ruined Castles and Abbeys of England (VALENTINE, 30 slides).

A series of views of castles and abbeys more or less renowned in bygone days, such as Kenilworth Castle, Haddon Hall, Warkworth Castle, Lindisfarne Abbey, etc.

542.—Ruined Castles of Scotland (WILSON, 45 slides).

Including Loch Leven, Craigmillar, Roslyn, Dumbarton, Bothwell, Rothesay, Carrick and Scalloway. This set of 45 different castles ought to give a very instructive history of Scotland in ancient times as told by its castles.

543.—English and Continental Cathedrals (NEWTON, 166 slides).

One hundred and thirty-four views in this set are in Great Britain and Ireland, the remaining 32 are chiefly in Belgium, France, Italy, and on the Rhine. The eight of *Wales* are made up of the exterior and interior of Bangor, Llandaff, St. Asaph and St. David's; *Scotland* is represented by the interior and exterior of the Glasgow Cathedral; for *England* are 16 views of Canterbury, nine of Westminster Abbey, six each of St. Paul's in London, Salisbury, Worcester and Gloucester, five of Ely, four each of Bristol, Carlisle, Durham, Lincoln, Peterborough, Rochester, Wells and York, three of Exeter and Winchester, two of Beverley, Chester, Chichester, Hereford, Lichfield, Manchester, Norwich, Oxford, Ripon, St Alban's, Southwell, Truro, and one of Liverpool, Newcastle and Wakefield Cathedrals; in *Ireland* we have an exterior and interior of St. Patrick's Dublin. Now taking the Continent, *Belgium* has two views each of Antwerp and Liege, *France* has one of Amiens (exterior), one of Tours (exterior), two of Lyons, and three of Notre Dame at Paris, on the *Rhine* we have one each of Strasburg and Bonn, two of Cologne; *Spain* has one each of Burgos and Seville; *Switzerland* has Basle, *Austria*, St. Stephen's, Vienna, while in *Italy* we have Catania, Florence, Milan (2), Palermo, Pisa (4), St Peter's, Rome (2), and St. Mark's, Venice (3). In all cases where one view only is given it is an exterior view. When two are mentioned, one is exterior and the other interior.

EUROPE.

544.—Austria (Wood, 29 slides)

The set gives nineteen views of Vienna, four of the Exhibition, four of Prague, etc., including the Imperial Palace, the Prague Suspension Bridge, Dalaas, Trieste, and Pesth.

545.—Austria (Newton, 21 slides).

The set gives nine views of Vienna, one of the Giessbubler Springs at Carlsbad, one of Lintz on the Danube, two in Hungary, at Prague and in Styria, and one each of Pesth and Salzburg.

546.—Belgium (York, 50 slides).

Reading by the Rev. J. Cowper Gray. There are two views of Courtray, three of Tournay, three of Dinant, six of Liege, four each of Brussels and Waterloo, eight of Antwerp, four of Ghent, six of Bruges, two of Ostend, etc.

547.—Belgium (Newton, 31 slides).

There are six views of Antwerp, four each of Bruges, Brussels, Ghent, and Liege, one of the Field of Waterloo, etc

548.—Cyprus (Newton, 23 slides).

The famous Island added to the British Empire by Lord Beaconsfield's Berlin Treaty of 1878 Five illustrations refer to Famagosta, seven to Larnaca, one to the Hoisting of the British Flag at Nicosia, the capital, etc.

549.—Denmark and Sweden (Newton, 30 slides).

The set shows the Battle of the Baltic, five views each of Copenhagen and Gottenburg, and six of Stockholm, two Maps, Reindeer Sledge Travelling, H R.H. the Princess of Wales.

550.—Views in France (Wood, 46 slides).

Caen is here represented by five views, Rouen by seven. We see also Boulogne, Calais, Dieppe (2), Havre (2), Lyons (2), Nice, and the interior and exterior of Fontainebleau, etc, etc.

551.—France and Paris (Newton, 85 slides).

Here are the cathedrals of Amiens, Bayeux, Lyons (4 views), Tours, two views of Caen, three of Dieppe, two of Marseilles, one of Cannes, etc. In Paris we have the Abbey of St. Denis, Tomb of Napoleon, three views of Notre Dame, Portrait of Napoleon III., Palace of Versailles.

552.—Wanderings in Paris (York, 50 slides).

From the railway station, after passing Trinity (Eglise de la Trinité) and other churches, we come to the gardens of the Tuileries and ruins of the Palace, then we shortly get to the Louvre (5 views) and pass Notre Dame (3), Luxembourg, the Panthéon, Palace of Versailles (5), St. Cloud (3), and St. Denis (2), both of the Cathedral

553.—The Riviera (50 slides).

From Marseilles to Genoa, passing Cannes (4 views), Nice (5), Monaco, Monte Carlo (5), Menton (6), San Remo, including the Villa Tzirio, the late Emperor Frederick's residence, etc , etc.

554.—Germany (WOOD, 42 slides)

Baden (3 views), Berlin (12), including the Royal Palace, Cathedral, Palace of Crown Prince, Dresden (6), Hamburg (2), Munich (5), Nuremburg (4), Potsdam (2 of old and 1 of new Palace), Stuttgart (3), Wiesbaden (3), Wurtzburg (1).

555.—Germany (NEWTON, 57 slides)

Berlin has nine views, Frankfort (4), Dresden (3), Worms (3), Homburg, Munich, Potsdam, Charlottenburg, etc. following, then leaving Germany proper, and entering Austria, we have views of Carlsbad, Gran and Presburg in Hungary, Vienna (9), including the Imperial Palace.

556.—Greece, Section I. (WOOD, 63 slides).

Prehistoric, Legendary, Classical, and Modern, including Map; Ruins of Troy; Constantinople, with the Seraglio, the Bospborus, and Entry of Mahomet II (history here seems rather confused, however), Cyprus, Ancient Athens; Costumes, Aphrodite in her Car, winding up with the Bay of Naples, which must have been in Greece, if ever, in *very* prehistoric times.

557.—Greece, Section II. (WOOD, 49 slides).

Prehistoric etc; Hades, Juno, Vesta, Ceres, and Diana grace this second section, together with a Grecian Banquet, a Marriage Procession, Storming a City, a Funeral Pyre, Modern Athens, a Greek Girl and Brigands after her.

558.-Picturesque Holland (YORK, 50 slides).

A view of Holland from the steamer starts the set, followed by seven scenes in Rotterdam, thirteen in Amsterdam, and others at Dordrecht (3), Delft (2), Scheveningen (4), Hoorn (2), and six in Marken. We also get an idea of the dog-drawn carts, the children, markets, canals, etc., two views of Utrecht completing the set.

559.—Iceland (RILEY, 48 slides).

Commencing with the Railway Station and view of Edinburgh, we have in passing a view of Leith Harbour, Bass Rock, Herring Fleet at Wick and Duncansby Head, we dive into Iceland. First seeing a map of that country, the majority of the pictures are views, but there are also illustrations of native life.

560.—The City of Florence (36 slides)

Florence shows us its Church of St Maria, three views of the Vecchio Palace, the Cathedral, the Pitti and Strozzi Palaces, the Villa Palmeria, etc.

561.—Venice (YORK, 30 slides).

After St Mark's (3 views), we have Doge's Palace (2), Bridge of Sighs, Canal Scenes, Fish Market, etc , etc., winding up at the Well, Campo SS. Giovanni e Paolo.

562.—Rome (YORK, 50 slides).

Rome is here shown from two points, we have views of St. Peter's and the Vatican, of the Ruins of Ancient Rome, including the Temple of Saturn, Arch of Constantine, etc

563.—Rome, Ancient and Modern (YORK, 50 slides).

Ancient Rome is represented by the Entry to Forum, Interior of the same, the Coliseum (2), Arch of Constantine, the Palatine, etc , and Modern Rome by the State Carriage of the Pope, Awaiting the Illuminations, Interior of St. Peter's, a Cardinal entering the Vatican, etc.

564.—Italy, Naples, Pompeii, and Venice (NEWTON, 193 slides).

Como is here shown by eight views, Florence ten, Milan six, Mount Vesuvius, going up, coming down, etc., in seven slides; Naples has twenty-two views, Pisa five, Pompeii eighteen (including a plaster-cast of human figure found in the Ruins), Venice fifty-three, including four of the Ducal Palace, and thirteen of St. Mark's, etc.

565.—Italian Lakes (YORK, 50 slides).

Lake Orto, Lake Como, Lake Lecco, Lugano from the Lake, Maccagno from the Lake, Luino from the Lake, Laveno from the Lake, Gravedona from the Lake, Tropical Plants in Garden of Villa Serbelloni, etc.

566.—Picturesque Italy (WOOD, 50 slides).

Here we have Venice (8), Florence (7), Naples (10), Pompeii (9), Genoa and Turin (3), Pisa (2), with Milan, Orto Lake, Lake Maggiore, Lake Como, Bologna, Arenzo, Palermo, and Lucca (one view of each).

567.—Rome and Roman History (NEWTON, 102 slides).

These are views of Ancient and Modern Rome, the former being illustrated by the Appian Way in the Olden Time, Britons in Rome, Alaric and the Councillors, Ruins of Old Bridge of Horatius, Excavations of Forum, etc., and the latter (Modern Rome) by views of St. Peter's, Portraits of the Popes Pius IX. and Leo XIII., etc.

568.—Views in Italy (WOOD, 60 slides).

These views embrace Turin (5), Palermo (5), Naples (2), Pompeii (8), Lac Majeur (11), Como (9), Venice (3), Verona, Bologna, and Loretto (two each) and Caserta, Padna, Chatillion, Milan, Brindisi, Ancona, Carpone, Domo d'Osola, &c. (one each).

569.—A Tour in Italy (WOOD, 50 slides).

The views are as follows :—Venice (9), Rome (14), Milan (3), Naples (3), Pompeii (5), Turin, Pisa and Florence (2 each), and Brescia, Terni, Tivoli, Ponzynoli, Amalfi and Messina (1 each), the set ending with two views of Palermo.

570.—Italy (YORK, 50 slides).

Turin has three views, Venice eight, Florence seven, Naples ten, two of Pisa, three of Genoa, nine of the world-renowned Ruins of Pompeii, etc , etc.

571.—Italy (THE SCIOPTICON CO , 92 slides).

Twenty-six slides are here devoted to Rome, not including the Vatican, which has seventeen to itself. Of Tivoli, Milan, Akani, Vietri, Verona and Palermo, we have one each; of Como and Sorrento, two each; of Pisa, three; of Florence, four, of Amalfi, seven; of Naples, seven; and of Pompeii eighteen.

572.—Norway (PUMPHREY, 47 slides).

This set gives views of a number of places, including Odde, Bergen, and Eide districts

573.—A Visit to Western Norway (YORK, 40 slides).

The set is made up of views at Nordtjord (7), Söndmore (6), Romsdal (6), Jotunheim (6), Sognefjord (4), Hardangerijord (5).

574.—Rambles in Norway (VALENTINE, 50 slides).

This set includes Christiania, Bergen (5 views), Odde (2), on the Rauma (3), Molde (2), Colonel Sinclair's Grave, etc., etc.

575.—Norway (NEWTON, 48 slides).

Nordtjord (7 views), Söndmore (6), Romsdal (6), Jotunheim (6), Sognefjord (4), Hardangertjord (5), Christiania (5), Geirangerijord (3), Trollhattan Falls (2), etc.

576.—The Hardangerfjord, Norway (41 slides).

Includes Bergen (2), mill near Voss, road near Seine, way to the Vöring Foss, two views of Odde, etc.

577.—The Rhine (YORK, 60 slides).

From Constance to Cologne. Of Cologne we have nine views, then we pass Bonn, Coblentz (4), Bingen, Wiesbaden (2), Mayence (2), Worms (3), Heidelburg (7), Baden-Baden, Strasburgh, etc., and thence through Bâle, Neuhausen, and Schaffhausen to Constance.

578.—The Rhine and Its Neighbourhood (NEWTON, 108 slides).

A comprehensive series of Rhine views too numerous to summarise. The "neighbourhood" is represented by Amsterdam, Hamburgh, Rotterdam and other places.

579.—From St. Petersburg to the Crimea (WOOD, 59 slides).

Of *St. Petersburg* thirty-three slides, including Peter the Great's Cottage, the Winter Palace, Assassination of Alexander II, a Droski Driver, etc. Of *Moscow* ten slides, including the Coronation of the Czar, the Great Bell, and a Great Gun; then we see peasants' houses, etc., and finish with a view of Odessa, of Sebastopol, and Balaclava.

580.—Dr. Lansdell's Tour through Siberia (WOOD, 36 slides).

Starting in Europe from St. Petersburg, we see Peter the Great's Cottage, thence we go *via* Kieff, Tobolsk, and other unpronounceable places, into Siberia, where we note Russian Prisoners, Siberian Gold Mines, Penal Colonies, etc.

581.—Through the Baltic, Russia and Finland (WOOD, 63 slides).

These views are to illustrate Dr. Lansdell's travels, and we have first a map referring to these, followed by Riga (9 views), Dorpat (6), Revel (10), Helsingfors (6), Wasa (8), etc, and we have also such scenes as a Russian convict prisoner chained for transport, etc.

582.—Russia and the Russians (NEWTON, 76 slides).

The principal views here are twenty-four of Moscow, twenty-six of St-Petersburg, and ten of Siberia, the latter including a view of the great central prison.

583.—Spanish Bull Fighting (HUGHES, 7 slides).

Illustrates Goading the Bull on Being Caught by his Horns, and Death of an Attendant.

584.—Spain and Portugal (NEWTON, 65 slides).

Barcelona (4) is here followed by Biarritz (2), Burgos (2), Cadiz, Cordova (3), Cuenca, Granada (8), Madrid (8), Seville (7), The Escurial (with a view of the palace and the Queen's chamber), etc , in Spain; and Alcabaca, Combra and Cintra are followed by six views of Lisbon before and after the earthquake, etc , the four latter towns representing Portugal.

585.—Spain (YORK, 50 slides).

Gibraltar has here six views, Seville seven, including two of a Bull Fight, Cordova four, Toledo two, Madrid four, Malaga three, and Granada, the capital in Spain of the last Moorish rulers, with its Alhambra or King's Palace, etc., thirteen, and then other well-known views in different parts of Spain.

586.—Spain (THE SCIOPTICON CO , 61 slides)

Commencing with Biarritz (4 views), we see Barcelona (6), Tarragona (10) Cordova, once the capital of the Calephate of the West (17), and end our journey at Seville (17), one of the most ancient towns in Europe, the capital of Spain under the Gothic Dynasty, and long the chief residence of Spanish monarchs

587.—Switzerland and the Tyrol (NEWTON, 192 slides).

Here we get a general view of Switzerland, twenty-six referring to Mont Blanc, eleven to Lucerne, fourteen to Righi, eight to Zermatt, other towns, villages, etc , having from one to six views devoted to them. In the Tyrol Innsbruck has three views, and Octythal thirteen.

588.—Mont Blanc (HUGHES, 28 slides).

Views of the Mountain and Mountain Scenery in Switzerland from special drawings by R. Beker, also such pictures as Sheep and Peasants at rest before the Avalanche, and the Destruction wrought by it, Dangers of crossing Natural Bridges, the Guides linked together, &c., coloured, price 4s. 6d. each.

589.—Ascent of Mont Blanc (WOOD, 17 slides).

Starting at Geneva, we pass the Monastery of St. Bernard, Dead House, Chamounir Valley and Village, a Snow Bridge, and see the top of Mont Blanc, &c., winding up by coming down again.

590.—The Engadine, Switzerland (50 slides).

We meet in a street in Coire, view the Cathedral, pass Silvaplana Village, Maloga, see St. Moritz from Johansberg, &c , a Diligence, then pass by the Weissenstein Valley and Bergun, back again to Coire. Goes a little out of the beaten track.

591.—A Swiss Tour (WOOD, 50 slides).

Starting from Schaffhausen we go *via* Zurich, Lucerne, Thun, the Matterhorn, of which we have three pictures, including a Disaster, Mont Blanc, Castle of Chillon, on the Lake of Geneva to Berne, passing many interesting views not here mentioned.

592.—The Forest Cantons of Switzerland (NEWTON, 50 slides).

We have here eight views of Lucerne, a neighbourhood especially celebrated for its beautiful scenery, and scenes at Goldau, Brunnen, Amsteg, Wasen, Hospenthal, etc.

593.—Views in the Engadine (WOOD, 24 slides).

These include Panorama de Silvaplana, Lac de Silvaplana, Lac de St. Moritz, Samaden, etc.

594.—Switzerland. Tour No. 1 (YORK, 50 slides).

The Northern Lakes and Bernese Oberland After Basle (5 views), Neuhausen (2), Constance (2), Zurich (2), Lucerne and its Lake (6), etc , etc , we come to Interlaken (3), Grindelwald (4), Thun (3), Fribourgh (3), Berne (3), passing a few unmentioned scenes on the way.

595.—Switzerland. Tour No. 2.

Here we have Geneva , Mont Blanc , the Matterhorn ; Amstag , Rhone Glacier, and Hotel , Vevay, Montreux, and Dent du Midi , the Castle of Chillon (2) , Lausanne (2) , and others

596.—Swiss and German Pictures (PUMPHREY, 227 slides)

These are a series of views from which lecture sets can be made up The chief items are Heidelberg Castle (23 views), Glacier Views (26), Lanterbrunnen Valley (9), Thun (16), Berne (12), and Basle (10)

597.—Turkey and the War (NEWTON, 35 slides).

Commencing with a map of Turkey in Asia, then Besika Bay, and three of the Bosphorus , the remaining thirty are devoted to Constantinople.

598.—The Turks, Past and Present (WOOD, 48 slides).

The views are of Trieste, Greece, Bosphorus, Constantinople, Widdin, Bedgrave, Kars, etc Other items are the Hour of Prayer, the Three Sultans, the Dogs, Battle before Plevna, etc.

599.—Continental Views (THE SCIOPTICON Co , 103 slides).

These are from negatives by Steven Thompson and others Among the principal are eighteen of Venice, thirty-nine of Switzerland, five of Rouen, three of Caen, eleven of Amiens, six of Versailles, seven in Berlin, and a few odd ones in Belgium and Germany.

600.— Costumes of Europe (WOOD, 18 views).

Europe is here represented by two Spaniards, two Russians, two Italians, two " Swissians," a Dane, a Hungarian, and a Swede, to which are added the Turkish idea of a Shopkeeper and a Lady, also a Belgian Milkmaid and a Dutchman.

601.—The Mediterranean (YORK, 50 slides).

Gibraltar (3), Barcelona, Marseilles (5), Cannes (4), Antipes, Nice (7), Corsica, Monaco (3), Mentone (2), Genoa (2), Naples (3), Sorrento, Capri (2), Amalfi, Messina, Malta, and Algiers (7).

AFRICA.

602.—Africa and the Africans (NEWTON, 52 slides).

We have here views of Cape Colony, of Waterfalls, of Tangiers, Ascension Island, Madiera, St. Vincent, and St. Helena, as well as Maps and a Portrait of Livingstone. The miscellaneous slides include Mumbo Jumbo, or Lynch Law, the Tomb of Napoleon, and one each of the capture of, and interior of, a Slave Ship.

603.—Morocco (WOOD, 10 slides)

Tangiers (3), Interiors of Moorish and of an European House, Group of Natives, etc.

604.—Algiers (YORK, 20 slides).

General Views, Palace, Museum, Street Scenes, and so on, including a Costermonger from the Arabian point of view.

605.—Algeria (NEWTON, 35 slides).

The set includes thirteen of Algiers, seven of Tlemcen, an Arab Inn, and the Women's Private Apartments in a Moorish House.

606.—The Egyptian War (YORK, 50 and 66 slides).

The reading in connection with this set refers to fifty slides; there are, however, sixteen extra slides for which there is no descriptive matter.

The first fifty show Arabi Pasha, leader of the Egyptians; a Massacre of Europeans on June 11th; Bombardment, its Effects; Sir Garnet Wolsley and his Generals; Highland Brigade at Close Quarters; Charge of the Horse Guards, etc., etc, winding up with the Review of Troups.

The remaining sixteen show Gibraltar, Malta, Cyprus, the English and French Fleets at Alexandria, the Charge at the Bayonet's Point at Tel-el-Kebir, Candidate for Victoria Cross under Heavy Fire, etc.

607.—The War in Egypt (HUGHES, 73 slides).

After suitable introductory slides, we see here Alexandria's fortifications, male and female Egyptians, H.M.'s ships "Dee" and "Don" for protection of Suez Canal, the English Consulate before and after the bombardment, Alexandria's principal street on fire and in ruins, an execution, Lang's daring exploit, blue jackets clearing the streets, charges, a rush for water after the battle of Kassassin, triumphant entry into Cairo by the British troops, the return of the Guards to London, etc.

608.—Egypt and the War (NEWTON, 67 slides).

British Ironclads ready for Action; H.M.'s ship "Inflexible" opposite Alexandria; a Night Scene, with effect slide—the electric light, flashing from vessel, illuminating the earthworks and batteries; an 81-ton gun; a Skirmish with Bedouins at Ramleh; the battle of Tel-el-Kebir; Charge of Bengal Lancers; Arabs on the Look-out from the Pyramids, and so on. The remaining twenty-three pictures refer to incidents in the Soudan war, scenery, portraits, etc, including three battles, three cataracts, portraits of General Gordon and Lord Wolsley, and a view of Khartoum, of unhappy memory.

In this set there are mechanical and effect slides, one referred to above, others in connection with explosions of shell on board one of H M 's ships, blowing up the enemy's fort, and Alexandria Grand Square on fire.

609.—Views of the Nile (PUMPHREY, 50 slides).

The views are of the country and river between Alexandria and the Second Cataract. One of the first views is of Cairo—by far the largest city of Egypt, if not in Africa. A little further on we have the Obelisk at Heliopolis, supposed to have been standing in its present position 1740 years B C., and, if so, it would be also there when Joseph was governor of Egypt. Continuing, we have the Pyramid of Gizeh, perhaps 200 years older even than the Obelisk; Karnac, a village on the site of Ancient Thebes, formerly the great capital of Central Egypt; various remains of Thebes; the Palm Grove on the Island of Philæ, remarkable for its magnificent temple ruins belonging to the last period of Egyptian prosperity, etc., etc.

610.—Egypt and the Nile, etc. (SCIOPTICON CO., 56 slides).

From negatives by Mr. F. M. Good, representing, at Heliopolis, the famous Obelisk and Fig Tree; at Cairo, among the fifteen views, the Mosque of Sultan Hassan, two women grinding corn, Shepherd's Hotel, etc. As types, we have an Arab, a Fellah, Turkish Ladies, and a regular Turk. There are also views on the Nile, and at Thebes, Philæ and Nubia; six at Constantinople, including the Golden Horn, and one at Smyrna.

611.—Egypt (York, 60 slides).

The Rev. J. Comper Gray's lecture is here illustrated by views of Alexandria (3), Cairo (14), Thebes (13), Philæ (7), Nubia (5), etc., etc. These include, at Cairo, the Prince of Wales's Palace, a Section of a Pyramid, the Suez Canal; and at Philæ, Pharaoh's Bed. There are also a Map of Egypt, views of the Nile, etc.

612.—Modern Egypt and its People (York, 50 slides).

The Rev. Dr Tremlett, a Fellow of the Royal Geographical Society, writes a description, which these slides are made to accompany, of Egyptian life, and objects of interest. The set shows Alexandria, Port Said, Suez Canal, Old Cairo, Pyramids, Rock Tombs, General View of Cairo, the Fellaheen (Peasants), a Medical Mission, a Fanatic preaching to the people, etc., etc.

613.—Central Africa (York, 51 slides).

The Rev. J Comper Gray's lecture is here illustrated by a Map, Portraits, Views of the Victoria Falls, of Lake Tanganyika, Pictures of Native Animals, People, Houses, Customs, Incidents of Travel, attention being drawn to slavery under the headings, "Hunters" and "Captives."

614.—Our West African Settlements (York, 48 slides).

This set contains five views and pictures of Maderia, shows the far-famed Peak of Teneriffe, views of Bathurst, River scenes, Sierra Leone ("The White Man's Grave"), Free Town, Kissy Road, School Matters, and the Effect of West African Fever.

615.—Livingstone and Stanley in Central Africa (Pumphrey, 49 slides).

A portrait of Livingstone, the Mill where he worked, &c., introduce the set, followed by Lions attacking a Buffalo, Reception of the Mission by Shinte, the Victoria Falls, Natives Weaving, Group of Slaves, Mrs Livingstone's Grave, a Mutiny, "Dr. Livingstone, I presume?" &c.

616.—Livingstone's Last Journey (Pumphrey, 22 slides).

This set shows the Doctor, Slavers revenging their Losses, a Group of Slaves abandoned to die of Starvation, last miles of the Doctor's Journey, Fac-simile of the last page of his Journal, &c.

617.—How I found Livingstone (Newton, 18 slides).

This little set includes the Mutiny on the Gombe River, "Dr. Livingstone, I presume?" Symptoms of a Fight, Portrait of Stanley, &c.

618.—Stanley in Africa (York, 29 slides).

We here find maps, illustrations of work with boats and canoes, pictures of battles, portraits of Stanley and other distinguished members of the expedition, an encampment in Cannibal land, etc.

619.—Madagascar and its People (Pumphrey, 50 slides).

This set starts with an explanatory map, and shows mode of travelling, bread-fruit tree, traveller's tree and mode of obtaining water from it, slave girls pounding rice, mode of punishing slaves, the slave market, royal reception at the palace, etc., etc.

620.—Madagascar (Newton, 13 slides).

Antananarivo, the capital, is followed by an audience at the palace, female slaves filling bamboos with water, travellers passing through great forest, etc.

621.—India (NEWTON, 162 slides).

Aden commences the set, then on arrival in India we have eleven views of Agra Proceeding, we have six pictures for Benares, we have eleven for Bombay, seven for Calcutta, four for Cawnpore Then we have Cashmere (7), Ceylon (18), with views in Colombo the capital) We behold a Daughter of the Land, Procession, with Budda's Tooth, etc We then see Delhi (13), H R H Prince of Wales receiving Native Princes, various Idols, Lahore (5), Lucknow (8, Madras 6), Maharajah of Rewal in State Costume, Missionary Preaching, Nizam (8), including Palm Tree Climbing, Fakir's Self-torture, etc, etc.

622.—Our Indian and Colonial Empire (WOOD, 73 slides).

Map of British Possessions, followed by Newfoundland (3), Nova Scotia, etc, Melbourne, Gold Crushing, Sydney, etc., Settler's Homestead, Hobart Town, Feji, New Guinea, Singapore, Queen Proclaimed Empress, Mandalay, Bombay, Madras, Ceylon, Cape Town, Diamond Fields, St. Helena, Jamaica, Malta, Gibraltar, and others.

623.—India (YORK, 360 slides).

A selection of these slides is made for the reading India is here divided into—Cashmere, 170 views (with 20 selected for the reading); Muree, 18 (with 2 selected), Lahore, 27 (with 6); Umritsur, 12 (with 3); Delhi, 42 (with 12); Delhi Camp Views, 15, Agra, 9 (with 4), Cawnpore, 5 with 3), Bittoor, 2 (with 1), Lucknow, 7, Sundry Places, 9 (with 2), Native Characters, 44 (with 1 selected).

Under the heading Cashmere, we have the Maharajah's Palace at Sreenugger, Ancient Temple at Pandretton, Glaciers (Bungalows), State Barges, Rope Bridge, Bridge of Shops, Ruins of Martund, Caves, Cascades, etc. In Muree we see a Bazaar, Barracks, etc. In Lahore, Runjeet Singh's Tomb, Lahore Railway Station, etc. At Umritsur, a view of the City with the Golden Temple, etc. At Delhi, the King's Palace, etc. At Agra, views of the Taj At Cawnpore, Suttee Chowra Ghat (scene of the massacre). At Bittoor, Nana Sahib's House on the river Under the heading Sundry Places, we have the Temple at Kurtespoor, Punjaub, and under Native Characters, a Fakir or Priest.

624.—Indian Tropical Scenes (PUMPHREY, 55 slides).

Views of the Ganges start the list, followed later on by a Jungle picture, Juggernaut Car, a Jungle infested with Tigers, Bungalow, an English Merchant at Breakfast, his Warehouse, a Pineapple Plantation, a Ricefield, Cocoa-nut Groves, etc.

625.—Rock Temples of India (SCIOPTICON Co., 13 slides).

The Seven Caves, Monastery, and nine others for Ajunta, the Entrance of Elephanta Cave, and a view of Indian Village Temples.

626.—India (APPLETON & Co, 459 slides)

These slides are from original negatives purchased by Messrs. Appleton. They make up a large series, under many headings, from which lecture sets are to be prepared from time to time, one being now ready under the title, "Two Months in India with a Camera"

627.—Life in India (WOOD, 43 slides).

The set shows a Dancing Woman, a Snake Charmer, the Basket Trick, a Man Tiger, Self-torture, Reading the Koran, Infants thrown to Crocodiles, etc., etc.

628.–Indian Native Characters (44 slides).

Native Soldiers and their Officers, Bankers, Ladies, Priests, a Postman, a Washerwoman, etc , etc.

629.–Two Months in India with a Camera (40 slides).

Views, People, and Objects in India, including Hindu Dancing-girls, a Group of High-caste Girls, views in Madras, Bangalore, Ahmedabad, Delhi, Bengal, Shevaroy Hills (3), and other odd places.

630.–Northern India (WOOD, 55 slides).

Cashmere, Coolies crossing Bridge, State Barge, Boatmen and Women, Lahore Railway Station, Umritsur (2), Muree (2), Delhi (9), Cawnpore, etc , etc.

631.–Southern India (WOOD, 51 slides)

Mysore, Seringapatam, Great Bull of Hallibeeb, Temple of Bobeneswar, Mumtapum, after which *we* may be mum.

632.–India (YORK, 50 slides).

This is a set taken from the larger series of Indian views It includes the Rope Bridge in Scindi Valley, Cashmere, an Ancient Temple, the Maharajah's Palace, Lahore Railway Station, the Golden Temple, Suttee Chowra Ghat, etc

633.–Mysore (YORK, 51 slides).

The lecture by Colonel Dixon. The views include the Deria Dowlut, where the Duke of Wellington resided after the taking of Seringapatam, the Old Palace of the Mysore Rajah, a gigantic Mango Tree, the Great Temple of Juggernaut, etc

634.–Burmah (YORK, 36 slides).

The slides refer to incidents of late War, and to Native Life. We see King Theebaw and his Wives, his Removal, and his Army, three views of Mandalay, and an illustration of the Capture and Shooting of Dacoits.

635.–Burmah and the Burmese (WOOD, 70 slides).

Buildings and Natives, six views for Rangoon, eight for Mandalay. eleven for Bhamo, one each of a Festival, Boy's Tricks, Priests, Scene in the Jungle, and the Interior of a Chinaman's House, etc.

636.–Mandalay, the Capital of Burmah (PUMPHREY, 60 slides).

A Plan of Mandalay, Wayside Food-sellers, King Theebaw and Wives, the Medical Staff, the Incomparable Pagoda, Residence of the Commander-in-Chief, Burmese Dacoits, etc , are examples of the scenes in this set.

637.–Prince of Wales's Visit to India (PUMPHREY, 50 slides).

A Map shows the Route of the Prince , then we see the Serapis and the Prince's Apartments (2), have a view of Athens, Suez Canal, Pyramids of Gizeh, and arrive at Bombay ; see an Elephant and a Rhinoceros Fight at Baroda, and reach Ceylon ; thence by Madras to Calcutta, Benares, Lucknow, Delhi, Cashmere, Nepaul (6), Bombay again ; and thence by the Mediterranean to Madrid.

638.–Hunting Expedition of the Prince of Wales (YORK, 30 slides)

Scenes of Indian Hunting, beating the Jungle, Bears, Tigers, the Prince in Danger, Elephant-hunting, Shipping the Prince's Menagerie, etc.

639.—Prince of Wales's Visit to India (YORK, 56 slides).

Maps showing routes, Malta, Acropolis of Athens, Aden, Bombay, and others start the set; then we see the Prince Dining in the Case of Elephants, State Elephants, Devil's Dance, Madras, Juggernaut, Calcutta, Cawnpore, Lucknow, Delhi, etc., etc., with lecture by Henry Gore, C.E.

640.—Prince of Wales's Visit to India (YORK, 80 slides).

From direct photographs and "The Graphic," a set from which the above is taken, with reading to 56 slides.

ASIA AND AUSTRALASIA.

641.—Persia and the Persians (NEWTON, 17 views).

The scenes include Persian Women, and views of Teheran (8), the latter dealing chiefly with the Shah's Palace.

642.—Java, etc. (SCIOPTICON CO., 142 slides).

There are 132 slides here referring to Java, including Scenery of the Tropics (4), the Emperor, Empress, Emperor's 32nd Child, his Spittoon Bearer, his Dancing Girls (2), Tropical Fruit, a Girl of the Lower Class, and various other Mortals, Places, and Objects, of which may be mentioned Singapore Harbour, four rustic views of Sumatra, a Natural Bridge at Sunda Straits, at Moluccas a Volcano, a Lake, and a Photographer's Studio.

643.—Australia (WOOD, 41 slides).

The set includes seven of Sydney, two of Melbourne, and the miscellaneous slides include views about Gold Diggings, and Snake-hunting by Natives.

644.—Our Colonial Empire (NEWTON, 80 slides).

Australia with its Map, its views at Adelaide, and at Sydney, its Kangaroo Hunt, Buck-jumping Horse, Sheep, Ox Waggons, Natives, etc., 23 pictures, *Canada* with Winnipegg the Emigrant Centre, a Forest Path in Winter, a Horse in Snow Shoes, etc., 22 pictures; *Cape Colony* with Cape Town, Kaffirs at Natal, etc., 7 pictures, *India* with Map, Snake-charmers, Native climbing Palm Tree, etc., 17; *New Zealand* with its Gods, Maoris, and Boiling Mud Springs, etc., 11 pictures.

645.—New Zealand. Part I. (YORK, 54 slides).

This part represents the North Island, with views of Auckland (14), one of which shows the Homestead System, the Bush (3), Tauranga (3), Maori Girls, Group of Maoris, Wellington (5), etc., etc.

646.—New Zealand. Part II. (YORK, 39 slides).

This part represents the Scuth Island, with views of Christchurch, Omarn, Gold Working, Canterbury (2), Dunedin (8), Otago (4), etc.

647.—New Zealand NEWTON, 78 slides).

Views of Towns and Villages, Natives, War Canoes, Clubs and Dances, White Terrace, Pink Terrace, Hot Springs, etc.

648.—New Zealand Volcanic Eruption (VALENTINE, 36 slides).

A set illustrating the Volcanic Eruption at the Hot Springs, principally consisting of various views of the White and Pink Terraces, concluding with representations of houses, etc., in ruins.

649.—A Tour in New Zealand (RILEY, 50 slides).

Views of the principal towns of both islands, and of the wonderful hot spring district, recently the scene of a fearful volcanic eruption. Several of the views give places before and after the catastrophe.

650.—China and the Chinese (PUMPHREY, 44 slides).

A series of pictures illustrative of the Cities, Dwellings, Trade, Religion, and Amusements of the Chinese. We begin with Hong Kong, and come to a Street in Canton, Great Wall of China, Silk and Tea Manufacture, Itinerant Barbers and Doctors, Marriage Performances, and Kite-flying on the 9th of the Moon, winding up with a little opium.

651.—China and the Chinese (YORK, 60 slides).

Starting with a Map of China, we get views of Pekin, and natives playing shuttlecock with the feet, we see the City of Nankin, and also Cat-sellers, Rice-sellers, a Frog-catcher, the Manufacture of Silk, Tea, and Tiles, and Opium-smoking.

652.—Scenes in China (WOOD, 19 slides).

Ballad-singers first draw our attention , we have to follow, mourning for the dead. We have five examples of Potters and Clay, etc., and terminate with Siam.

653.—Japan (73 slides).

The set shows the People, their Life, Religion, and Country, including a Master giving Orders to his Servant, a Girl to show the general style of Dress, How the People Sleep, a Daimio entering his Turimono (conveyance), surrounded by Retainers, and so on.

654.—Japan and the Japanese (42 slides).

A Carrier with a load on the back, Cart Coolies, Cleaning and Preparing Rice, Domestic Life, Lady making Tea, a view of Jeddo, a Portrait of the Spiritual Emperor, etc , give one an idea of the life of these enterprising Asiatics.

655.—South Sea Islands (NEWTON, 24 slides).

Of Hawaii, the largest of the Sandwich Islands, we have here seven scenes, and of Honolulu, in Oaku, the capital of the Isles, we have six views, and so on.

656.—Fiji and the Fijians (RILEY, 51 slides).

These are from a series of negatives by the Rev. Joseph Nettleton, who, for thirteen years, laboured there as a missionary. It represents the People, perhaps more than the Country, giving many portraits of Natives, Chiefs, Native Ministers, etc.

657.—Social Life in Oriental Lands (WOOD, 40 slides).

Except in the case of Syria, we have no special land mentioned, and the scenes are such general ones as House tops, Removing the Shoes, a Mosque, a Wine Press, etc.

AMERICA.

658.—North America (SCIOPTICON Co., 83 slides).

This series give 46 views connected with Niagara, three with California, five with Washington, 14 with Philadelphia. The three of California relate to Mammoth Trees. There is a view of Chicago, and also another of the Interior of a Mississippi Steamer, while at Washington City we have the Tomb of Washington, the great General and Patriot.

659.—North America (NEWTON, 175 slides).

In the set are ten pictures of California, 40 of New York, 27 of Niagara Falls, seven of Philadelphia, 16 of Washington, six of Yellowstone Park, 17 of Yosemite, four of Chicago, a general view of Sierra Nevada, one of the Mammoth Cave of Kentucky, etc.

660.—North American Scenery—Niagara in Winter (YORK, 59 slides).

This set commences with the Ice Statue on Luna Island, continues with American Falls and Ice Mounds, gives a snow effect on Luna Island, and another snow effect on the grove above Horse Falls. There are slides illustrative of Frozen Spray, of the Horse Shoe Fall, of Terrapin Tower, and of the Rapids, in which the famous English swimmer, Captain Webb, lost his life

661.—North American Scenery—Ithaca and Vicinity, N.Y. (YORK, 7 slides).

Big Waterfalls, including the lower one of the Wallaupaupack.

662.—North American Scenery—The Glens of the Catskills (YORK, 31 slides).

Views of Mountains, Falls, Ferry Boats, the Hudson River, and "All quiet on the Ramapo."

663.—North American Scenery—Central Park, New York (YORK, 9 slides).

Views in the Park, of the Lake, Wonderful Arbor, etc.

664.—North American Scenery—Portage and Vicinity, N.Y. (YORK, 6 slides).

Views of Waterfalls.

665.—North American Scenery—Washington City (YORK, 12 slides).

Views of the Capitol, and of famous Statues and Monuments.

666.—North American Scenery—Public Buildings, New York (YORK, 19 slides).

This set includes a bird's-eye view of New York, views of the Academy of Design, Mary Stewart's New Store, Grand Opera House, Broadway, Streets, Falls, Rivers, and Woods.

667.—South Pacific Railway (HUGHES, 60 slides).

Illustrating a Tour across the American Continent, by the Central Pacific Railway.

668.—American and Canadian Views (PUMPHREY, 150 slides).

Twelve views on board the Vancouver, 2 of Quebec, 2 of Winnipeg, 35 of Yellowstone Park, 4 of Chicago, 7 of Boston, 5 of New York, the miscellaneous pictures including "In a Pullman Car going West," the Cap of Liberty, an Emigrant's Waggon, the Whirlpool and River.

669.—Three Weeks in the United States and Canada (VALENTINE, 30 slides).

Starting with a view of s s. Parisian, we are shown Halifax (Nova Scotia), Boston (3), New York (6), Philadelphia (2), Baltimore (2), Capitol and White House at Washington Then with single glances at St. Louis, Chicago, and Niagara Falls, we pass into Canada, where we have six of Montreal, and four of Quebec.

670.—Washington to the North-West, U.S.A. (YORK, 50 slides).

Commencing at the Capitol, Washington, we have a picture of Baggage Smashers, Interior of a Pullman Car, views of Washington (1), Philadelphia (4), New York (7), Albany (3 , Niagara (5), Detroit (2), Chicago (4), St. Paul's (2), Minneapolis (4), Ascending the Rocky Mountains (2), ending with the Great Falls at Yellowstone Park.

671.—Washington City (YORK, 50 slides).

In this set we see the Capitol (6 views), the Senate, House of Representatives, Interior, President taking the Oath of Office, the Interior and Exterior of White House, the Treasury, British Legation, etc , ending up with a Negro Sermon.

672.—New York to the White Mountains (YORK, 60 slides).

Starting from the New York Central Station, we pass through Newhaven (4), Hartford (5), Worcester (2), thence to Boston, of which we have 24 views, passing on to Cambridge (7), and Portland (3), to the White Mountains (11), including Profile Rock, Echolake, and the summit of Mount Washington in summer and in winter.

673.—New York (YORK, 50 slides).

This set gives views of the principal Buildings, Streets, Monuments, etc., including Broadway, Brooklyn Bridge, Wall Street, the Treasury, Morse's Monument, and the Fifth Avenue Hotel.

674.—Niagara (YORK, 41 slides).

The lecture by the Rev J. Comper Gray, starting with Diagrams, showing the cause of the receding of the Falls, and followed by the usual views.

675.—Canada, from Quebec to the Rocky Mountains (YORK, 60 slides).

There are 10 views for Quebec, 12 for Montreal, 10 for Ottawa, 3 for Hamilton, 8 for Niagara, 4 each for Toronto, Lake Superior, and Winnipeg, and 5 for the Rockies.

676.—Canadian Snow Scenes (PUMPHREY, 6 slides).

The first three views show a railway station under 4ft. of snow, the remaining three show wintry scenes in the streets of Montreal.

677.—Canada (NEWTON, 79 slides).

Thirty-two views of Montreal, Ottawa, Quebec, the Rockies, Toronto, and Winnipeg, 20 of Indians and Indian Life, including an Indian Chief, now a clergyman of the Church of England, etc., and the Arrival of a Missionary at Camp.

678.—Canada and British North America (HUGHES, 87 slides).

Commencing with a Map of British America, we get a view of the (Maine) Eastern Terminus of the Grand Trunk Railway, as also of Quebec, Montreal, Halifax, and a fine view of Fraser River. The miscellaneous slides include a view of the Rocky Mountains, pictures of a Horse in Snow Shoes, a Farm Snowed Up, Going to Church during a Flood, Portrait of Prime Minster, Sir John Macdonald, Indian Burial on the Plains, Beavers at Work, Shooting the Rapids.

679.—Through South-western Texas with a Camera (VALENTINE, 36 slides).

Commencing with a view of an American Liner, and a Map of Texas, and passing several places, we have Capote Farm (6) and San Antonio (7); the other slides principally represent Life in Texas, Cattle, Cowboy, Mexicans, Settlers, etc.

680.—West India Islands (20 slides).

St Vincent, Tortola, Antigna, Jamaica, St. Thomas, with Earthquake and Natives.

681.—South America (WOOD, 60 slides).

The set is divided into 7 views for Bahia, 9 for Buenos Ayres, 3 for La Plata, 2 for Monte Video, 6 for Pernambuco, 22 of Rio Janerio, 4 for San Domingo, 4 of s s. Tagus, and 1 of Santiago. The slides refer to Natives, Bridges, Wonderful Trees, Sugar Mills, etc.

682.—A Year Within the Arctic Circle (YORK, 50 slides).

An account by the Rev J. Comper Gray of Arctic travel and episodes, starting from Portsmouth Harbour, on May 29th, 1875, with pictures of the ships "Alert," "Discovery," and "Pandora," some of the officers, such as Captain Nares, and a series of events connected with sledge-travelling on the ice.

683.—American Franklin's Search Expedition (YORK, 20 slides).

The events and scenes of the expedition sent to discover traces of Franklin, as described by a Fellow of the Royal Geographical Society, in which the party learn Eskimo life, hunt, fish, and make important discoveries.

684.—Arctic Views No. 1 (HUGHES, 50 slides).

Life in the Arctic regions, illustrating amusements, adventures, and sorrows, together with the Discovery of Greenland, by Eric the Red, views of ships, men, and surroundings, loss of the ship McLellan, and a view of the perilous position of H.M.S "Assistance" and "Pioneer" on the 12th Oct., 1853, in Disaster Bay.

685.—Arctic Views No. 2 (HUGHES, 19 slides).

Shows life in the Arctic regions as concerns winter dwelling, pitching tents, hunting bears, Franklin's winter quarters, funeral on the ice, with effect slides of Sir J. Franklin's winter quarters, and one of the Aurora Borealis.

686.—Arctic Views No. 3 (HUGHES, 12 slides).

Two views of the Arctic, as also two ports, a glacier, native animals and human inhabitants.

687.—Arctic Expedition, 1875-6 (PUMPHREY, 50 slides).

Sir John Franklin left England, accompanied by Captain Crozier, in the ships "Erebus" and "Terror," in the year 1845. Nothing being subsequently heard, as many as 17 expeditions from England, and three from

America were sent for the relief. In 1855, the British Government satisfied of the death of Sir John, desisted from sending other expeditions. Franklin's idea had been to force a passage either to east or west, or sail *across* the Pole to India In 1870, however, an attempt was made to *reach* the Pole, and the American Government fitted out a wooden gunboat, under Captain Hall's command, but the Captain died, and disaster overtook his followers. The British Government thereupon resolved in 1875, to send out an expedition extremely well equipped, with instructions to approach as near as possible to the Pole. Captain Nares was chosen Commander, and the ships "Alert" and "Discovery," were fearfully and wonderfully fitted up for the purpose. The Prince of Wales and his brother inspected the ships on the 21st May. On the 29th the ships left Portsmouth, and in course of time, sighted Cape Farewell The ships, however, got separated after leaving England, and did not meet till just before reaching Disco Island. Here the travellers met Eskimo friends. The ships touched at Upernavik, the most northerly abode of civilised man, one ship grounding in Discovery Bay, the other in Rodmore Harbour. After staying in winter quarters, during which a push was made for the Pole, and many interesting incidents occurred, the ships safely arrived again at Portsmouth 29th October.

688.—Arctic Discovery and Adventure (PUMPHREY, 50 slides).

Gives a portrait of Sir John Franklin, and shows the opening of the cairn containing Franklin's record, and also the record of his death We see also in this series items of interest touching the Eskimos, and we encounter bears, seals, the walrus, and Arctic dogs, whilst No. 27 gives a portrait of the devoted Lady Franklin.

689.—Arctic Regions (NEWTON, 50 slides).

Show the "Erebus" and "Terror" vessels accompanying Sir John Franklin in his last voyage, icebergs, Greenland whalers, Eskimo village, Aurora effect, chase of the walrus, winding up with a picture of the Franklin relics.

General.

690.—To the North Cape and Back (WOOD, 66 slides).

Norway, Sweden, and Denmark; places, people, travelling, etc.

691.—Ceylon to Labuan (WOOD, 50 slides).

Starting with the Harbour of Galle, Cocoa-nut Groves, an Alligator's Retreat, etc., etc., after encountering Arabs and Hindoos, we arrive at Penang. Of this latter town we get 15 views, including Malay houses and Siamese boys. Thence, passing Malacca, Singapore shows nine, and Labuan fourteen scenes, one with the police.

692.—Plymouth to Java (WOOD, 57 slides).

Starting with the Plymouth Dockyard, H.M. Ships, etc., we come to Madeira; in Ascension we then get seven scenes, in St. Helena seven, the Cape three, Keeling Islands five, etc , etc., concluding with thirteen views of Javan joys, including the Banyan Tree and Javanese servants.

693.—London to Rome (YORK, 50 slides).

Passing Rouen, Paris, Berne, the Matterhorn, Milan Cathedral, Pisa, (8 views), Florence (4 views), etc., etc., we have nineteen views of Rome, including the Coliseum, and sixteen views in the Vatican.

694.—From London to the Falls of Niagara (YORK, 46 slides).

Starting from Euston Square Station, we reach Queenstown and the Cove of Cork, we examine the Saloon of Steamer, and take a bird's eye view on arrival of New York. We then note Broadway and the *Herald* Office, enter the Central Park. After noting Catskill Mountain House, Steamboat on Lake George, etc., we get a view of the Rapids, Suspension Bridge, etc., winding up with five views of Washington.

695.—London to Mount Vesuvius (RILEY, 42 slides).

Commencing at London Bridge, the tourist passes through Paris (6), Mont Cenis Tunnel, Milan, Intra, Venice (8), Bologna, Florence (3), Rome (11), Naples (4), to Vesuvius and Pompeii, ending at Genoa (5).

696.—Through Turkestan to the Afghan Frontier (50 slides)

Starting from Omsk, we get *inter alia* a view of the famous Thian Shan Mountains, one of Khiva, one of Merv, and one of Herat. We see here local life, local buildings, portraits, including the Khans of Khiva, and can form from this set an idea of the neighbourhood in which Russia approaches our Indian Empire. The lecture to accompany the slides is written by the Rev. Dr. Lansdell.

697.—To the Antipodes and Back (WOOD, 51 slides).

A tour round the world. Leaving Southampton, we make for the West Indies, thence along the Coast of South America, past the Straits of Magellan (near the extreme southern point of the continent), through the Pacific Islands, to Australia; continuing to South Africa, and thence by Napoleonic St. Helena and the Canary Isles back to England. The set shows the places of interest, and gives such illustrations of local matters as Slave Emancipation in West Indies, Lassoing Wild Horses in South America, Napoleon's Tomb at St. Helena, Human Sacrifice, etc.

698.—Overland Route to India (WOOD, 32 slides).

Women Pounding Rice, a Confectioner's Shop, Domestic Servants, a Group of Afghans, Hindoo Pilgrims, Sweet-sellers, etc. Though called an overland route, the places met with, including Malta and Gibraltar, are rather few.

699.—Overland Route and Egypt (NEWTON, 157 slides).

The set commences with the usual introduction type of slide, the route is shown by a map, the vessel is shown leaving Southampton, there is a wave effect slide of the Needles; we see the ship at sea, full sail; then the ship is moving in a storm, and suddenly gets struck by lightning, followed by a fire. The fire having been extinguished, we reach Oporto by moonlight, and thence proceed to the Port of Lisbon. At Cape Trafalgar Nelson obligingly comes on the scene, gives his last signal, and in the battle dies once more. We see Gibraltar in eight views, by moon and daylight, having a siege while we stop. At Malta, in eight views, St. Paul destroys the viper, and a couple of Her Majesty's ironclads appear immediately afterwards. We have eight views of Alexandria, and one of the Rosetta Stone Inscription. In the 37 pictures for Cairo, we see a caravan of Mecca pilgrims, Arab woman and child, Bedouin Arab, Turkish lady, and fellahs and guides, as well as shops, tombs, Pyramids, and the Sphinx. We witness a simoon and a sandstorm in the desert, as also a mirage. Of the Nile we have a map, a water-raising wheel, oxen corn treading. Of Thebes we have eight slides, including the Palace of Rameses at Medinet Aboo. Then three views of Luxor, five of Karnak, six pictures for Philæ, with a crocodile on sandbank. We have Rameses restored, Pharaoh meeting Moses, water turned into blood. We see the Suez Canal four times, land at Brindisi, and then return to London.

700.—Round the World in 100 Minutes (RILEY, 60 slides).

Commences, of course, at London Bridge, thence by Paris and other European places, through Palestine, Egypt, etc., to India, China, New Zealand, and America, coming back *viâ* Dublin to Liverpool.

701.—A Voyage Round the World (NEWTON, 100 slides).

Making the start at Portsmouth, with effect slides of the Eddystone Lighthouse, the voyage proceeds *via* Lisbon, Gibraltar, Algiers (4), Malta, Suez Canal, Egypt (9), to India (10). Then the sight-seer has a rest, and is treated to Crossing the Line, Sunset at Sea, &c. Resuming, Java (5) and China (12) are visited, and another rest "for refreshments" in a few effect slides of Ship on Fire, etc After that San Francisco, Yosemite Valley, Salt Lake City, New York, Niagara and Montreal, and on the way home to Queenstown two whaling pictures. The whole concluding with H.M. the Queen.

702.—Round the World with a Camera (YORK, 60 slides).

The route is first to Gibraltar, thence by the Mediterranean to Naples, Constantinople, by Egypt and the Red Sea to New Zealand; here we have about twenty-seven views. Thence crossing the Pacific Ocean, and visiting Honolulu, the capital of the Sandwich Islands, we reach San Francisco, Niagara, of which we have three views, New York two, and Washington one. Thence we cross the Atlantic to England, arriving in the Mersey after encountering an Iceberg on the way.

703.—Round the World in a Yacht (YORK, 45 slides).

This set is associated with Mrs Brassey, the first views showing the famous "Sunbeam" and apartments. We pass *via* the Canary Isles to South America, and thence by the Pacific to Japan, China, Ceylon, and the Red Sea home.

704.—A Visit to the World's Great Capitals (RILEY, 54 slides).

An evening spent in the principal cities of the globe, our own country being represented by three of London, two of Edinburgh, and one of Dublin; the rest of Europe by twenty-eight pictures, Asia by nine, two in Africa, six in Australasia, and three in America.

History.

705.—Historical and Poetical (NEWTON, 29 slides).

These slides include the Death of Moses (C F. Alexander), Battle of Marathon (Byron), Bernardo del Carpio (Mrs. Hemans), Battle of Lake Regillus (Macaulay), Death of Marmion (Scott), etc

706.—The Romance of History (YORK, 36 slides).

Written for an entertainment. We see five views of the Tower of London, two of Windsor Castle, three of Westminster Abbey, Hogarth's Laughing Audience, etc., Canterbury Cathedral, Kenilworth, Stratford, etc.

707.—English History (NEWTON, 109 slides).

This set presents subjects of historical interest from the time of the Druids to the Battle of Inkerman in 1854, taken chiefly from the pictures in Cassell's History of England.

708.—Places of Interest in English History (WOOD, 37 slides)

The places include Stonehenge, Salisbury and Canterbury Cathedrals, Windsor Castle, Alnwick, Kenilworth, Warwick, Skipton, and Howard Castles, Anne Hathaway's Cottage, etc.

709.—The Great Civil War (PUMPHREY, 29 slides).

Commencing with a Portrait of Charles I., we have the Arrest of Stratford, the Queen at Bridlington, Storming of Bristol, Montrose discovered in disguise, Battle of Naseby, the King's attempt to escape discovered, Queen Henrietta interceding for the King, Cromwell viewing the dead body of the King, and so on.

710.—The Church in Britain before Augustine (NEWTON, 38 slides).

Stonehenge Plain, the remains of a Roman Villa, St. Alban's Martyrdom, the Apostles Preaching, Bangor and other Cathedrals and Churches are sample examples of this set, which comes under the general heading of Church History, and is called Lecture No. 1.

711.—The Growth of Papal Authority in England (NEWTON, 50 slides).

Gloucester Cathedral, Tintern Abbey, Lincoln Cathedral, Carthusian and Gilbertine Monks, Murder of Becket, John surrendering Royal Insignia, Fac-simile of Magna Charta, House of Lords, House of Commons, etc., are included in this set, which makes up Lecture No. 4 of English Church History.

712.—The English Church during the last 200 Years (NEWTON, 48 slides).

The Landing of Prince of Orange, William III. asseating to Toleration Act, "No Popery" Rioters, Interior of a Typical Church fifty years ago, such a Church as now restored, the Throne of England, various Churches and Cathedrals (21), etc., are included in this the seventh Lecture of Church History.

713.—Eighteen Centuries of Church History in Britain (NEWTON, 60 slides).

Twenty views of Churches and Cathedrals, ruined and otherwise, Typical Church 50 years ago, such now, Portrait of Wycliffe and Lutterworth Church, Charles I. at Carisbrooke, etc. This set is a kind of epitome of the seven others on the same subject, and makes the eighth Lecture of Church History.

714.—The Struggles of the Church against Romanism (NEWTON, 60 slides).

Portrait of John de Wycliffe, his Church at Lutterworth, Popular Sympathy with Wycliffe, Lollard Preaching, Wolsey's Palace in Fleet Street, his Dismissal, Cranmer and Cromwell, Bristol and other Cathedrals, portraits, etc., are included in this the fifth Lecture of Church History.

715.—The Struggles of the Church against Puritanism (NEWTON, 47 slides).

John Knox Preaching in Saint Andrew's, Puritans Embarking for America, Strafford going to Execution, Signing the Covenant, Charles I. at Carisbrooke, his Trial, etc , are included in this the sixth Lecture of Church History.

716.—The Conversion of the Heptarchy (Newton, 45 slides).

The Angli in Rome, Map of the Heptarchy, King Edwin and his Witan, Saxon Missionary Preaching at a Cross, Bede and his Scribe translating St. John's Gospel, Monks in Scriptorium, with Melrose Abbey, Lichfield, Dorchester and other Cathedrals and Churches, are included in this the second Lecture of Church History.

717.—Influence of the Danes and Normans on the Church (Newton, 48 slides).

Early Danish Invaders, Alfred the Great in Camp, Cnut and the Waves, Cnut's Palace, Southampton, Norman Thanksgiving for Victory, Saxon Doorways and Church Towers, Norman Naves, are included in this the third Lecture of English History.

718.—The Jubilee of Her Majesty, 1887 (Newton, 91 slides).

Seven views of Buckingham Palace, six of the Houses of Parliament, Westminster Abbey (4 views), Windsor Castle (24), views in London, Maps, Her Majesty going in State to the Abbey, also with the Imperial Crown and Robes of State as Empress of India, may be taken as specimen pictures.

719.—Kenilworth (9 slides).

Leicester's Banquet, Elizabeth's interview with Amy, Leicester's visit to Amy, Amy's Murder, etc., illustrate this subject.

720.—Ireland: Coercion at Work (Riley).

This lecture and set of slides is intended to be published at the commencement of this season We cannot therefore give a list, but the secondary title is "The First of Lord Salisbury's Proposed Twenty Years of Firm Government," and this will give a sufficient idea of it.

721.—Ireland: Its Beauty and its Blight (Riley, 66 slides).

Starting with a few general views of Ireland to represent its beauty, we have representations of evictions, emergency men, destroyed cottages, the "Grand Old Man," and various portraits of Irish members, concluding with the Tullamore farce of Balfour stealing O'Brien's breeches.

722.—The Irish Question (Riley, 101 slides).

Views of labourers' dwellings, Glenbeigh; official buildings, Mitchelstown, Bodyke, Ponsonby, Vandeleur. Concluding with many portraits of persons intimately connected with the campaign.

723.—Irish Political Cartoons (Riley).

This series has special reference, of course, to the Irish Question, and is a copy from the *Freeman's Journal* supplements.

724.—Politics of the Day (Wood, 8 slides).

Cartoons and Portraits from the *Liberal and Radical* Newspaper. The set to be increased.

725.—Cleopatra's Needle (York, 47 slides).

The History of Egypt as told by its Monuments, the Lecture being written by Mr. H. Gore, C.E. We commence with the Pyramids, and follow with Abydos, Heliopolis, Alexandria (at one time the second city of the Roman Empire), Thebes, the Rosetta Stone, etc., etc., and terminate with the tale of Cleopatra's Needle, now standing on the Thames Embankment in London.

726.—Primitive Man (WOOD, 34 slides).

Scenes in the Mammoth Period, in the Reindeer Period, in the Stone Age (5), Bronze Age (8), Iron Age (4).

727.—The Fugitive Slaves (HUGHES, 15 slides).

A set illustrating Slave Life in South America, from original drawings by Frank Weeks. The scenes of the Slave Lovers in the Cotton Plantation—"Ye black cuss, the gal is mine", "Strike her, massa, if you dare."; "Take dat, you white debil—take dat" will give an idea of the set.

728.—The Coronation of the Czar (HUGHES 61 slides).

With views of Russia, St. Petersburg and Moscow (from nature), scenes of the Coronation of the present Czar Alexander III., and Assassination of the late Czar Alexander II., the views including 19 of St. Petersburg, 10 of Moscow.

729.—The Victoria Cross and our Soldiers (NEWTON, 18 slides).

Here we we see the Deeds for which some of our soldiers obtained the Cross.

730.—Afghan War (YORK, 36 slides).

A Fort in the Khyber Pass, a group of Generals, the Army awaiting the order to advance, a view of Jellalabad, Conflict with theAfghan Cavalry, a view of Candahar, Disaster to 10th Hussars, views of Cabul and Herat, the Ameers, Charge of the Bengal Lancers, etc.

731.—The Soudan War (HUGHES, 42 slides).

The scenes here presented include Spoils from the Desert, a Fight between Egyptians and Arabs, Black Troops, Escape of Baker Pasha, Battle of El Teb, British Square at El Teb, British Tars to the Front, etc.

732.—Soudan War (YORK, 57 slides).

Views of villages, pictures of Dervish preaching a Holy War, a Fight with the Arabs, the Mahdi, Zebehr Pasha, the Charge of the Hussars, King John (of Abyssinia) at home, Steamer passing the First Gate of the Second Cateract, battles, etc , are here represented.

733.—The Zulu War (HUGHES, 37 slides).

The subjects are taken from the *Illustrated London News* and *Graphic*, and include the Defence of Rorke's Drift, the King's Kraal, Zulu Attack on a Waggon Escort, Struggle inside a Laager, Battle of Ginghilova, etc.

734.—The Zulu War (NEWTON, 18 slides).

Here there are no fights or battles, but we see Cetewayo, his Kraal, his Cooks, and his country, winding up with Sir Garnet Wolseley.

735.—The Restoration of the Jews (WOOD, 7 slides).

People mourning over the destruction of Jerusalem, Cyrus restoring the vessels, etc , make up this set.

736.—Martyrs and the Reformation (NEWTON, 54 slides).

After the noble army of martyrs, we see Alban, the first British martyr, seven references to Wycliffe, twenty-two references to Martin Luther, Cranmer, Ridley, Latimer, John Knox, and so on.

737.—John Wycliffe (HUGHES, 12 slides).

Wycliffe, Portrait, at home, his Church, before Convocation, Scene with Monks, before Prelate at St. Paul's, Trial, etc.

738.—Life and Times of John Wesley (PUMPHREY, 62 slides).

Here we see portraits of John Wesley himself, his father and his grandfather, the escape from the fire, John and the German Moravians, his preaching to Creek Indians, dealing with Wednesbury rioters, his father's grave, mother's grave, etc.

739.—The Life of Martin Luther (YORK, 12 slides).

Includes the School, Poverty, Call from God, Diet of Worms, Domestic Life, etc.

740.—Scenes from Life of Luther (PUMPHREY, 26 slides).

Luther taken to school, begging, his friend killed by lightning, appears before the Emperor, beside the coffin of his daughter, his obsequies, etc.

741 —Lady Jane Grey (YORK, 24 slides).

The scenes depicted include Sion House, the Crown offered to Lady Jane, her interview with Dudley, under Royal Canopy, a Prisoner, Mary signing the Death Warrant, Execution, etc

742.—Mary Queen of Scots (YORK, 24 slides).

View of Holyrood, Rizzio slain, Loch Leven Castle, her neck is bared, the blow is struck, etc.

743.—Mary Queen of Scots PUMPHREY, 12 slides).

The history of the unfortunate Queen is here given in verse.

744.—Mary Queen of Scots (NEWTON, 16 slides).

This set includes a portrait of the Queen.

745.—Mary Queen of Scots (HUGHES, 17 slides).

The last three slides of this set are effect slides, producing the instant effect of the head being severed from the body The set gives the usual scenes in the Queen's life.

746.—Life of Bunyan (YORK, 111 slides).

These slides illustrate Routledge's Pilgrim's Progress. There are eleven slides as an introductory set, showing portraits of Bunyan, his wife persuading him to read his books, his parting with his wife and children, Bedford Gaol, etc. Then in Parts I. and II. we have pictures referring to Pilgrim's Progress

747.—Life of General Gordon (WOOD, 24 slides).

General Gordon is introduced by a portrait, followed by his School at Taunton, his work before Sebastopol, in Armenia, China, Khartoum, and Abyssinia, the Fall of Khartoum, etc.

748.—George Stephenson (WOOD, 30 slides).

We see here the stage coach to start with, and after viewing George's birthplace, various engines, and celebrated bridges, we wind up with an express train.

749.—Portraits of Celebrities (NEWTON, 51 slides).

Here are members of the Royal Family, Napoleon Bonaparte, Napoleon III., Gladstone and Beaconsfield, the Shah, Bismarck, Popes, Nelson, Wellington, Livingstone and Stanley, Garibaldi, Mary Queen of Scots, Tennyson, Sir Walter Scott, Charles Dickens, and others.

750.—Portraits (York, 6 slides).

These are portraits of half-a-dozen leading members of the Royal Family.

751.—Portraits (York, 33 slides).

There are some of the Royal Family, two Popes, Cardinal Manning, Archbishop of Canterbury, Emperor Napoleon III. and his wife Empress Eugénie, and fourteen English Statesmen.

752.—Notable Men and Women (Wood, 163 slides).

Life and scenes connected with Gladstone (6), Napoleon (11), Wellington (8), Nelson (6), Wycliffe (12), Bunyan (15), Tyndale (11), Knox (9), Milton (2), Hogarth (7), Garfield (14), Burns (11), Longfellow (4), Tennyson (4), Byron (5), Shakespeare (3), Southey and Wordsworth (4), Goldsmith (3), Dickens (4), etc.

753.—Portraits (Wood, 90 slides).

These include members of the Royal Family, Empress Eugénie, Emperor of Germany, Prince of Germany, Prince Bismarck, Moltke, Emperors of Russia and Austria, Stanley, the Shah, Carlyle, Spurgeon, Baroness Coutts, Sir C. Dilke, Cardinal Manning, Darwin, etc.

Educational.

754.—Natural Phenomena (Newton, 62 slides).

The sub-title is "Wonders of the World," and the wonders include Lunar Rainbows, Mock Suns, Aurora Borealis, Will-o'-the-Wisp, Meteoric Showers, Simoons, Water Spouts, Whirlpools, Avalanches, Petrifying Springs, Earthquake, etc.

755.—Natural Phenomena (Pumphrey, 26 slides).

The "Phenomena" include Avalanches, Coral Reef, Water Spouts, Aurora Borealis, Will-o'-the-Wisp, the Geyser, etc.

756.—Remarkable Trees and Plants (Newton, 47 slides).

Here we see the Bread-fruit Tree, the Butter Tree, Carniverous Plants, Cedars of Lebanon, a Cotton Plant, Lemon Tree, an Olive Tree, Orange Tree, a Palm Grove, Tea and Coffee Plants, etc.

757.—Physical Geography Wood, 19 slides).

The slides accompany Keith Johnstone's "Physical Geography," and illustrate the Mountains, Plains and Valleys of each Continent, the Action of Rain, Sea, Ice, etc., etc.

758.—Sunday Schools, and What they Have Done (Wood, 26 slides).

Boys as they Were, a Cottage School, an Arab Sunday School, an Indian Mission School, etc.

759.—Highland Dancing (4 slides).

Two of the Ghillie Cullum, and two of the Reel o' Tulloch.

760.—Wild Sports of the World (Newton, 36 slides).

These slides introduce us to Bears, Bisons, Hippopotami, Lions, Tigers, Snakes, Wolves, and so on, illustrating hunting, general incidents connected therewith, etc.

761.—Children's Sports (Wood, 33 slides).

Children are here represented Making a Cat's Cradle, Minding Baby, Titivating Dolly, etc.

762.—Art of Printing (Wood, 12 slides)

The Papyrus Plant, Monks Copying Manuscripts, specimen of Old Printing, Early Wood Engraving, etc., are shown.

763.—Paper Making (Wood, 4 slides).

Rags are cut, and we see Washing and Beating engines, and a Paper-making Machine.

764.—Art of Pottery (Wood, 6 slides).

The Grinding Wheel, Potter's Wheel, Stacking Kiln, etc.

765.—Architecture (Wood, 55 slides).

There are here given examples of Celtic, Egyptian (3), Greek (4), Asiatic (2), Roman (8), Oriental (7), Gothic (3), Italian Renaissance (4), and other styles.

766.—Railway Waggon Factory (Sciopticon, 5 slides).

The Iron Foundry, Lathe Room, Forge, etc , are here shown.

767.—Medals of the British Army (Newton, 15 slides).

Indian War, Waterloo, Indian Mutiny, and other medals, as well as the Victoria Cross, are here shown.

768.—The Indian and Colonial Exhibition (Wood, 76 slides).

From this set we get an idea of what was to be seen in the famous "Colondries" Exhibition at London, and see the products of the Colonies and India.

769.—Art Collections of the South Kensington Museum (York, 59 slides).

Monuments, Gateways, Statuary, Pulpits, etc , from renowned spots.

770.—British Museum (York, 129 slides).

In this series we have twenty-eight pictures touching Egyptian antiquities, nine Assyrian, nineteen Grecian, thirty Græco-Roman, eighteen devoted to Zoological Saloons and Galleries, ten to Geological Galleries and Specimens (2), and amongst the odd slides we have two of the Ethnographical Room, four of the Royal Library, and one of the Reading Room.

771.—A Visit to the British Museum (York, 50 slides).

We enter the Egyptian Rooms, Helleric Rooms, Geological Galleries, six Geological Rooms, the Royal Library, the Reading Room; we see Mummies, Assyrian Bas-reliefs, the Head of Julius Cæsar, etc.

772.—The History of a Cotton Bale (York, 10 slides).

Showing the processes of cultivation and manufacture.

773.—Bale of Cotton (Wood, 8 slides).

The growing and manufacture of Cotton is depicted.

774.–Our Bread and How We Get It. (VALENTINE, 14 slides).

The cultivation of the grain, and subsequent processes till it reaches the Breakfast table (and Gipsy Camp) in the form of a loaf.

775.–The History of a Quartern Loaf (YORK, 10 slides).

The course of cultivation of the grain, and the subsequent dealing with corn to produce the loaf.

776.–History of a Tea-cup (HUGHES, 13 slides).

Flint Grinding, Slip-maker, Sagger-maker, the Glost Placers and Sorters, etc , are here shown.

777.–The History of a Pound of Tea (YORK, 10 slides).

The cultivation, and subsequent drying, sorting, transport, etc., are here shown.

778.–Our Tea and How We Get It (RILEY, 14 slides).

Here we see the cleaning, pruning, picking, firing, pressing, etc.

779.–The History of a Pound of Sugar (YORK, 10 slides).

The cultivation and manufacture are here shown.

780.–History of a Grain of Rice (7 slides).

The cultivation and manufacture of rice are here shown.

781.–The History of a Golden Sovereign (YORK, 10 slides).

The Gold found, being Weighed, Gold Cheeses, the Cheese Vault, Coining, etc.

782.–Gold Mining in Central America (STEWARD, 30 slides).

The views are taken in the mountains of Chontales, in the Republic of Nicaragua, and embrace scenes of the Country, Officers' Dwellings, native Huts and Villages, etc.

783.–The History of a Scuttle of Coals (YORK, 10 slides).

Commencing with a view of Newcastle, we see the Colliery, the Ship on its way, etc.

784.–Our Coal and How We Get It (WOOD, 60 slides)

We are here shown fossils, a view of Newcastle, the Critical Situation of an Engineer, a Consultation in a Mine, Pony Descending, the Story of Jim the Collier (6 scenes) with an effect slide, Pit on Fire, Interior of a Miner's Cottage, and so on.

785.–Coal and Coal Mining (NEWTON, 43 slides).

A set illustrative of Underground Life with the Miners, and depicting the boring for coal, sinking a shaft, descent of a horse, falling in of roof, cross sections showing imprisoned men and the rescuers, etc.

786.–Scenes in a Coal Pit (PUMPHREY, 31 slides).

These scenes are from photographs taken in a coal pit by artificial light, and illustrate the Cage at Pit-top, Bottom of Pit, Examining for Gas at a Breakdown, Smithy Underground, Furnace for Ventilation, etc.

787.–Mines and Miners (YORK, 130 slides).

Ninety of the slides are in connection with Messrs. Chapman and Hall's "Underground Life," illustrating the history, modes, and difficulties of procuring coal and other minerals. *Inter alia* we have Fossils in Coal (14), Explosions, Engineer's Critical Situation, Prisoners Condemned to the Mines in Mexico. The latter 40 slides show Wigan Colliery Girl at Work, the Same Girl on Sunday, Section of Pit on Fire, Boring the Channel Tunnel, Prince of Wales descending Botallack Mine, Gold-digging in Australasia, etc.

788.—Mines and Mining (YORK, 41 slides).

This is a selection from the larger set.

789.—Mines and Miners (PUMPHREY, 34 slides).

The set is illustrative of some of the dangers and difficulties in mining, and the subject is taken from Chapman and Hall's "Underground Life; or, Mines and Miners." The slides show Coal with Impression of a Fern, etc; Boring Operations, Explosion of Fire Damp, Prisoners Condemned to the Mines at Mexico, etc., etc.

790.—Explosion in a Coal Mine (HUGHES, 23 slides).

The Suspense, Explorers Searching, Bringing up a Body, Volunteers to the Rescue, Working in the Pit, Pet Stable Horses, Wigan Colliery Girl, etc., are here depicted.

791.—Fire Engines and Firemen (WOOD, 9 slides).

Here we see pictures of fire brigades past and present.

792.—The Lifeboat and its Services (WOOD, 28 slides).

In this set we see a Wreck Chart, Section of Lifeboat, the Cork Jacket, the Light Ship, Section of a Lighthouse, Rocket Apparatus (and effect), the Rescue, and the Return, etc.

793.—Shipping (24 slides).

Views of Herring Fleet, Yawl Race, and Liners at different Ports.

794.—The Construction of a Dry Dock (HUGHES, 5 slides).

The Excavation, Cofferdam, Launch of the Cassion, Dry Dock views, etc.

795.—Artistic Studies (SCIOPTICON Co., 80 slides).

Seven from Ulleswater, 13 from Patterdale, 15 are Hedgerow studies, 6 Cattle, and the remainder principally of country scenes and character.

796.—Studies of Trees (8 slides).

Birch trees in the Trossachs, Fir trees in Scotland.

797.—Landseer Series (SCIOPTICON Co, 22 slides).

These slides are from engravings after Sir Edwin Landseer, "There is no place like Home," "Young Roebuck and Rough Hounds," "Dog of St. Bernard" are in the series.

798.—A Peep into Nature through the Microscope (YORK, 54 slides).

We here see magnified Fleas, parts of Flies, and of other insects, a Cat's Tape-worm, the stem of an Exogen (cross section of Clematis) annual rings of Growth, Medullary Rays Stomata, Volvox Globator, Diatoms, Stem of Plant in Coal, and some Foraminifera.

799.—Instantaneous Groups of Animals (WOOD, 39 slides).

Here we see Cattle in the Water, Horses, Sheep, Deer at Rest, Pigeons, Swans, Mountain Ponies, Dogs, including Pug Dog and Kitten, etc.

800.—Animals at the Zoological Gardens (NEWTON, 127 slides).

The descriptive book used in connection with this set of slides is Routledge's "Illustrated Natural History."

801.—A Walk in the Zoo (YORK, 48 slides).

We commence this set with a Plan of the Gardens; we see an interesting selection of animals, and wind up with Whit-Monday and the Zoo. Some of the slides are no doubt taken from the larger set.

802.—Animals at the Zoological Gardens (YORK, 215 slides).

Out of these pictures fifty-three have been taken for a lecture set. It would serve no good purpose to specify the animals.

803.—Prince of Wales's Animals at the Zoo (YORK, 18 slides).

These include Indian Elephants, Zebras, Goats, Sheep, Antelopes, Wild Dogs, etc.

804.—The Life of a Knat (6 slides).

Showing the Eggs, Larva, Birth, Eye, Digestive Apparatus, and Manhood.

805.—The Beehive (WOOD, 10 slides).

These pictures show the Cells, Arrangement of Comb, Tongue, and Leg of Bee, etc.

806.—Favourites of the Farm Yard (STEWARD, 6 slides).

These are the farm steading, and a horse, dogs, fowls, turkey, and cat.

807.—Round the Farm Yard (VALENTINE, 27 slides).

Pictures of the ordinary animal denisens of a farm.

808.—Beautiful Birds from Far-off Lands (VALENTINE, 18 slides).

Birds of Paradise, Parrakeets, Cockatoos, Coquettes, etc., are good examples of these birds.

809.—Animal Sagacity (NEWTON, 24 slides).

Showing cat getting milk out of narrow jug, twelve wonders touching dogs, and others relating to horses, lion, rat, etc.

810.—Animal Sagacity (YORK, 24 slides).

These pictures are for a lecture on Kindness to Animals. We see music-loving horses, a donkey gatekeeper, cow working a pump, and so on.

811.—Animal Sagacity (PUMPHREY, 10 slides).

Including Tommy the Cat's Clever Trick, Danco (the lion) and his keeper, etc.

812.—South African Animals, with Tropical Scenery (NEWTON, 12 slides).

The animals include Crocodile, Hippopotamus, Kaffir Village (!), Rhinoceros, and "Lion and Kaffir," the latter depicting the fate of Hendrick

813.—People we Meet (RILEY, 38 slides).

A new series of slides photographed in the streets from life, and illustrating the various characters met with every day. Many of them are most amusing, and all are exceedingly realistic. We see the Lamp-lighter, Gas Inspector, the Hansom Cabby, two Old Fruit Women, the Quack Doctor, etc., etc.

814.—Astronomy (YORK, 300 slides).

Physical, Observational, and Instrumental. The slides are selected and arranged by the Rev. Philip Sleeman, F.R.A.S, F.R M.S, &c. The series is specially adapted to meet the requirements of public lecturers and teachers of science. Under Part I, "The Sun" (75 slides), the relative sizes of the sun and planets, sun spots, the solar spectrum, solar eclipses, solar prominences, are the chief items dealt with. Under Part II., "The Solar System" (115 slides), the Ptolemaic and other systems, the comparative sizes of planets, the seasons, the tides, sidereal and solar day, thirty-eight moon views, and views of Mars, Jubiter and Saturn, are the chief items. Under Part III, "The Sidereal System" (70 slides) we have the constellations, star spectra, comets, and fourteen ancient cosmographies. And under Part IV., "Astronomical Instruments" (40 slides), we have three equatorial telescopes, an alt-azimuth, reflecting telescopes, transit instruments, micrometers, the sextant, and various observatories.

815.—Astronomical (PUMPHREY, 45 slides).

This set shows illustrations of the Solar System, Comparative Sizes of Sun and Planets, the Moon-Craters, etc.; Eclipses, Tides, Venus, Mars, Mercury, Jupiter, Saturn, Double Stars, Spectra of the Light of the Sun, Stars, etc., etc.

816.—Astronomical (NEWTON, 86 slides).

Here we have two lectures. In No. 1 we have 33 illustrations of the Sun, Planets, Systems, Tides, Comparative Sizes of Sun and Planets, etc. In No. 2 we have 54 illustrations of Planetary Motion, Comets, Refraction, Constellations (5), Solar Eclipses, Solar Spots and Prominences, Forms of Clouds, etc

817.—Astronomical (THE SCIOPTICON CO., 22 slides).

Views of the Moon, etc.

818.—The Solar System Illustrated (YORK, 50 slides).

With a Glimpse at the Stellar Universe. The Relative Sizes of the Sun and Planets, Sun Spots, the Seasons, Eclipses, Moon views, Saturn, Neptune, Comets, Part of Constellation "Gemini" as seen with the Naked Eye, and with Telescope, etc, etc., are here shown.

819.—Botanical (SCIOPTICON CO., 10 slides).

Illustrations of a Banana Tree with fruit, Tree Fern, Lemon Tree and fruit, Orange Tree and fruit, the Date Palm, the Aloe, etc.

820.—Entomological and Botanical (NEWTON, 29 slides).

These are real insects, etc, mounted in balsam on 3¼in square glasses, to show in the ordinary magic lantern. They include ants, centipedes, fern leaves, glow-worms, grasshoppers, lantern flies, locusts, birch tree leaves, sea weeds, skin of snake, etc.

821.—The Fructification of Flowers (PUMPHREY, 55 slides).

The Formation and Conjugation of Zoospores, Head of the Bee, illus trations showing the difference between the Humble Bee and Hive Bee, the Spatul of the Cuckoo Pint, the Bungalow of an English Merchant at Singapore, an Opium Clipper Ship, Penang with Cart, and Natives, etc, etc., will give an idea of this varied set.

822.—Flowers of the Year (PUMPHREY, 54 slides).

The flowers include Anemones, the Garden Bean, Digitalis (Foxglove), Delphinium (Larkspur), Linaria Vulgaris (Toad Flax), Lilium Auratum Roses, Snapdragon, Globe Thistle, Veronica, etc.

823.—Life of a Plant (NEWTON, 39 slides).

Various Selections of Stems, Scalariform Tissue, Raphides Stomata, Pollen, Ovaries, and other microscopical matters are here represented.

824.—Voyage of the "Challenger" (YORK, 42 slides).

This set represents a few specimens and items of interest to naturalists arising out of the scientific expedition of H.M.'s ship "Challenger" We see, *inter alia*, the "Challenger," the Arrangements for Dredging and Sounding, Foraminifera (15), Radiolarians (5), Siliceous Sponges, Chalk and Flint (6), Crinoidea (3), and Echinodermata and Crustacea (5). The reading is by Wm. Lant Carpenter, B.A., B.Sc., etc.

825.—Photo-Micrographs (PUMPHREY, 65 slides).

These are in two sets; the first, of 31 slides, shows transverse sections of teak wood, etc., stellate cells, parts of spiders, snails, sponges, etc.; the second set, of 34 slides, shows human fleas, a bed bug, crickets, spiders, parts of flies, beetles, etc. These slides reproduce in a popular manner the wonders of the microscope, showing markings which can only be otherwise seen with high powers of the microscope and with expensive instruments, and are such as cannot be rendered by microscopes attached to the lanterns. from insufficient light and consequent loss in definition, and it is intimated that starch granules already magnified on the slide 200 diameters are capable of enlargement to 4,800 diameters, if made to fill a disc of 6ft. diameter, and so on.

826.—Microscopical Objects (YORK, 150 slides).

We are here shown well-known insects, with members and development, Foraminifera, Sections of Whalebone, of Horns, of Roots, and Stems of Stems, Geological Sections, etc., etc., photographed as slides. The subjects are magnified eight diameters for some objects and up to about one hundred diameters for others.

827.—Vegetable Kingdom (SCIOPTICON CO., 31 slides).

Transverse sections of stems of sugar cane, burdock, poppy, etc., are shown in the first twelve illustrations; then we see fourteen Diatoms; and lastly, five specimens of parasitic algae.

828.—Mineral Kingdom (SCIOPTICON CO, 5 slides).

Four of these slides illustrate Crystallisation.

829.—The Animal Kingdom (SCIOPTICON CO., 117 slides)

The nomenclature and classification of Dr. C. Claus has been here followed. Under *Group* 1, Protozoa (animals of the simplest constitution and smallest size), we have the class Rhizopoda, with fourteen representatives of the order Foraminifera, and six of the order Radiolara. Under *Group* 2, Cœlenterata (Zoophytes), we have four specimens of the class Spongia. Under *Group* 3, Echinodermata, we have nine specimens of the class Asteroidea (Star-fishes), eight of the class Echinoidea (Sea Urchins), and one of the class Holothuroidea. Omitting group 4, we have *Group* 5, Arthropoda, embracing the class Crustacea, with one specimen (Water Flea) of the order Phyllopoda, and one (Barnacle) of the order Cirripedia, the class Arachnida, with four specimens (Spiders) of the order Araneida, and eight (chiefly Itch-Mites) of the order Acarina; the class Insecta-Hexapoda, with two specimens (House Cricket, etc.) of the order Orthoptera, one (Scorpion Fly) of order Neuroptera, six (Lice, Aphides) of order Rhynchota-Hemiptera, twenty-six of the order Diptera, including Blow Flies, Gnats, Horseflies, in the sub-order Pupipara (Lice Flies), parasites of Bees, Bats, etc., and in sub-order Aphaniptera, various Fleas; eight of the order Lepidoptera (Butterflies); four (Water Beetles) of the order Coleoptera; and three (Fairy, Oak Apple, and Saw Flies) of the order Hymenoptera. Omitting group 6, we have *Group* 7, Mollusca, with three specimens (Ear Shell, Pinna Shell, and Whelk); and omitting group 8, we have seven items (sections of Human Scalp, of Shark's Tooth, etc.) coming under *Group* 9, Vertebrata.

830.–Microscopical Gems (YORK, 50 slides)

From the three kingdoms of nature Amongst these Gems we see Trichinæ in Human Muscles, and in Tongue of a Rabbit, sections of Bones and Horns, of Stems of Plants, Platino-cyanide of Magnesium polarised, etc.

831.—Parasites and their Hosts (YORK, 40 slides).

To mention these Parasites by name would be to convey small matter of interest to the popular mind. As an example, however, we may mention the head and neck of Bothriocephalus Proboscideus. It is a matter of interest, however, to know that the reading is by such a rare good dealer with pôpular science as Dr Andrew Wilson.

832.—Human Physiology Popularly Explained (YORK, 52 slides).

The sub-title is "The House we Live In." The subject is divided into Skeleton (10 illustrations), Muscles (2), Digestive Organs (7), Circulatory Organs (4), Respiratory Organs (11), Nervous System (5), Senses—touch, smell, taste, etc. (12), and the lecture is by Mr. William Furneaux.

833.—Magnetism and Electricity (YORK, 460 slides).

This is an extensive series, from which lecture sets may be made up. It enters with thoroughness into both subjects, and it is useless for us to specify any particular items.

834.—Magnetic Curves (YORK, 33 slides).

This set shows Curves of Bar and Horse Shoe, Magnets, Curves of Attraction and Repulsion, Lines of Force, etc , etc.

835.—Mechanics, Hydrostatics, etc. (YORK, 110 slides).

This set is selected and arranged by the Rev. P. R. Sleeman, F.R A.S , F.R M.S., etc. We have illustrated the Parellelogram of Forces, Pulleys, Levers, Windlass, Elasticity, Crystallisation, Centre of Gravity, Centrifugal Force, Pendula, the Principle of Hydraulic Press, Liquid Pressure, and other items touching the various properties of Solids, Liquids, and Gases.

836.—Sound (YORK, 177 slides).

These slides are illustrative of the larger text-books on the subject. We may mention the slides illustrative of the Reflection of Sound, the Vibrations and Nodes in a string shown by Damping, Marloye's Harp, Manometric Flames, graphic method of representing a sound composed of several sounds, etc.

837.—Spectrum Analysis (YORK, 100 slides).

These slides show representations of the various instruments and apparatus made use of in recent Spectroscopic researches, and are descriptive of the results of the application of prismatic analysis to the study of terrestrial substances, and to the physical constitution of the celestial bodies. There are included Charts of Radiation Spectra, Charts of Absorption Spectra, Spectra of Stars, Spectroscopes, etc.

838.—The Telephone (YORK, 135 slides).

The full title is "The Telephone, Microphone, Phonograph, and New Inventions of 1878, also Lamps, Magneto and Dynamo Electric Machines, and Applications of the Electric Light." The illustrations embrace Reis's Electro-magneto, Say's musical, Bell's Articulating and other Telephones, Edison's Inventions, Hughes's Microphones, etc., etc.

839.—Volcanoes of Italy (PUMPHREY, 48 slides).

These slides are to illustrate a paper read before the British Association for the Advancement of Science, at Bath, 1888. These are illustrations of Vesuvius (9), of Pompeii (3), of Postum and Etna (3 each), of Stromboli and Vulcano (4 each).

840.—Heat (YORK, 260 slides).

The slides are selected and arranged by the Rev. P. Sleeman, and full illustrate the whole subject. We may mention those illustrative of the expansion of gases, the tension of vapours, the liquefaction of gases, snow crystals, conduction, the radiation, reflection and refraction of heat.

841.—Geological Subjects (HUGHES, 17 slides).

This set illustrates Upheaval, Subsidence, Denudation, Footprints and Raindrops (on stone), Lake Dwellings, etc.

842. - Geological (NEWTON, 36 slides).

Igneous Rocks, Old Red sandstone, the Triassic Period, a Fish Reptile Deinotherium (a terrible wild beast), etc.

843.—Six Days of Creation (WOOD, 60 slides).

The Six Days of Creation are here looked at from a geological point of view, and ideal landscapes and scenes are given of the various geologica periods.

Scriptural and Religious.

844.—Bible Manners and Customs (YORK, 30 slides).

The Manners and Customs include Tent Life, the Wine Press, Flat Roofs of Houses, Putting off Shoes, etc.

845.—Bible Animals (PUMPHREY, 46 slides).

The set is taken after Wood's Bible Animals (Longman). We may mention the Dog of Palestine, Camel going through the Needle's Eye, the Locust, the Scorpion, and the Ox treading out the Corn.

846.—Bible Pictures (PUMPHREY, 188 slides).

One hundred and fifteen of these pictures are from the Old Testamen and we have represented a large number of the most prominent scenes in Bible History, from the account of the Creation to that of Daniel, and from the Nativity to Paul's farewell to the Elders of Ephesus.

847.—Bible Pictures (PUMPHREY, 125 slides).

These pictures are from engravings selected from the works of the old and from modern masters in the Imperial Family Bible (Blackie). We note Martin's Fall of Man and his Deluge, Telbin's Damascus, Martin's Belshazzar's Feast, Warren's Jerusalem, Schadow's Wise and Foolish Virgins.

848.—Ancient Bible Cities (WOOD, 47 slides).

Eight scenes are here shown for Jerusalem, twelve for Nineveh, one of the Ruins of Babylon, and we also see Bethlehem, Samaria, and Siloam.

849.—History of the English Bible (Wood, 53 slides).

The Lost Roll Found, Rosetta Stone, Samaritan Pentateuch, Codea Vaticanus, Alfred the Great Studying, Wycliffe's Bible, Fac-simile Page or Tyndale's Testament, the Bible made Free, Bible Society's House, etc., are amongst the illustrations of the History.

850.—The Book and its Story (Newton 40 slides).

The Rosetta Stone Inscription, Portrait of Wycliffe, of Martin Luther, and of Tyndale, Reading the Bible in the Streets of Old London, Bible Society's Old House, the Colporteur on his journey with Bibles, etc., are here represented.

851.—Dore's Bible Illustrations (Pumphrey, 102 slides).

The first set of 50, belonging to the Old Testament, commences with the Formation of Eve, and we have scenes touching Abel, the Deluge, Jacob, Joseph, Moses, Samson, David, Solomon, Elijah, Job, Daniel, and so on. In the New Testament we have Wise Men guided by the Star, Sermon on the Mount, Miracles, the Annunciation, the Ascension, and other scenes taken from the illustrations by Gustave Doré in Cassell's Holy Bible.

852.—The Dore Bible (York, 250 slides).

One hundred and forty-six of these pictures refer to the Old Testament, 23 to the Apocrypha, and the remainder to the New Testament. As the pictures deal with almost all the prominent point of Bible History, it is not necessary here to enter into detail.

853.—Old Testament History (Hughes, 12 slides).

These scenes refer to Adam and Eve, the Deluge, the Brazen Serpent, Abraham, David, Elisha, Joseph, etc.

854.—Old Testament Scriptures (York, 40 slides).

The scenes refer to Paradise, the Deluge, Jacob, Joseph, Moses, Samson, Saul, David, Elijah, Daniel, Oppression of the Israelites, the Firstborn Slain, the Death of Eli, etc. The Lecture is written by the Rev. F. Hardwick.

855.—The Old Testament (York, 154 slides).

These slides illustrate in a very full manner the events of the Old Testament History. The Lecture is by the Rev. F. Hardwick.

856.—The New Testament (York, 77 slides).

A companion series to the larger set on the Old Testament, and they also illustrate in a full manner the events of the New Testament.

857.—The New Testament (Hughes, 18 slides).

These illustrate the Nativity, Flight into Egypt, Christ Blessing little Children, the Crucifixion, Ascension, etc.

858.—Scripture Texts (York, 34 slides).

This set shows such texts as "Grow in grace," "The Lord is my Shepherd, I shall not want."

859.—Scriptural Natural History (Newton, 12 slides).

The animals include the Ass, Badger, Bears, Dogs, Locusts, etc.

860.—Scripture Figure Scenes (Newton, 162 slides).

There are here scenes of the Deluge (9), scenes relating to Abraham (8), to Isaac (4), Joseph (18), Moses (34), Joshua (9), David (9), Samuel (5), Solomon (4), Elijah (13), etc., etc.

861.—Scripture Manners and Customs (PUMPHREY, 15 slides).

The manners and customs are here illustrated by Washing the Hands, Bottles, Mills, Beds, Threshing Corn, Flat Roofs of Houses, etc.

862.—Collects (WOOD, 15 slides).

These slides show such collects as that for the Sixth Sunday after Epiphany, according to the Ritual of the Church of England.

863.—Life of Christ (PUMPHREY, 20 slides).

Commencing with the Announcement to the Shepherds, we see the Clearing of the Temple, Christ Blessing Little Children, etc.

864.—Life of Christ (MASON, 50 slides).

The Annunciation of the Birth, Adoration, Worship of the Wise Men, the Flight into Egypt, the Sermon on the Mount, Jesus Asleep during the Tempest, Christ and the Woman of Samaria, the Betrayal, Crucifixion, etc., are here represented.

865.—Life of Christ (NEWTON, 89 slides).

In this set we see the Annunciation, Adoration, Massacre of the Innocents, Christ with the Doctors, Baptised, Walking with His Disciples in the Corn Fields on the Sabbath Day, the Sermon on the Mount, Performing Miracles, the Parables, Peter betraying Christ, the Crucifixion, Resurrection, Ascension, etc.

866.—Life of Daniel (WOOD, 21 slides).

The events of the life here shown include Nebuchadnezzar's Dream, Daniel introduced to the King, the Siege of Babylon, the Lions' Den, etc.

867.—Life of David (WOOD, 23 slides).

We see in these pictures David slaying Goliath, David showing Saul that he had spared his life, David chosen King, etc.

868.—Elijah and Elisha (WOOD, 28 slides).

There are here shown the well-known scenes in the lives of Elijah and Elisha, such as Elijah fed by the Ravens, the Call of Elisha, etc.

869.—Queen Esther (WOOD, 12 slides).

In the scenes depicted we see Esther at the Palace, Mordecai discovering the Plot, Esther accusing Haman, etc.

870.—Life of Joseph (WOOD, 16 slides).

We see here such scenes as Joseph sold into Bondage, his Prison, his Appearance before Pharaoh, his Meeting with his Father Jacob.

871.—Life of Joseph (HUGHES, 16 slides).

We see in this set Joseph in the Pit, the Baker and Butler telling their dreams to Joseph, Embalmment, etc.

872.—The Judges (WOOD, 26 slides).

The scenes show Jericho Destroyed, Joshua commanding the Sun to stand still, Samson slaying a Lion, Boaz and Ruth, etc.

873.—Scenes from the Life of St. Paul (PUMPHREY, 12 slides).

The scenes include the Conversion of St. Paul, St. Paul at Athens, his Shipwreck, in danger from Mobs, etc.

874.—Life and Journeys of St. Paul (NEWTON, 63 slides).

Tarsus, Jerusalem, and Damascus are here represented, while the 11 scenes of St. Paul's First Journey include Paul and Barnabas at Antioch, People wishing to worship Paul, etc, 13 scenes of the Second Journey, are represented by St. Paul preaching to the Thessalonians, and views of Athens, Corinth, Ephesus, etc.; 7 scenes of the Third Journey show the Ephesians burning Books, the Ruins of Tyre, etc.; and 21 scenes in the Voyage to Rome show Paul before the Council, Destroying the Viper, etc

875.—The Life and Epistles of St. Paul (PUMPHREY, 21 slides).

The views shown include Damascus, Jerusalem, Philippi, Athens, Corinth, Miletus, Rhodes, Syracuse, and Rome.

876.—The Life of St. Paul (HUGHES, 20 slides).

The scenes depicted show Paul's Conversion, Ananias sent to him, Paul Healing the Lame, his Preaching at the School of Tyrannus, etc.

877.—Life and Jouneyings of St. Paul (WOOD, 44 slides).

The views include Tarsus, Damascus, Antioch, Jerusalem, Ancient Athens, Tyre, Mount Carmel, Cæsarea, Rome, etc, and the scenes depict Peter being rebuked by Paul, Paul and the Jailer, the Storm and Wreck, Paul in Prison at Rome, etc.

878.—Life of Moses (RILEY, 23 slides).

The scenes in the life represent the Finding of Moses, Moses before Pharaoh, Destruction of Pharaoh, Worshipping the Golden Calf, Consecration of Aaron, Moses viewing the Promised Land, etc.

879.—The Patriarchs (WOOD, 40 slides).

The Creation, Adam and Eve Hiding, Cain and Abel, Noah Cursing Canaan, Abraham and Isaac, Jacob's Dream, Job hearing of his Ruin, etc, are here represented.

880.—The Patriarchs (MASON, 22 slides)

The Creation of Birds and Fishes, the Expulsion from Paradise, Cain and Abel, Noah leaving the Ark, scenes connected with Abraham (7), Isaac (2), Jacob (6), and Job (3) are here shown

881.—The Prodigal Son (NEWTON, 12 slides)

We see here the Prodigal Son a boy at Home, his Leaving Home, his Begging for Work in a far country, his Return, etc.

882.—The Prodigal Son (HUGHES, 13 slides).

In this set the Prodigal Son is represented asking, "Father, give me the portion that falleth to me," going into a far country, Living Riotously, a Famine coming, Returning to the Father, etc.

883.—The Prodigal Son (PUMPHREY, 12 slides).

This set is taken from Hamilton's Work, illustrated by Selous and published by Nisbet.

884.—Solomon (WOOD, 8 slides).

We here see the Judgment of Solomon, the Building of the Temple, Marriage with Pharaoh's Daughter, the Visit of the Queen of Sheba, and Solomon in all his glory.

885.—Samuel (WOOD, 18 slides)

The scenes represent Hannah's Prayer, Samuel in the Tabernacle, the Message to Eli, Samuel's Sacrifice, the Witch of Endor, etc.

886.—Joseph and his Brethren (PUMPHREY, 12 slides).

Joseph the Young Shepherd, Joseph cast into the Pit, in Prison Interpreting Pharaoh's Dream, the Cup found in Benjamin's sack, etc.

887.—The Rich Man and Lazarus (HUGHES, 6 slides).

The Rich Man, the Beggar Lazarus, the Beggar dies, carried away by the Angels, the Rich Man dies, Hell.

888.—Parable of the Ten Virgins (HUGHES, 5 slides).

Here we see the Five Wise, Five Foolish Virgins, the foolish saying " Give us of your oil," etc.

889.—The Parables of Our Lord (HUGHES, 16 slides).

The Parables include those of the Sower, the Tares, the Unjust Shepherd, the Unmerciful Servant, etc.

890.—The Parable of the Sower (PUMPHREY, 5 slides).

No 1 represents the Sower going out to sow, No. 2 " As he sowed some fell by the wayside," and so on.

891.—Miracles and Parables of Christ (MASON, 12 slides).

These scenes include the Casting out Devils, Healing the Palsy, Raising the Widow's Son, Restoring Sight, Parables of the Good Samaritan, Pharisee and Publican, etc.

892.—The Good Samaritan (HUGHES, 5 slides).

Representing a man falling among thieves, people passing by, and a Good Samaritan, who helps him, etc.

893.—Israel's Wanderings in the Wilderness (WOOD, 46 slides).

We witness the Departure from Rameses, Israel Pursued, the Crossing of the Red Sea, Mount Sinai, the Tabernacle and Court, the Most Holy Place, the Pillar of Fire, Death of Moses, Passage over Jordan, etc.

894.—The Tabernacle in the Wilderness (YORK, 15 slides).

This set shows us the Encampment by Day overshadowed by a Pillar of Cloud, the Laver, the Holy Place, the Incense Altar, etc.

895.—The Tabernacle in the Wilderness (PUMPHREY, 12 slides).

Here we have general views, the Altar of Burnt Offering, the Laver, Holy of Holies, Sin Offering, the Poor Man's Sin Offering, etc.

896.—The Tabernacle in the Wilderness (HUGHES, 18 slides).

These pictures are from special drawings by R. Beker, and include the Outer and Inner Coverings, the Outer and Inner Curtains, the Holy Place, Brazen Altar, High Priests in Robes, etc.

97.—Bunyan's "Holy War" (NEWTON, 25 slides).

We have such scenes in the "Holy War, or Siege of Mansoul" as Diabolus Addressing the Inhabitants of Mansoul and Inciting them to Arms, Mansoul Deaf to the Appeals both of Mercy and Judgment, Hoisting the Flag of Defiance, and Barricading Ear-gate, etc.

898.—Bunyan's "Holy War" (PUMPHREY, 41 slides).

These scenes are taken from H. C. Selous and D. F. Friston's illustrations to Cassell's edition of the " Holy War," and include Bunyan Listening to the Old Women at Bedford, the Arrest of Bunyan, etc., and also Diabolus and his Companions seeking Mansoul, First Attack upon Ear-gate, Understanding and Conscience sent to Prison, Godly Fear, and Carnal Security, Feasting, etc.

899.—The Catholic Martyrs (HUGHES, 20 slides).

The martyrs here represented are John Finch, Henry Heath, Mary Queen of Scots, Margaret Clitheroe, William Viscount Stafford, etc.

900.—Book of Martyrs (PUMPHREY, 52 slides; YORK, 52 slides).

These views are taken from Cassell's Illustrated "Book of Martyrs,' and represent Latimer and Ridley at the Stake, Roman Christians worried by Dogs, Torture before the Inquisition, the Rack, Lady Jane Grey sees her husband taken to execution, Cranmer brought to trial, etc.

901.—The Child Martyr (PUMPHREY, 4 slides).

A true story in verse of a child bound to a stake to die with the influx of the tide, and her heroic martyrdom.

902.—The Gospel in England (WOOD, 44 slides).

This subject is illustrated by the Martyrdom of St. Alban, Portrait of Wycliffe, Burning Bibles, the Pilgrim Fathers, the Rack, Latimer and Ridley, Bunyan Preaching, Wesley, etc.

903.—The Holy Land (SCIOPTICON CO., 46 slides).

Of Sinai we have here seven views, of Jerusalem fourteen, of Athens four, in addition to views at Bethlehem, Samaria, Nazareth, Damascus, and Baalbec.

904.—The Holy Land (YORK, 60 slides).

We have scenes here showing Mounted Travellers, Mount Sinai, Women of Bethlehem, Gethsemane, the Virgin's Tomb, Old Street in Jerusalem, the Rich Man's House, Ploughing in Palestine, the Sea of Galilee, Cana, Damascus, etc.

905.—Views in the Holy Land, Asia Minor, &c. (NEWTON, 160 slides).

There are eight slides for Bethlehem, eight for Damascus, six of Galilee, forty-five for Jerusalem, fifteen for Mount Ararat. There is a view of the Interior of the Cave of Adullam, and one of Babylon; also one of Tripoli in Syria, and two for Tyre, etc., etc.

906.—Views of the Holy Land (PUMPHREY, 50 slides)

Of the views here shown twenty-one are in Jerusalem, six refer to Baalbec, and we also see Joppa, the Tomb of Rachel at Bethlehem, the Dead Sea, Nazareth, etc.

907.—India (PUMPHREY, 46 slides).

The Lecture is arranged by Mr. W. Yorke, a resident member of the Christian Vernacular Society of India. The set opens with the hymn "All people that on earth do dwell," and subsequently shows a group of Street Coolies, Hindoo Dancing Women, Hindoo Swallowing a Sword, Infants thrown to Crocodiles, Hindoo Ascetic, various forms of Self torture, etc., etc.

908.—Missionary Scenes (NEWTON, 140 slides).

Of these scenes 17 are laid in America, where we see the Arrival of a Missionary at Camp, an Indian Chief now a Clergyman of the Church of England, Indians going to Church, etc.; 21 scenes are in Africa, where we see a Missionary received by a Chief, Slaves Chained and Marching to the Coast, the Interior of a Slave Ship, etc.; 36 scenes are in China, where we see a Furnace for burning Imperial Sacrifices in the grounds of the Altar of Heaven, Temple of 500 Gods, Temple of Heaven, etc. In India we have 42 scenes, including a group of Fakirs at Cashmere, a group of Gods and Priests, various Idols, Missionaries preaching at Ganges Peshawur, methods of Self-torture, the Devil Dance in Ceylon, etc. And we have scenes in Madagascar (4), New Zealand (10), and Japan (7).

909.—Missionary Enterprise in Many Lands (Wood, 58 slides).

After the introductory slide we have a Missionary Map, followed by the Hindoo Trinity, Temple of Juggernaut, Christian Converts, Opium Smokers, Temple of Ten Thousand Idols, a Maori Chief, an African God, a Slave Market, an African Alphabet Class, Dr Livingstone, a Christian in Chains, etc.

910.—Bethlehem to Olivet (Wood, 67 slides).

The secondary title is "In the Footsteps of the King" We start from the Vale of Moreb, we have next Moses on Nebo, with views of Bethlehem, Jerusalem, the Sermon on the Mount, the Centurion's Servant Healed, Feast of the Dedication, etc , etc.

911.—Home Mission Subjects (Hughes, 12 slides).

We have here views of St. Peter's Mission, Hoxton, 700 London Poor at Tea, a Scene on Emigrant Ship, Thames Church Mission Ship, Eel Boats at Billingsgate, etc.

912.—The London City Mission Work and Workers (Wood, 7 slides).

Here we see an Attack on a Missionary, a Conference with Thieves, Bethel Service, Open-air Preaching, etc.

913.—The Rock of Ages (Newton, 3 slides).

Illustrative of the well-known hymn.

914.—The Cross and its Story (Wood, 16 slides).

Jesus before Pilate, Led Forth, Falling Beneath the Cross, at Calvary, the Crucifixion, etc., are here depicted.

915.—Horrors of War (Sciopticon Co., 11 slides).

The scenes include the Conscription, Departure for War, Famine, Treason, Human Jackals, etc.

916.—The Heart and its Inmates (Pumphrey, 8 slides).

We here see "The natural heart possessed by evil passions, indicated by wild and unclean beasts," "The evil passions free and unrestrained," "The result—remorse, despair, etc."

917.—The Passion Play (York, 30 slides).

The Lecture is written by the Rev. Canon Wilson, and the set shows Ober-Ammergau, the Entry into Jerusalem, Joseph Maier, the Last Supper, Joseph sold by Judas, the Crucifixion, the Resurrection, etc.

918.—The Lost Lamb (Pumphrey, 6 slides).

These slides illustrate the story in verse, published by the Religious Tract Society, commencing, "See the little lamb, Near it is its dam."

Stories, Tales, Illustrations of Authors, &c.

919.—Amateur Photographer (PEXTON, 12 slides).

A tale in verse of two youths who, "doting on art," find inconveniences to the practice thereof, and think of photography. They buy a lens and try the cat, but fail, and take a romantic bridge, subsequently dealing with the fixing process They next try a bull dog, who, however, does damage, and then Aunt McCuddy, who is not pleased with the result. Trying landscapes, they quarrel and fight with themselves, and then with a stray bull. At last, with clothes torn, hopes dashed, their apparatus smashed, they're told to "shun what can't be done."

920.—Æsop's Fables (PUMPHREY, 12 slides; YORK, 50; NEWTON, 63).

The Miller and his Son, Wind and the Sun; Dog in the Manger; The Man and Satyr, etc., etc.

921.—Adventures of Briggs with a Bull (YORK, 4 slides).

The bull very much objects to the colour of Mr. Briggs's waistcoat, and puts Mr. B. in a fix.

922.—Adventures of Brown, Smith, Jones, and Robinson (YORK, 4 slides).

The four go fishing and land a monster lobster, which they take to town to be cooked

923.—Androcles and the Lion (YORK, 6 slides)

The old story of the Roman slave who helps a lion in distress, the lion in turn helping him.

924.—Adventures with a Flea (YORK, 12 slides).

A Mr. Morpheus aroused by a flea, for which he searches, even in his night-cap He gives up once, but tries again, in the bed, up the bed. At last it is found in his stocking, and after operations with a wet finger, a pin, and a candle, he sleeps like a top.

925.—Arabian Nights Entertainments (PUMPHREY, 50 slides).

Illustrations to some of the Arabian Nights' stories.

926.—Alone in London (Newton, 12 slides).

A story of children taken from the streets.

927.—Army and Navy Drolleries (NEWTON, 24 slides).

A humorous Army and Navy A.B.C., by Major Seccombe.

928.—Aurelia's Unfortunate Young Man (PUMPHREY).

Illustrations to the most humorous story by Mark Twain, in which after the engagement the young man becomes "pitted like a waffle-mound" with small-pox, loses the use of his limbs, his eyes, and his hair, and when M.T. isappealed to as to whether A. ought to marry what is left of the Y.M.: "Try it," says M.T., "it would have been a happy conceit if he had started with his neck and broken that first, but since he has tried to string himself out as far as possible you can't upbraid him for it, the best must be done and you must try and not feel exasperated at him."

929.—Baby's Sugar Bag (YORK, 12 slides).

The tale of a baby and a sugar bag, the dogs who stole it, and the hornet whose sting the dogs are made to feel, and the general adventures of the baby.

930.—Bells, The (STEWARD, 6 slides).

A set to illustrate the poem by E. A. Poe, The Sledge Bells, Moving Sledge, the Marriage Bell, the Alarm Bell and fire effect, the Iron Bell and effect of Goblins.

931.—Boys and Ravens (YORK, 8 slides; NEWTON, 8).

Two boys attempt to capture the nest, the ladder slips and lands them into a filthy pool. A dog fetches one boy out, the dog's master has to fetch the other, and all march away, to the raven's joy.

932.—Brown and the Bear (YORK, 12 slides).

Brown out Bear Hunting, whilst arranging plans receives from a grizzly sudden notice to move, which (after waving his unmentionables in his foe's mouth) he does up a tree, and so on.

933.—Brown and the Mouse (YORK, 12 slides).

The disturbed slumberers give chase to a mouse that thereupon seeks refuge in B.'s nightcap, from whence it is conveyed in triumph to a tub with a hole in, but being re-caught, Mr. and Mrs. B have a leap and downfall, during which Mr. Mouse wishes Mr. and Mrs B. good morning.

934.—Billy's Rose (YORK, 10 slides).

Illustrations from Life Models to G. R. Sims's well-known poem

935.—British Workman (PUMPHREY, 6 slides).

Specimens of him; for instance, He who mends the Gutters—and undoes the Slates; he who mends the Slates—and spoils the Garden.

936.—Barber and his Clever Dog (YORK, 12 slides).

This clever dog welcomes customers on account of the morsels that fall from accidental slip of the barber's razor, and, of course, watches the shaving operations with an intelligent face.

937.—Barred Window (YORK, 8 slides).

On the strength of a door-knock Smith puts his head between the window-bars, and being stuck, his visitor pays him his respects till his cries bring the wife and the blacksmith.

938.—Beautiful Snow (YORK, 5 slides).

An American Poem.

939.—Blunderbore, Mr., in Society (YORK, 12 slides).

The clumsy, blundering man who always gets into trouble.

940.- Bookworm, The (YORK, 12 slides).

The bookworm wipes his spectacles, lights his cigar, and is absorbed in his book, when the cat springs on to his shoulders, an annoyance for which he kicks her down stairs. Continuing the reading he spills some paraffin, is blown up into the air, and, finally lies in bed a pitiable object, but the victim of his own folly.

941.—Bashful Man (PUMPHREY, 7 slides).

Showing the scrapes and misfortunes he gets into at a dinner.

942.—Boy who did not get the Prize (PUMPHREY, 6 slides).

A boy sat down to get up home lessons, when two little feathers float just before his nose, he finds rare fun in puffing them about, and, of course, does not get his work done.

943.—Brook, The (PUMPHREY, 6 slides).

Illustrations to Tennyson's poem commencing "I come from the haunts of coot and hern, I make a sudden sally."

944.—Burns, Robert (NEWTON, 20 slides).

A portrait, views of his cottage, monuments to his memory, illustrations of "Tam o'Shanter," "Auld Brig o'Ayr," etc., are here shown.

945.—Byron (NEWTON, 7 slides).

A statue of Byron, a scene of the night before Waterloo in the ball room: "And there were sudden partings," with other scenes referring to Waterloo.

946.—Ballad Singer (6 slides).

From life models. A drunken father reformed by the ballad singing and illness of his child.

947.—Baron Brag (PEXTON, 28 slides).

As the name implies, the tale of a braggart's supposed adventures, principally with animals, such as the bear, walrus, and lion.

948.—Cat and Serpents (YORK, 8 slides).

Puss looks curiously into a basket, and is seized by snakes therein. After vigorously disputing possession, one snake dies of suffocation and the other of indigestion.

949.—Chimes, The (YORK, 24 slides).

Illustrating the tale by Charles Dickens.

950.—Careless Maggie (PUMPHREY, 6 slides).

The results are shown of careless and untidy habits.

951.—Christie's Old Organ (HUGHES, 24, YORK, 24).

The Religious Tract Society's well-known story.

952.—Clever Nephew (YORK, 12 slides)

We here have a discourse on pudding, "Stickification" and "Catch-'em-alive-oh" exemplified, etc., etc.

953.—Conceited Pig (PUMPHREY, 8 slides).

The pig who dresses and tries to act as a man, with the woful failure.

954.—Colonel Slasher and his Neighbour's Cat (6 slides).

The would-be annihilation of the tom-cat, which, however, was not a success.

955.—Crotchet's Musical Recital (YORK, 12 slides).

Illustrations of the professor's motions and positions of body corresponding with the musical movements, such as "Capricuoso," "Fuga del, diavolo," etc.

956.—Cruikshank's Works (YORK, 28 slides).

These illustrations of Cruikshank's works include "Breaking Up," "Home from School," "Black Monday, or the End of the Holidays," various illustrations of "London Nuisances," "The Blue Devils," "Raining Cats, Dogs, and Pitchforks," etc.

957.—Curfew Must Not Ring To-night (York, 10 slides).

Rose Hartwick Thorpe's very attractive poem of how Bessie saved her lover.

958.—Crab-catching Extraordinary (Hughes, 5 slides).

This Crab-catching was designed by Mr. Weeks. Joe tries to catch a crab, but the crab catches him.

959.—Colonel Slasher and the Mosquito (6 slides).

A fight to the death with the insect, but the colonel gets the worst of it.

960.—Cat Battery (Pumphrey, 6 slides).

A satirical and humorously written lecture on electricity.

961.—Christmas Carol (York, 25 slides; Newton, 24; Pumphrey, 12).

Dickens, Scrooge, Marley's ghost, and the nephew, Bob Crotchett, are represented, also the visitors and the three ghosts.

962.—Cruikshank's Comic Illustrations (Pumphrey, 24 slides).

The illustrations are of a Gentleman Intended for the Bar, a Tale of a Tub, a Drop of Comfort, a Vane Man, etc., etc.

963.-Comical Acting Cats (Pumphrey, 6 slides).

Scenes taken from life "Over the Garden Wall," the Serenade, the Duet, etc.

964.—Cat and the Mouse (Underhill, 5 slides).

Miss Pussy tries to catch mousie in a boot, but fails.

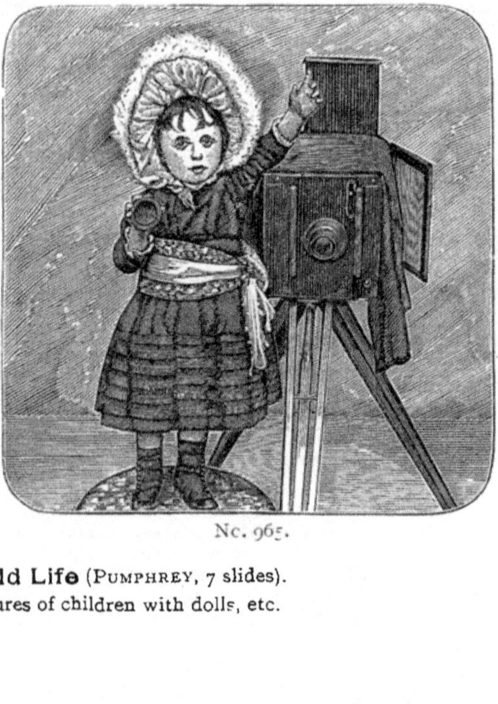

No. 965.

965.—Child Life (Pumphrey, 7 slides).

Pictures of children with dolls, etc.

966.—Cotter's Saturday Night (PUMPHREY, 8 slides; YORK, 9).

Burns. Slides to accompany the poem.

967.—Cowper (NEWTON, 6 slides).

Portrait, Cowper's House at Olney, Weston Lodge, and other views.

968.—Contrasts for Juveniles (PUMPHREY, 18 slides).

Short moral tales for children, such as the Cowardly Boy, the Polite Boy, the Mischievous Boy, the Tidy Girl, etc.

969.—Corporal Smart (STEWARD, 8 slides).

Illustrations of the Ups and Downs of the Corporal—on a Donkey.

970.—Christiana and her Children (NEWTON, 18 slides).

Illustrations to the second part of Bunyan's Pilgrim's Progress.

971.—Dame Perkins and her Grey Mare (YORK, 8 slides; NEWTON, 8).

Tells how a poor old grey mare, who could hardly carry her dame to market, wakes up at the sound of the hounds, clears gates and hedges, smashing eggs, etc , and wins the brush.

972.—Death of the Bluebottle (YORK, 12 slides).

A bluebottle pitches upon a sensitive part of Dr Snoozum after dinner. The Doctor wages war and mounts upon a chair, from which he unexpectedly descends, but after a desperate conflict he witnesses the total annihilation of the fly.

973.—Don Quixote (PUMPHREY, 44 slides).

These are copies of Gustave Doré's illustrations to Cassell's edition

974.—Dante's Inferno (PUMPHREY, 17 slides).

The subjects are taken from Cassell's Dante's Inferno, illustrated by Gustave Doré. We are here shown Dante lost in the Wood, the Punishment of Gluttony, of Heretics, of Seducers, of Hypocrites, etc.

975.—Dilbury Diddle, Sir (PUMPHREY, 6 slides).

The gallant Captain, who goes forth to fight for his country, his queen, and his fair lady.

976.—Dog of St. Bernard (UNDERHILL, 9 slides).

Illustrating the Dog, his Work of Mercy, and giving views of the Hospice. The reading is in verse.

977.—Dinah's Fiddlestick (WOOD, 12 slides).

Shows Dinah paying the Rent, Barnaby in Bed, Dinah Lecturing, the Fiddlestick, the Wedding, Barnaby's Happy Home, etc.

978.—Diogenes and the Boys of Corinth (PUMPHREY, 7 slides; YORK, 12).

From life models in appropriate costume.

979.—Dr. Spiritus and the Moon (YORK, 8 slides).

Illustration of the various fears of a Dr. Spiritus, who believes the moon is coming nearer the earth.

980.—Dreams at Sea (YORK, 5 slides).

Fancy pictures of "The tender limbed and timid boy," "The husband sees the harbour," etc, by the Rev T Greenway.

981.—Ducks and Frog (YORK, 12 slides; NEWTON, 12)

A duck and drake quarrel over the swallowing of a frog, putting him to a variety of tortures In the end the frog escapes where they can't follow, but where they do succeed in getting their heads stuck till a poulterer carries them off. The frog was laid up for three weeks, but now, thank goodness, he is able to smoke his pipe.

982.—Dick's Dive into Duck Pond (NEWTON, 12 slides).

Dick, home for holidays, falls asleep and dreams he becomes very small, and is introduced by a frog to High Life below Water. He sees hydras, water fleas, caddis flies, water scorpions, gnats, and gnat larvæ, etc.

983.—Domestic Tragedy, A (PUMPHREY, 2 slides).

A gentleman on duty as special constable is invited into his own house by his own cook to have some cold mutton and pickles, the cook falling aghast down the area steps on discovering her mistake

984.—Elephant's Revenge (YORK, 12 slides).

A native has a shot at an elephant. The animal, giving chase, catches him by the ear, shakes him, throws him into a lake near a crocodile, gives him a shower bath, shakes him by his drawers, and finally throws him away on the prickles of a cactus bush.

985.—Equestrian Courtship (PUMPHREY, 6 slides).

Illustrations of Tom Hood's poem, commencing "It was a young maiden went forth to ride, And there was a wooer to pace by her side."

986.—Enoch Arden (NEWTON, 23 slides).

Illustrates the poem by Tennyson, by a Seaside Village, a Scene in the Wood, where Philip, "like a wounded life, crept down into the hollows of the wood," etc., etc.

987.—Elephant and the Tailor (12 slides).

An elephant pays a friendly call to a tailor, who pricks trunk with a needle. The elephant goes and fills his trunk with water, and gives the tailor a regular swamper.

988.—Eugene Aram (16 slides).

Illustrations of Hood's well-known poem.

989.—Five Senses (YORK, 5 slides, HUGHES, 5).

Comic pictures of Seeing, Hearing, Smelling, Tasting, and Feeling.

990.—Foolish Toper, The (YORK, 12 slides).

Has a beery quarrel, meets an insulting stranger, eventually slaughters the stranger, is troubled with the stranger's ghost, becomes his country's care, etc.

991.—Friendless Bob (NEWTON, 18 slides; YORK, 18).

The story from the "Children's Friend" of November, 1879, in which Bob makes acquaintance with a Missionary, makes a friend of Jerry, meets Ethel, etc., and in which Mr. Fortescue chastises the donkey boy.

992.—Funny Faces (UNDERHILL, 18 slides).

The funny faces include those of Ally, Mrs , Miss, and Master Sloper. a Clown; Miss Mary Ann Spriggs, a servant; Miss Crabstick, an old maid' Mr. Jollyboy, an old gentleman , Ching Ching, a Chinaman, and so on.

993.—Frizzlebrush, the Barber (HUGHES, 7 slides).

Shows how a barber's brushing machine did not work properly.

994.—Factory Chimney (PUMPHREY, 4 slides).

Relates how a man left on the top of a factory chimney escapes by unwinding his worsted stocking, tying a piece of mortar to the end of the worsted, dropping it down, pulling up string and then a rope, which he fastens, and with its aid descends.

995.—Fine Art Gallery (PUMPHREY, 12 slides).

Humorous pictures of an Old Shoe Sole, representing "A poor old soul," a music stool with an egg upon it, "The lay of the last minstrel," etc.

996.—Fairy Gift (PUMPHREY, 5 slides).

A maiden, who thinks there is too much to do, asks a fairy for a magic wand to get her work done, but fairy gives her ten little workmen, who get all the work done, and then she promises to leave them if they are never allowed to be idle (the workmen are the maiden's own fingers and thumbs).

997.—Faust and Marguerite (9 slides).

Illustrates the play by Goethe, and shows Faust in his Study, Faust seeing Marguerite enter the Cathedral, Mephistopheles introducing Faust to Marguerite, the Love Test, etc.

998.—Fiddler and Crocodile; or, The Power of Music (PUMPHREY, 6 slides; YORK, 6).

The power of music. The fiddler overcomes a crocodile by the beauty of the strains from a fiddle.

999.—Gabriel Grub (NEWTON, 18 slides; YORK, 17).

Charles Dickens's tale. The story of the sexton who was stolen by the goblins.

1000.—Golden Goose (YORK, 12 slides).

Illustrates the tale in which Sawney departs for the forest, discovers the golden goose, departs from the inn with the innkeeper's three daughters, parson, and clerk, call on the wood-cutters to help them, etc., etc.

1001.—Gossips, The (YORK, 12 slides).

Mrs. Chatterbox and Mary Dawdle, enjoying a gossip, get separated by barrow-men, town criers, the military school, geese, rain, by a pig, and finally, by thunder and lightning.

1002.—Greatest Plague of Life (PUMPHREY, 12 slides; YORK, 12).

The adventures of a lady in search of a good servant. Sketches of the *genus domesticus*.

1003.—Great Gun Trick (HUGHES, 4 slides).

A story about a clown sending a bobby to the moon by means of a 100-ton gun.

1004.—Grindlestone, The (PUMPHREY, 3 slides).

Lancashire dialect verses by Edwin Waugh on the folly of stopping to help an ungrateful fellow.

1005.—Gemze Fawn (PUMPHREY, 6 slides).

An orphan, chamois fawn is petted by a young girl, but on hearing the voices of other fawns, darts after them, leaving her disconsolate.

1006.—Giant Bugaboo Bill (PUMPHREY, 4 slides).

Describes the way in which the tyrant was overcome by a good dinner and sent up in a balloon, from which he did not return in a parachute *à la* Baldwin.

1007.—Gray's Elegy (PUMPHREY, 12 slides; NEWTON, 14).

Illustrating the poem by Gray commencing, "The curfew tolls the knell of parting day."

1008.—Given in Charge (WOOD, 30 slides).

Here we see the Refuge, visions of the past, the present, the future; the little boy's appeal; his being cast adrift, detected in crime, taken in charge, etc.

1009.—Gipsy's Revenge (PEXTON, 32 slides).

The story of Eric, Ella, Bertram, Albert, &c. From life models.

1010.—Greedy Boy and the Witch (10 slides).

Sammy steals jam, a bait is laid for him, he is caught by the nose, a witch changes him into a pig, sharpens a knife to take his life, and he is just going to be killed when Jane comes to the rescue.

1011.—Hare and the Robbers (UNDERHILL, 8 slides).

Two robbers after a hare, by means of a rope, are pursued by an officer of justice, whom they manage to tie up aloft, and thus escape.

1012.—Heathen Chinee (YORK, 9 slides).

The tale by Bret Harte, in which we see "that Heathen Chinee," Bill Nye, "Ah Sin astonishing the pair," etc , etc.

1013.—Hogarth (YORK, 30 slides).

These slides illustrate Hogarth's paintings, such as "Morning," "Noon," "The Laughing Audience," "The Sleeping Congregation," "The Analysis of Beauty" (No. 1 and No. 2), etc.

1014.—Harlot's Progress (YORK, 6 slides).

Hogarth here illustrates the country girl arriving in town, quarrelling with the Jew, in lodgings, in prison, her death, etc.

1015.—Honey Stealers (YORK, 8 slides).

An account of two lads who go honey stealing, and forget that bees have stings. After trial of the water cure, and the discovery of inability to swallow potatoes, they resort to the apothecary, and in bed resolve never to steal honey again.

1016.—Hostile Neighbours, The (YORK, 12 slides).

An artist and a musician live with only a wall and a grudge between them The musician practises much, and the artist conceives a plan to stop such *vile* performances, thereby rousing the fiddler's fury. On the artist resuming work, life suddenly springs into his picture, the sequel having a picturesque and profitable termination.

1017.—How to Catch a Hippopotamus (HUGHES, 3 slides).

The Little Piccaninies' Picnic on the Nile. They coolly have a "picnic in state," whilst they "for old hippopoppy wait," and accomplish their business of catching him in rare old style.

1018.—How Geese are Plucked (PUMPHREY, 6 slides).

Verses. The geese are overcome by some brandied cherries thrown away by Gretchen. Deeming them to be poisoned, they are plucked for the sake of the feathers, but afterwards come round and run about. The farmer winds up with a wise remark—

> "There's many a human goose is plucked,
> When overcome like you."

1019.—Homes Without Hands (NEWTON, 38 slides; PUMPHREY, 70).

Illustrations of the nests of various birds, etc.

1020.—Humorous Sketches (PUMPHREY, 106 slides).

Humorous illustrations of the various vagaries of the British workman, the British domestic, our gas supply, etc , etc.

1021.—Horse and Oysters (PUMPHREY, 3 slides).

A cold traveller's device to get a seat near the fire by sending out oysters to his horse, the occupants of the seats rushing to see him eat so strange a meal.

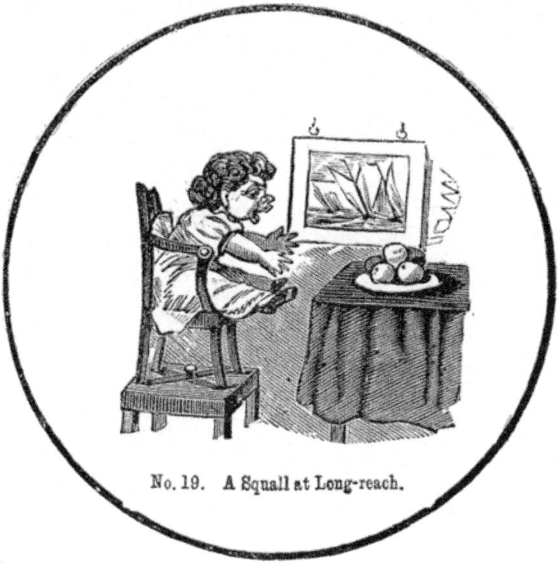

No. 19. A Squall at Long-reach.

No. 1022.

1022.—Hood's Comicalities (PUMPHREY, 24 slides).

We see here Hood's rendering of "A Note of Admiration," "Getting a Hole Holiday," "Preparing a Hot Bed," etc.

1023.—Hans in Luck (NEWTON, 10 slides).

Hans changes a lump of gold for a horse, a horse for a cow, a cow for a pig, etc , and finally becomes a wiser man.

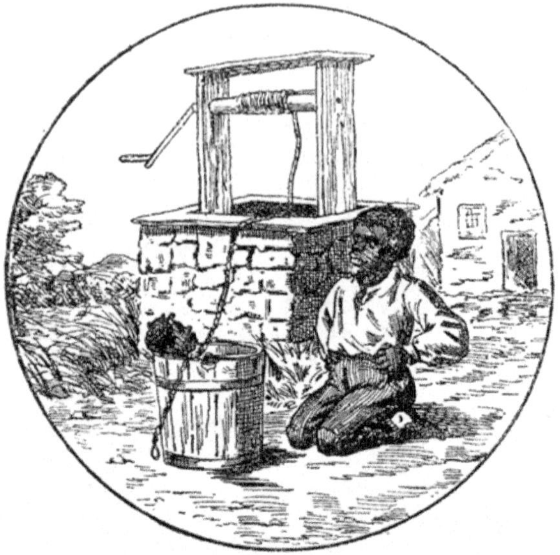

No. 1024.

1024.—Hieronymus Pop and the Baby (PUMPHREY, 6 slides).

A very humorously written story of a little nigger boy minding a baby. To quieten the fretful baby, he pops it into the well bucket, where it is discovered later on by the anxious parents, Hieronymus having meanwhile gone off to a dog fight.

1025.—Ideal Pictures from Life (YORK, 24 slides).

Chiefly pictures for children, such as "Minding Baby," "Admiring Baby," "Dressing Dolly," "Look at yourself in the glass, dolly!" etc., etc.

1026.—Industry and Idleness; or, The Two City Apprentices (NEWTON, 18 slides; PUMPHREY, 12; YORK, 12).

Hogarth here represents one apprentice going wrong, and finally being taken to Tyburn for execution; the other going to church, being taken into partnership, marrying the master's daughter, and finally becoming Lord Mayor of London.

1027.—Inchcape Bell (STEWARD, 8 slides).

A set of slides to illustrate the well-known poem, with effects.

1028.—Impulsive Gardener, The (YORK, 8 slides).

The dog barks, and lo! those confounded pigs are in his garden again. He tries to "give that black one a lesson this time," but "pigs will be pigs," and he doesn't meet with success.

1029.—Inverted World (PUMPHREY, 5 slides; YORK, 6).

Things generally are turned round the other way, such as the tree cutting up the woodman.

1030.—Instantaneous and Life Pictures (Sciopticon Co., 47 slides).

River views with vessels, specimens of dogs, cats, rabbits, etc.

1031.—Inez de Castro (Hughes, 4 slides).

Don Pedro, son of Alphonso, King of Portugal, privately married the beautiful Inez de Castro. On this becoming known to the king, he endeavoured to annul the marriage, but failing to obtain the co-operation of the parties, Don Pedro was imprisoned, and Inez savagely butchered. When Pedro at last came to the throne, he punished with death all who were in any way connected with his wife's death, and, disinterring his late wife, he had her crowned with him as queen.

1032.—I Love You, Mother (Pumphrey, 6 slides).

A story in verse of three children who all loved their mother. One loved, and helped her, too, and this was the one who perhaps liked her best

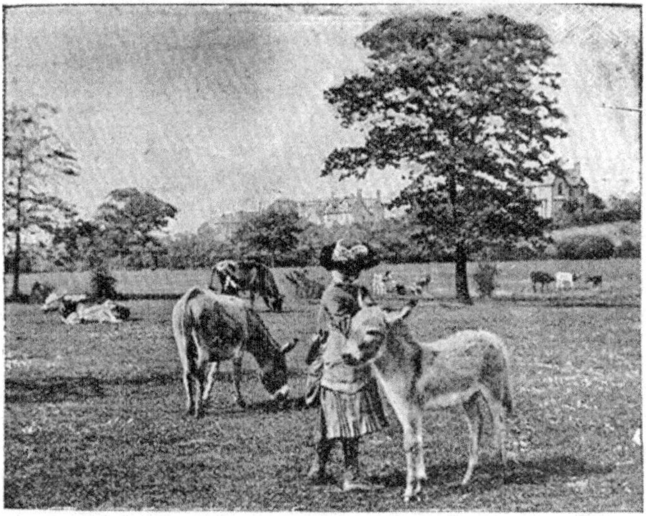

No. 1032A.

1032a. Instantaneous Pictures (Pumphrey).

This maker has a large number of these very interesting instantaneous pictures of animals.

1033.—Irish Life and Wit (Wood, 48 slides).

We have here illustrations of "Going wid the Pigs," "Tay or Whiskey?" "Donnybrook Fair," "Inconvaynience of Married Life," "Inconvaynience of Single Life," "Bewildherin' the Tourists'," etc., etc.

1034.—Isaac Newton and the Apple (York, 8 slides).

Sir Isaac Newton, pacing up and down, notices the fall of an apple from a tree, and after careful consideration, discovers the wonderful law of gravity.

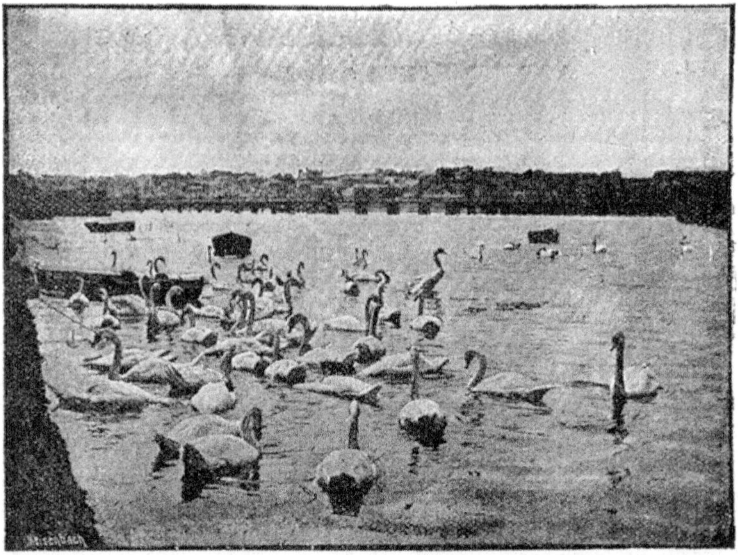

No. 1032A.

1035.—Jack the Conqueror (PEXTON, 26 slides; PUMPHREY, 12; YORK, 12).

A tale for juveniles of perseverance and its success.

1036.—Jackdaw of Rheims (PUMPHREY, 12 slides; YORK, 13).

Illustrations to this Ingoldsby Legend.

1037.—Jane Conquest (NEWTON, 16 slides; PUMPHREY, 8; YORK, 16).

Originally from the *Methodist Family Budget* and by Dr. Croft, but now known everywhere as a lantern set. Relates how she rang the bell and saved the crew of a ship.

1038.—Jessica's First Prayer (HUGHES, 12 slides; PUMPHREY, 12; YORK, 10).

An affecting story, originally brought out by the Tract Society.

1039.—Jocko, the Brazilian Ape (YORK, 14 slides).

Jocko is here shown cutting capers in his native forest, then with pitch-lined boots, with two Angola cats, stealing brandy, etc., etc.

1040.—John Ploughman's Pictures (YORK, 38 slides; HUGHES, 40).

Picture hymns, illustrating the God of Eternity, the wonderful power of the Gospel, a temperance song, etc.

1041.—Jones's Baby, Origin of (YORK, 4 slides).

Scenes in connection with the ordering of a baby.

1042.—Juvenile Smokers (YORK, 6 slides).

Tom and Harry buy two cigars to smoke like men. The smoke overpowers them, and on sneaking home they get lectured for their folly.

1043.—Jim the Collier Boy (HUGHES, 7 slides; YORK, 6).

A touching story, in verse, of the death of a young collier boy.

1044.—Jacob Goodheart (HUGHES, 21 slides).

A tale illustrating "the good spirit to forgive and forget."

1045.—Jap the Butcher (HUGHES, 6 slides).

Showing the ups and downs of a piggy's life, as invented, written, and illustrated by Frank Weeks.

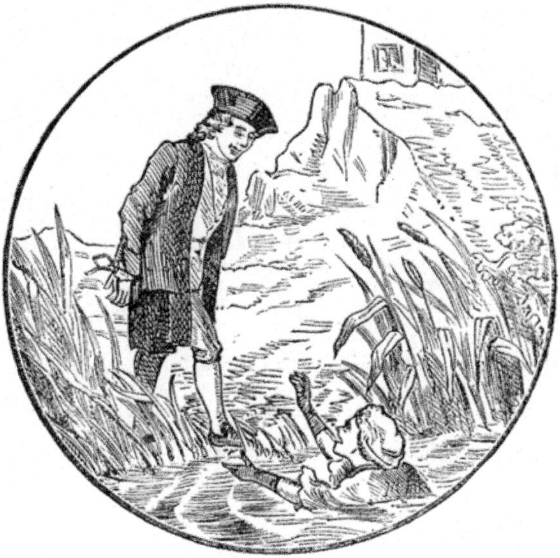

No. 1046.

1046.—Johnny Sands (PUMPHREY, 4 slides).

Illustrations to the old-fashioned song.

1047.—Jack Holiday (PUMPHREY, 14 slides).

A story for juveniles, by the late Albert Smith; a mixture of prose and verse. Some most amusing skits on school life and school books, sure to be relished by boys.

1048.—Jungle to Let (PUMPHREY, 4 slides).

An elephant, "Old Punjaub," sets his back against the innovation of a railway train, with the result of nailing up a ticket, "This jungle is to let."

1049.—Knight and the Lady (PUMPHREY, 6 slides; YORK, 6).

The Ingoldsby Legend of Sir Thomas and his spouse, who was very fond of eels.

1050.—Kate Maloney (York, 6 slides).

By G. R. Sims. A powerful Irish poem of the woman who, lying at death's door, refused her husband's offer to earn a hundred guineas reward by betraying a fugitive from justice. The husband, however, steals away to do the deed, and is shot as a traitor by his wife.

1051.—King and Clown (Pumphrey, 2 slides).

A little tale, in verse, of a king who acted so like a clown that, on a real and cunning clown getting hold of mace and crown, people hailed *him* monarch of the town.

1052.—La Fontaine's Fables (Pumphrey, 36 slides; York, 85).

These are some of the well-known fables, such as the Robbers and the Ass, the Town Rat and the Country Rat, the Sick Lion and the Fox, etc.

1053.—Lazy Traveller (York, 12 slides)

After a second rousing this traveller begins to dress; then, being hurried, he puts on his boots before his trousers, and his trousers the hinderside before, finally, of course, missing his train.

1054.—Life Boat (Newton, 9 slides; York, 7).

A tale in verse, by G. R. Sims, in which the husband of a woman, who is lying dying of grief for the loss of a son, is begged by the captain of the lifeboat to join in a daring rescue. The dying woman begs her husband to go; he obeys, he seizes a body in the water, gets a terrible blow, is laid up at home, and at last discovers that he rescued his son, and that his wife has recovered her health at the joyful sight of her long-lost son.

1055.—Little Artist and Large Portfolio (York, 8 slides).

The L. A. uses his L. P. as a shelter from rain, as a raft, a tent, and a sledge. At last, in prosecuting his studies in high art, he rises in the world, as does the wind, the portfolio, acting as a balloon, coming down badly.

1056.—Little Town of Weinsburgh (York, 6 slides; Pumphrey, 6).

The story of the wives who carried their husbands out of the besieged town on their backs.

1057.—Lodgings to Let (Pumphrey, 6 slides).

An account, in verse, of Will Waddle's rapid decrease in rotundity in new lodgings by sleeping over a baker's oven

1058.—Labour-saving Appliance (Pumphrey, 3 slides).

Blotherwich is induced to fix up a L S A., for which "All you have to do is to open the window and down comes the basket," but a midnight cat-howl results in B. bringing down the basket first—on his own head.

1059.—Longfellow (Newton, 4 slides).

A portrait and illustrations of Evangeline, Maidenhood, and the Wreck of the Hesperus.

1060.—Lost in Snow (Wood, 10 slides).

Showing Willie's home in the fishing village, Willie sent out, food refused him, his despair, etc.

1061.—Little Meg's Children (Wood, 12 slides)

Showing the mother's death, Meg and her neighbour, a friend in need, etc.

1062.—Lovechase and his Dog Tray (York, 12 slides).

Lovechase commences the education of Tray by placing a sausage before him to teach him self-denial Tray hunts cats and foxes, the latter in a peculiar manner, and at last meets with a tragic end by Mr. L.'s own clumsy shot.

1063.—Level Crossing (9 slides).

The poem, by G. R. Sims, commencing "The eight o'clock up's just gone, sir."

1064.—Lord Bateman (Pumphrey, 12 slides; Newton, 12).

These are illustrations, after Cruikshank, to the well-known verses.

1065.—Little Black Monkey (Pumphrey, 5 slides).

The man who tried to shoot a monkey because he grinned.

1066.—Marriage a la Mode (York, 6 slides).

Hogarth's "à la mode" represents the husband staying out at night and getting drunk, and the wife being inconstant, and dying by poison.

1067.—Mad Umbrella (York, 12 slides).

Showing the little vagaries of an umbrella under the influence of the wind.

1068.—Mother Tabbyskins (Pumphrey, 4 slides).

The cat falls ill and calls in doctor mouse, whom she gobbles up, but is in turn gobbled up by doctor dog.

1069.—Mary the Maid of the Inn (York, 10 slides).

Southey's poem, illustrated by life model pictures, Mary engaged to Richard, to fulfil a wager visits an old abbey, where she sees Richard and another villain burying a corpse secretly. The sight turns her brain.

1070.—Man and the Calf (York, 12 slides).

Mr. Bah resolves to sell his calf, but meets with many difficulties over attempted removal. At last, happy thought! he ties the cow's bell round his neck, and calf immediately follows him.

1071.—Meg and her Brother Ben (York, 12 slides).

A story for the young of a brother's wish to be a chorister, and his sister's efforts to help him. "Carotty Meg" is assisted by the Dean's brother, and Ben gains the coveted position in the choir.

1072.—Miller and the Sweep (York, 12 slides).

A sweep finds Nancy in love with a miller, and wages war on his bold rival, who, unable to hide, gets a hiding. After which the sweep has to retreat up the chimney, and is followed by the miller. On both descending, we see the confusion that results from not knowing whom you love.

1073.—Mistletoe Bough (Newton, 7 slides; Pumphrey, 4; York, 7).

The very well worn story of Lovell's Bride.

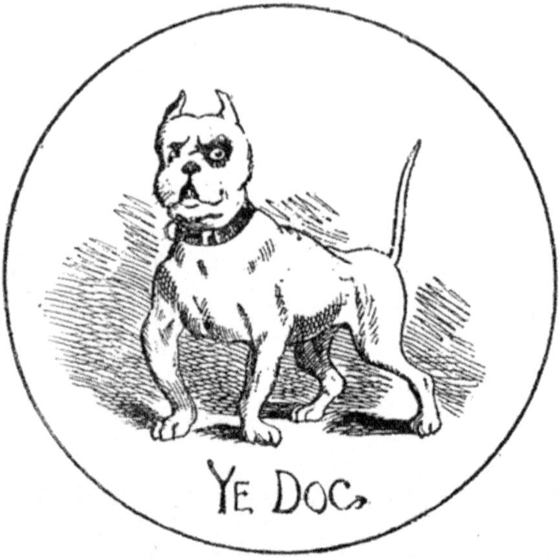

No. 1074

1074.—Mrs. Somebody Swallowed a Fly (PUMPHREY, 7 slides).

Firstly, Mrs. Somebody swallows a spider to catch the fly, then a bird to catch the spider, and so on *ad lib.*

1075.—Mischievous Boys (3 slides).

They put a stone under an old hat which a masher kicks, afterwards, of course, wishing he hadn't.

1076.—Mother's Last Words (HUGHES, 13 slides; NEWTON, 19; PUMPHREY, 15; YORK, 6; YORK, 12).

Illustrations to Mrs. Sewell's ballad of two boys left to fight their way as crossing-sweepers. Their temptations to be dishonest are finally overcome, though with the battle against cold and hunger one succumbs.

1077.—Munchausen, Baron (PUMPHREY, 17 slides; YORK, 25).

Taken from Cassell, Petter & Galpin's edition, with Gustave Doré's illustrations.

1078.—Months of the Year (HUGHES, 12 slides).

From drawings by R. Beker, representing the months with their respective features, such as June, haymaking; November, Guy Fawkes, etc.

1079.—Malle Scrub, The (PEXTON, 24 slides; HUGHES, 24).

An Australian story of two children sent out by their mother to gather firewood. They get lost, are searched for and finally discovered.

1080.—Mrs. Giles's Run with Hounds (PUMPHREY, 8 slides).

A farmer's wife returning from market with butter and eggs follows the hounds against her will, owing to the old recollections of the mare Peggy

1081.—Men and Dogs (PUMPHREY, 8 slides).

Eight designs after Harrison Weir, such as the Low Dog, the Fast Dog, the Old-fashioned Dog, etc.

1082.—Musicians of Bremen (PUMPHREY, 6 slides).

The donkey, dog, cat, and chanticleer, who by a concert frighten away a robber band and secure a good supper.

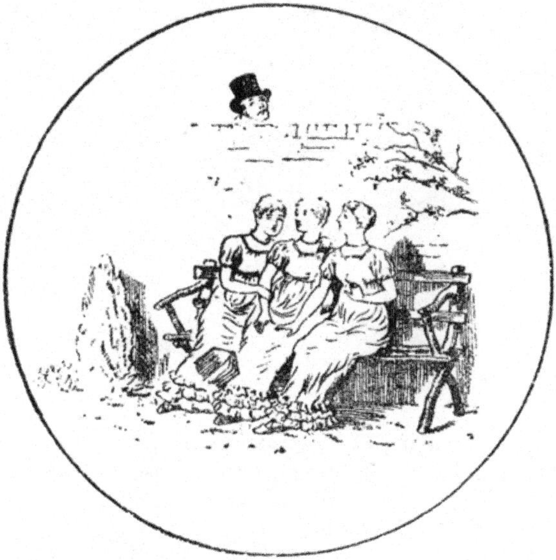

No. 1083.

1083.—Maids of Lee (PUMPHREY, 12 slides).

F. E. Wetherby's well-known verses of the three maids who had so many faults to find with their suitors, that in the end they remained the three old Maids of Lee.

1084.—Men of Ware (PUMPHREY, 11 slides).

Verses by F. E. Wetherby, well known as a companion song to the Maids of Lee. The three young men want special extra superfine wives, and of course are sold.

1085.—Miss Popps's Pet YORK, 12 slides).

The tale of a greedy dog to whom Miss Penelope Popps was devotedly attached, with the moral "to escape being made a sausage, never grow too fat."

1086.—Morrow of the Carouse, The (YORK, 8 slides).

Poor Jones has been getting into bed for the last half-hour when we first see him, and next day the pains in the head being inexpressible, he tries various odd remedies.

1087.—Miller's Daughter, The (WOOD, 12 slides; PUMPHREY, 12).

Illustrations to Tennyson's poem.

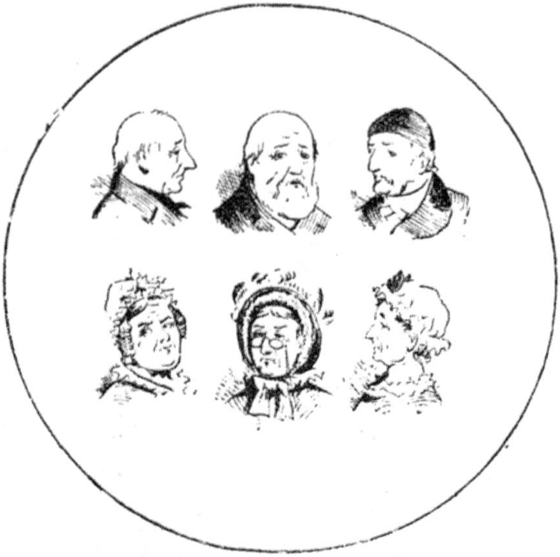

No. 1084.

1088.—May Flower, The (Wood, 16 slides).

The history of the pilgrimage of the Puritans in the "May Flower," with portraits of Sir Walter Raleigh and Miles Standish.

1089.—Misguided Elephant, The (Steward, 20 slides).

The adventures of an elephant who gets decidedly on the loose, with his repentance

1090.—Mr. Vanity's Valentine (Underhill, 12 slides).

Verses. His adventures after receiving a valentine, as he thinks, from Miss Spriggs.

1091.—Nine Lives of a Cat (Pumphrey, 12 slides, York, 12).

The cat is hung, burnt, shot, poisoned, flattened, etc, but always comes out alive.

1092.—New Hat (York, 9 slides).

Very proud of his new hat, a man forgets to take the precaution of carrying an umbrella, rain coming on, he rushes for shelter to a tree, which is struck by lightning, and the hat takes quite another shape With humble pride he learns the necessity of precaution.

1093.—Nursery Tales—

Aladdin (Pumphrey, 6 slides, Newton, 16; Sciopticon, 13).
Aladdin and the Magic Lantern (York, 8 slides).
Alice in Wonderland (York, 42 slides).
Blob the Spider (Newton, 6 slides).
Children in the Wood (Pumphrey, 7 slides, York, 8; Newton, 8).
Cock Robin, Death of (Newton, 14 slides).

Cinderella (YORK, 8 slides; PUMPHREY, 6).
Cock Robin, Marriage of (NEWTON, 11 slides).
Dick Whittington (YORK, 24 and 8 slides; NEWTON, 27, PUMPHREY, 6).
Five Little Pigs (NEWTON, 8 slides; HUGHES, 12).
Gulliver's Travels (PUMPHREY, 31 slides).
Grimbeard the Wolf (NEWTON, 9 slides).
House that Jack Built (PUMPHREY, 10 slides).
Hop o' My Thumb (9 slides).
Jack and the Beanstalk (PUMPHREY, 7 slides; NEWTON, 12; YORK, 8).
Jack the Giant Killer (NEWTON, 12 slides; YORK, 8).
John Gilpin (PUMPHREY, 13 slides; YORK, 12, NEWTON, 15).
Little Woodman (PUMPHREY, 12 slides).
Old Mother Hubbard (YORK, 8 slides).
Puss in Boots (NEWTON, 12 slides).
Red Riding Hood (PUMPHREY, 7 slides; NEWTON, 8; YORK, 8).
Robin Hood (NEWTON, 12 slides).
Robinson Crusoe (PUMPHREY, 17 slides; NEWTON, 22; YORK, 17).
Sinbad the Sailor (NEWTON, 18 slides).
Three Bears (PUMPHREY, 7 slides, YORK, 8).
Three Kittens that Lost their Mittens (YORK, 9 slides).
Tom Thumb (8 slides).

1094.—Neddie's Case (PEXTON, 25 slides).

Or suffer the little children. The story of two poor destitute boys, Neddie and Dickey

1095.—Newsboy's Debt (6 slides).

A gentleman trusts a newspaper boy to get change, which he was to bring to his office. The lad, however, is run over, but sends his brother to say that when he gets well he'll work to refund the money lost at the time of the accident.

1096.—Newcastle Apothecary (PUMPHREY, 6 slides).

The old humorous poem—"Before taken, to be well shaken."

1097.—Oiled Feather, The (YORK, 12 slides).

A moral story by the Rev. P. B Power, M.A, on the power of love, kindness and good temper, representing an oiled feather.

1098.—Old Curiosity Shop (YORK, 24 slides, NEWTON, 24).

Slides to illustrate Dickens's well-known novel, and the wanderings of Little Nell and her grandfather.

1099.—Old Man and his Ass (YORK, 8 slides).

A man and boy drive the ass to market, the boy rides and the men grumble, then the old man rides and women jeer. Finally, they carry the ass and manage to kill him.

1100.—Outcast London (YORK, 40 slides).

This set illustrates a London court, a furnished apartment, getting dinner ready, a brave girl, the tart shop, and other scenes amidst the London poor.

1101.—O'Toole and the Umbrella (YORK, 9 slides)

O'Toole takes a walk in Africa with umbrella for a sunshade. While napping a lion appears, but is astonished at the umbrella, which keeps him at bay. Still more astonished at the umbrella when opened the animal retreats. O'Toole now advises everyone to carry a large umbrella.

1102.—Old Oak Chest (HUGHES, 8 slides).

Giles Jumble's terrible adventure with a nightmare. An empty (M.T.) legend.

1103.—Ocean Waif (HUGHES, 12 slides; WOOD, 12).

A wreck is discovered, and upon examination a tiny baby is found in a bunk. It, however, dies soon after.

1104.—Our Penny Reading (HUGHES, 25).

A most comical history of the village penny reading entertainment, who performed, and the funny effects.

1105.—One-legged Goose (PUMPHREY, 8 slides).

. Verse. The cook's young man eats one leg of a roasted goose, leaving her to get out of the scrape with the squire at dinner time as best she can.

1106.—Owl and Pussy Cat (PUMPHREY, 4 slides).

Verse. The owl marries the pussy cat, and borrows a ring from the piggy-wig's nose for the ceremony.

1107.—Owl Critic (PUMPHREY, 2 slides).

Verse. A man finds fault with the stuffing of an owl, which turns out to be alive.

1108.—Our Father's Care (NEWTON, 8 slides).

The story of little Nellie who sells watercresses whilst her father is in the hospital.

1109.—Old Prickles (NEWTON, 6 slides).

An elf steals a supper from a hedgehog, who recovers the supper and gives the elf a very uncomfortable ride.

1110.—Old Coaching Days (10 slides).

From life models. The journey by the coach to the old-fashioned country house, and adventures there.

1111.—Photographer and the Dog (6 slides).

Mr. Hypo attempts to photograph a bull-dog, but does not manage it. The dog, however, manages Mr. Hypo.

1112.—Photographer's Perplexities (YORK, 12 slides).

Details a poor photographer's tremendous wrestle with a baby, in which "the mother and the photographer are willing, but Bobby is not."

1113.—Precious Couple on the Road to Gretna Green (YORK 12 slides).

From Dickens's tale of The Boots at the Swan.

1114.—Passions, The (YORK, 12 slides).

Examples of love, joy, despair, jealousy, pity, etc, illustrated by a tale of a policeman and two servant girls.

1115.—Peasant and his Ass (YORK, 12 slides; PUMPHREY, 9).

Humorous verses on the miller, the peasant, the ass and the peasant's wife. The ass is tied to the mill, which, set going by the miller, kills it. The peasant saws the mill down in revenge.

1116.—Pilgrim's Progress (PUMPHREY, 36 slides; NEWTON, 48; YORK, 40·

Illustrations of Bunyan's great work.

1117.—Pied Piper of Hamelin (NEWTON, 12 slides; PUMPHREY, 12; YORK, 12).

Robert Browning's verses on the Pied Piper who, deprived of reward for bringing out all the rats, revenges himself by leading all the children into the mountain.

1118.—Portraits from Nature, Humorous (HUGHES, 11 slides).

These are the well-known studies of priest laughing, etc., etc.

1119.—Positive Electricity (PUMPHREY, 3 slides).

The humorous adventures of Professor Dynamo with a lamp-post.

1120.—Peep Behind the Scenes (36 slides).

The Religious Tract Society's well-known story.

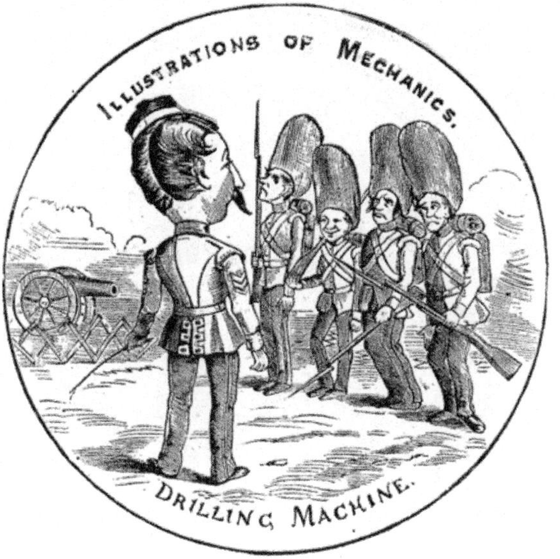

No. 1121.

1121.—Puns (PUMPHREY, 12 slides).

Puns and jokes upon mechanical subjects, such as " A Boring Machine" (lecturer), " Drilling Machine " (drill sergeant).

1122.—Penny Readings, Our (NEWTON, 14 slides).

A very funny account of the different persons taking part in a village "penny readings," with illustrations of their manners, sayings and songs.

1123.—Progress of Mr. Lambkin (24 slides).

Cruikshank's pictures of his adventures in the pursuit of pleasure and amusement, and also in search of health and happiness.

1124.—Pegasus in Harness (PUMPHREY, 8 slides).

The poet brings Pegasus to the fair, where he is bought by a farmer, and goes through various trials.

1125.—Pilgrim and the Peas (PUMPHREY, 3 slides).

Illustrations to Peter Pindar's well-known verses.

1126.—Please Give me a Penny (PUMPHREY, 4 slides).

The little boy begging, "Please give me a penny, sir, for bread."

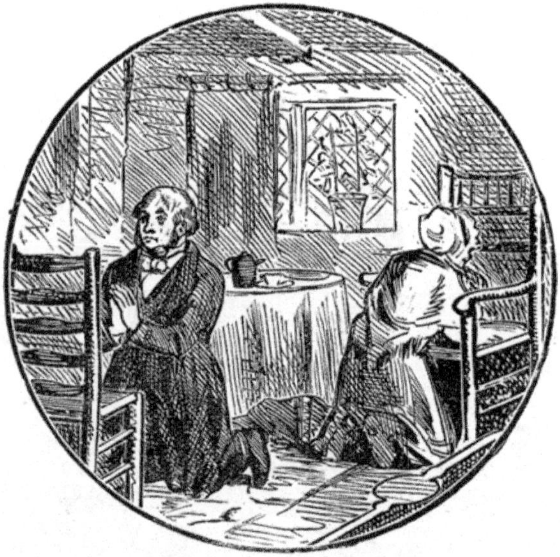

No. 1127.

1127.—Prayer and Potatoes (PUMPHREY, 9 slides).

The widow sends for the Deacon, with the idea that he will provide her with potatoes, but he prays for her spiritual good, whilst she wants the potatoes.

1128.—Priest and Mulberry Tree (PUMPHREY, 3 slides).

The Priest stands on the back of his good steed reflecting on its docility while he gathers the mulberries, but the animal at the sound of his voice moves and alters the situation of affairs.

1129.—Phaeton Junior (PUMPHREY, 6 slides).

The sad misfortunes of a would-be Jehu fresh from school.

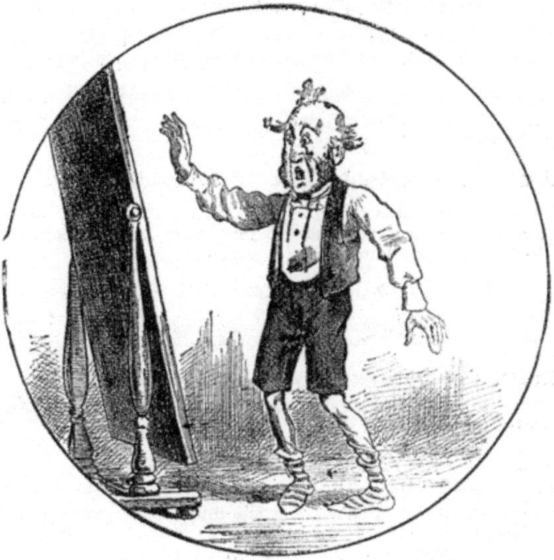

No. 1130.

1130.—Poor Pa's Trousers (Pumphrey, 9 slides).

His trousers are by mistake cut shorter by each member of his family, with astonishing results. A slight hit, too, at the higher education of women.

1131.—Painters, An Evening with (Wood, 137 slides).

Illustrations of Landseer's, Callcott's, Raphael's, Ruben's, Turner's Gainsborough's, Grottgen's, and other painters are here shown.

1132.—Pair of Studies in Black and White(wash) (2 slides).

Sambo is at work on a ladder whitewashing, a negress coming out of the door upsets Sambo, etc.

1133.—Pussy's Road to Ruin (York, 12 slides).

Pussy, after being taught to work, to spin, to catch mice, etc., takes midnight rambles, and follows these by stealing and killing indiscriminately, thereby eventually getting into prison.

1134.—Philosopher's Escape, The (Pumphrey, 3 slides).

An amusing story of the very wise man who could not escape from prison because the door was barred, whilst the window was fully open all the time

1135.—Quarrelsome Dog, The (York, 12 slides).

His biting and snarling, and the reward he received in a new home.

1136.—Quarrelling (Wood, 10 slides).

The result of a quarrel between Tom and Kate, a snowball, which rolls them away.

1137.—Quartette Party (YORK, 9 slides).

The story of Charley Crochet, Simi Breves, little Jerry, and Peter Piper and their quarrel. Verses fairly bristling with puns.

1138.—Quaker and the Robber (PUMPHREY, 4 slides; YORK,

The old story of Jimmy Barlow disguising himself as a lady, but nevertheless bowled out by the Quaker.

1139.—Rake's Progress, The (YORK, 8 slides).

Hogarth here illustrates a youth just in possession of property, then going to excess, drunk, arrested for debt, marrying a rich old maid, gambling, in a debtor's prison, and finally a hopeless lunatic.

1140.—Reward of Cruelty (YORK, 4 slides).

Torturing animals for sport, leading to murder and execution. The reward shown is the cruel man being dissected by the doctors.

1141.—Rumours of an Election (YORK, 4 slides).

Hogarth here illustrates the feast, canvassing, polling, and chairing the member in the "guid old times."

1142.—Retaliation (YORK, 12 slides).

Dan starts the fun by tying a string to Will's knocker, who retaliates with a garden syringe, and merrily the retaliation proceeds till both determine to leave the house.

1143.—Reward of Covetousness (YORK, 8 slides).

A hunting Prince of France often took a meal of radishes and water with a peasant at the latter's humble home in the forest. When the Prince became King the peasant went to him and presented a fine radish, for which he was given a thousand ducats. A courtier, noting how presents were received, presented a horse, for which he received the radish, that, as the King said, cost 1,000 ducats.

1144.—Reynard the Fox (NEWTON, 13 slides; PUMPHREY, 31; YORK, 12)

Reynard the Fox is impeached before the Court of Noble King of Beasts, to which all animals were summoned, but it is discovered that Mr Reynard is a very artful dodger, and a match for all.

1145.—Rival Lovers, The (YORK, 12 slides).

Nina, refusing to listen to the Sultan's addresses, is assisted to escape by Arthur, who supplies the gaoler with rum. A slave tells the Sultan, and both nearly reach the escape ladder when A. and N. pull it away, leaving the Sultan and slave to settle their own affairs.

1146.—Robbers, The (YORK, 12 slides).

The robbers ask for charity from an old gentleman. He treats them sumptuously. At night they break into his house, but the old man catches them neatly, nails them to the wall, belabours them, and sends them home wiser men.

1147.—Road to Heaven (6 slides).

Pictures from life models of a poor little London waif and his death.

1148.—Rip Van Winkle (YORK, 12 slides; HUGHES, 15).

Illustrations to Washington Irving's story of Rip, his adventures in the Katskills, and return to the village 20 years after.

1149.—Richard Doubledick (HUGHES, 20 slides).

Dickens's story from "The Seven Poor Travellers."

1150.—Ralph the Rover, Sir (PUMPHREY, 4 slides).

Southey's poem of the Inchcape Bell.

1151.—Rationalistic Chicken, The (PUMPHREY, 2 slides).

S. J. Stone's verses on the chicken who says, "what I can't see I won't believe in."

1152.—Roger Ploughman's Excursion to London (PUMPHREY, 12 slides).

These are pictures in illustration of the renowned Roger's travels to and about London.

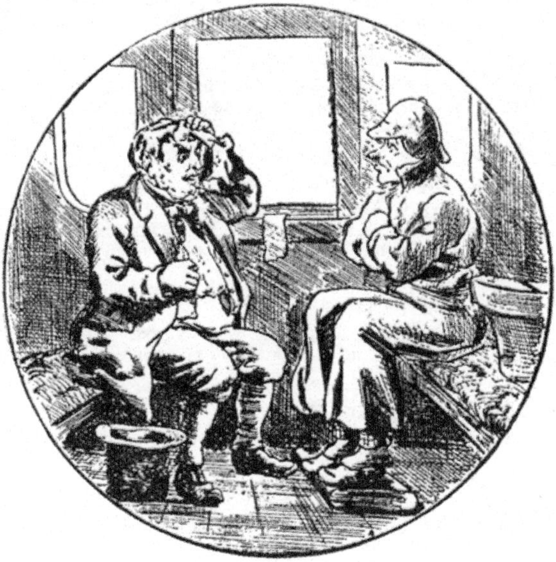

No. 1153.

1153.—Railway Story (PUMPHREY, 8 slides).

Two men quarrel about the window, one wants it open and the other shut, till eventually both tumble out of the door.

1154.—Raw Gorbs (PUMPHREY, 6 slides).

The negatives in this set are taken from life, and illustrate the snaring of birds by boys to prevent the loss of the farmers' corn. One little boy collects the unfledged birds, hence the little "Raw Gorbs."

1155.—Scrub, the Workhouse Boy (PUMPHREY, 11 slides; YORK, 11).

Alfred Pamperfield becomes apprenticed, goes to the theatre, and enjoys his glass, and nearly gets burned to death. Scrub, in the kitchen, declines beer, resists temptations, is promoted to serve behind the counter, and finally becomes a Sunday-school teacher.

1156.—The Seasons (UNDERHILL, 4 slides; YORK, 4).

Pictures of spring, summer, autumn, and winter, represented by landscapes, figures, and comic studies.

1157.—Settler and the Savages (YORK, 2 slides).

Showing a ludicrous effect of Indians trying to rob a Settler's Shanty in the Backwoods.

1158.—Seven Ages of Man (PUMPHREY, 8 slides; YORK; HUGHES).

Shakespeare's soliloquy commencing "All the world's a stage."

1159.—Signal Box (YORK, 6 slides).

By G. R. Sims. The Stationmaster's story of doing his duty when in the signal box at the expense, he believes, of his own child's life, but, as it turns out, really saving both the child and its mother.

1160.—Simon and his Pig (YORK, 12 slides).

Simon buys a pig and takes a drop to drink. The pig wakes to a savoury smell in the inn, and rushing inside meets many obstacles, but overturns them all. Rushing out he carries Simon into a pond. Finally, the butcher finishes him, Simon exclaiming, "You're done now, my boy!"

1161.—Sledge Party (YORK, 12 slides).

Master Fritz taking a ride downhill in the sledge, trips up a schoolmaster, puts a hunter and his dog into the schoolmaster's lap, compels an old market woman to join the party, and so on, till they all go over a precipice—on to soft snow.

1162.—Snow White (YORK, 7 slides).

A fairy tale of a maiden who was obnoxious to a Queen, the latter's endeavours to kill her, and the usual ending, married to a prince, and bad end of the Queen.

1163.—Soldier's Dream (YORK, 8 slides; PUMPHREY, 3).

Illustrates Campbell's verses commencing "Our bugle sang truce."

1164.—Songs of the Brave (NEWTON, 24 slides).

Soldier's Dream, Battle of the Baltic, Burial of Sir John Moore, Ye Mariners of England, Waterloo, Charge of the Light Brigade, &c.

1165.—Scott, Sir Walter (NEWTON, 29 slides).

A portrait; views of the exterior and interior of Abbotsford (13), Melrose (4), Dryburgh (3), and various monuments to Scott are here shown.

1166.—Shakespeare, William (NEWTON, 11 slides).

Views of Shakespear's House, Stratford Church, Warwick Castle, etc., are here shown.

1167.—Southey (NEWTON, 12 slides).

A portrait, views of Keswick and Bristol, and illustrations of the Death of Nelson, Nelson on the "Victory," Battle of Trafalgar, Nelson's Last Signal, etc.

1168.—Soldier's Trumpet, Notes on (NEWTON, 26 slides).

An army ABC, by Surgeon-Major Scanlan, commencing "A was an Adjutant, Ride a Cock Horse," "B was a Brigadier, Crusty and Cross."

1169.—Shrove Tuesday in the Backwoods (STEWARD, 4 slides).

Pompey pulls the tablecloth and gets totally eclipsed (in the batter), so that his own father wouldn't know him.

1170.—Sleepy Hollow (6 slides).

Ichabod Crane and his scholars ; his marvellous stories.

1171.—Soldier's Return (4 slides).

The return home of a son to his father after the war, and recognition by the dog Tray.

1172.—Stolen Sausage (YORK, 9 slides).

Fritz and his relatives buy a sausage, which, while being brought home, is seized by F.'s little dog. F. in pursuing falls down the family cellar's trap door on to the cook; he is followed by pa, and the cook is crushed positively flat, by the sister who presses her comparatively flatter, and by the uncle who squeezes her superlatively flattest.

1173.—Story of a Toadstool (NEWTON, 9 slides).

The story of two little elves who decide not to work, but to dance on a toadstool, with their exciting adventures.

1174.—Suspicious Travellers (YORK, 12 slides).

Two travellers in a train are mutually nervous of each other being a robber, but end up good friends. Story in verse.

1175.—Scaramouche and the Mouse (HUGHES, 3 slides).

The spoilt supper, revenge, the trap, caught, the sentence and doom.

1176.—Sailor's Apology for Bow Legs (PUMPHREY, 6 slides).

Tom Hood's humorous verses describing the sailor's ride on a brown mare and how his legs got bowed.

1177.—Scaramouches (PUMPHREY, 12 slides).

The amusing misadventures of four children.

1178.—Story Without an End (PUMPHREY, 8 slides).

A turbulent monarch offers his daughter's hand to the man who invents a story without an end. This is accomplished by one who tells about locusts pilfering grain.

1179.—Sambo's Five Senses (5 slides).

Comic pictures of a nigger's five senses.

1180.—Saved from the Sea (10 slides).

From life models. A child saved from the sea is cared for by a good wife instead of being sent to the union house, with a happy ending.

1181.—Stubbs's Seaside Adventures (30 slides).

Adolphus takes his wife and twins to the seaside, where they (and he in particular) pass through many adventures, such as meeting a Count at billiards, losing "Julius Cæsar," etc.

1182.—Sally in our Alley (PUMPHREY, 8 slides).

Sally's sweetheart here tells us in verse that Sally is the darling of his heart, how he loves to walk with her on Sunday dressed in his best, and looks forward to the happy day that will see them united.

1183.—Seaside Pictures (PUMPHREY, 12 slides).

Illustrates sand scenes at the seaside, riding, racing, the donkey's home, the coming storm, etc.

1184.—Sweep and the Whitewasher (10 slides).

Tormenting each other, they change colour entirely by the aid of whitewash and soot. Reading on each slide.

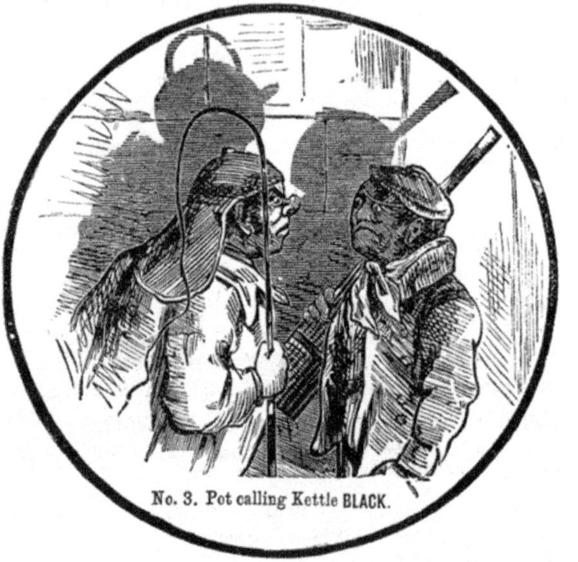

No. 1185.

1185.—Shadow Pictures (PUMPHREY, 18 slides).

Sketches by C. H. Bennett in which the figures are represented as casting appropriate shadows.

1186.—Shadows on the Blind (PUMPHREY, 3 slides).

Pictures that require no explanation, the meaning being self-evident.

1187.—Sir Hotspur Cock (PUMPHREY, 2 slides).

How he got henpecked for not taking his wife's advice.

1188.—Story of an Apple (PUMPHREY, 3 slides).

Two boys quarrel about the division of an apple, with the result that a cow gets the whole of it.

1189.—Story of Eyam (CUBLEY, 24 slides).

Views of Eyam, a village in Derbyshire, showing places connected with the unselfish and heroic conduct of the villagers, led by two noble ministers, during the great plague of London. A very interesting set.

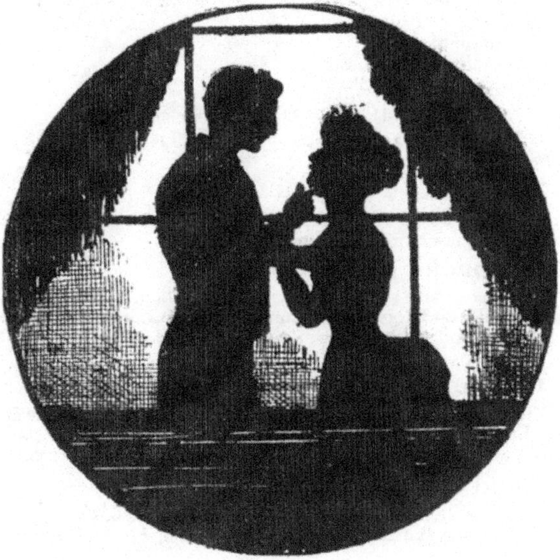

No. 1186.

1190.—Trap to Catch a Sunbeam (York, 15 slides).

A story to illustrate the blessings of peace, contentment, and self-help.

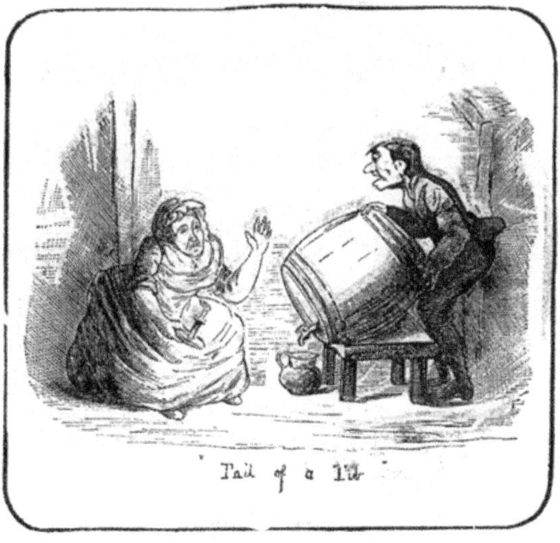

No. 1191A.

1191.—Tale of a Tub (YORK, 7 and 12 slides; NEWTON, 8).

"The tale of a picnic,
The tale of a tub,
The tale of two gentlemen
Eating their grub."

1191a.—Tale of a Tub (PUMPHREY, 6 slides).

A distinctly different tale, to which the illustration on the preceding page will be a fair guide.

1192.—Tipsy Geese (YORK, 6 slides).

How they became tipsy by devouring cherries used in making cherry brandy, and their comical actions.

1193.—Toothache (YORK, 12 slides).

Grimshaw can't eat his dinner for toothache, and after trying various remedies calls on Professor Twister, for whom G. hesitates to open his mouth, but at last is induced to do so, is soon rendered comfortable, and at last eats up his dinner.

1194.—Two Housebreakers (YORK, 12 slides).

These men break into a pawnbroker's, the pawnbroker awakes and seizes pistol, but is gagged, tied, and hung up out of the way. Finally policeman appears, and an artful use of umbrellas proves the robber's ruin.

1195.—Timid Man (HUGHES, 4 slides).

A timid man hopes the bedroom is not haunted, but wakes up with a fright to see two glaring eyes, the fright ending when puss slinks under the bed

1196.—Three Rogues of Bagdad (PUMPHREY, 4 slides).

Fun is made out of the way a countryman is treated by three thieves.

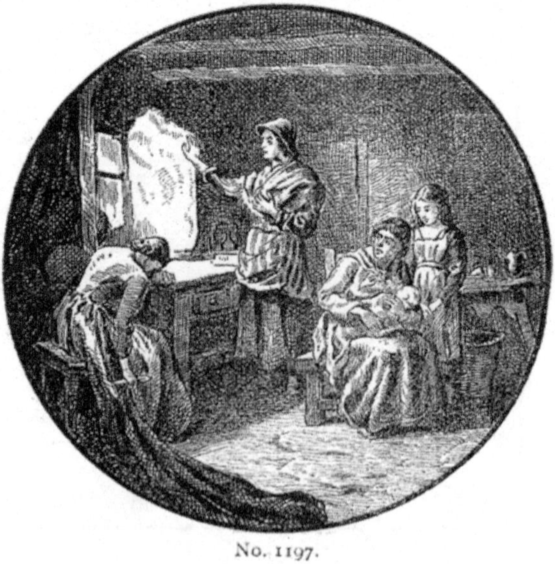

No. 1197.

1197.—Three Fishers went Sailing (PUMPHREY, 3 slides).

Slides to illustrate the Rev. C. Kingsley's beautiful verses, one to each verse.

1198.—Tommy Wilful (PUMPHREY, 6 slides).

What happened to him in the train.

1199.—Tommy Tait (PUMPHREY, 9 slides).

Illustrating how Tommy dawdles his time away when sent out on errands, playing marbles, bird-nesting, etc.

1200.—Two Crossing Sweepers (PUMPHREY, 4 slides).

From life models. A story of two friendless boys, with the moral of "Do unto others as you would they should do to you."

1201.—Tennyson (NEWTON, 5 slides).

A portrait, Tennyson in his study, and illustrations of the "Charge of the Light Brigade."

1202.—There's Help at Hand (28 slides).

Another of Mrs. Sewell's instructive stories of the poor.

1203.—Troublesome Security (12 slides).

An editor demands security for the insertion of a circus advertisement, and takes a bear in charge, with most awful results to the editor personally, and the *Fieldshire News* office.

1204.—Troubles of an Artist (6 slides).

How his work is spoilt by animal critics.

1205.—Unskilful Ratcatchers (YORK, 12 slides).

The clumsy attempt and misadventures of three dogs rat-catching. At last the rat thinks, "just in time;" the dogs, "just too late," and the latter thereby lose the day.

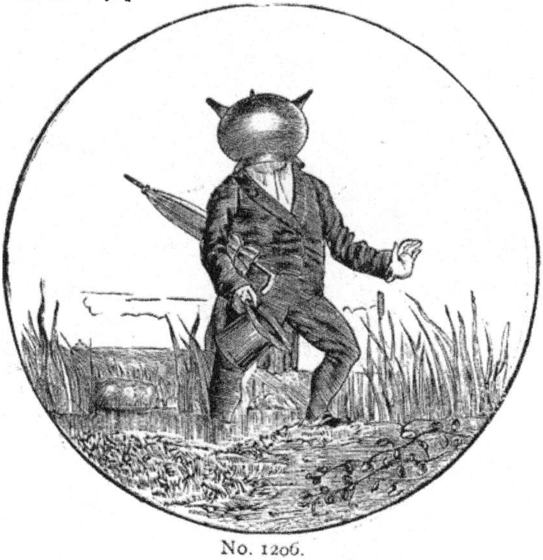

No. 1206.

1206.—Unlucky Present (PUMPHREY, 6 slides).

The story of a covetous parson who obtained an iron pot from a widow. Carrying it home on his head, it slipped down over his nose, and he has fearful adventures until relieved by a blacksmith.

1207.—Uncle Tom's Cabin (PUMPHREY, 20 slides; HUGHES, 30).

Uncle Tom's Cabin. Uncle Tom and Eva.

1208.—Ugly Duckling (PUMPHREY, 8 slides).

A story with the moral that "It matters not being born in a duck-yard when one is hatched from a swan's egg."

1209.—Undine (11 slides).

The German legend of the beautiful knight who marries Undine, her disappearance beneath the Danube, and death of the knight.

1210.—Valentine and Orson (HUGHES, 16 slides).

Ye ancient legend of the Wild Man of the Woods.

1211.—Vagabonds, The (YORK, 10 slides; NEWTON, 7).

An American poem of the Beggar and his Dog Roger.

1212.—Vulgar Little Boy, The (YORK, 6 slides; PUMPHREY, 6).

The amusing Ingoldsby Legend of the charitable but simple man and his misadventures at Margate.

1213.—Village Blacksmith (UNDERHILL, 10 slides).

This is a very fine set to illustrate the well-known poem of Longfellow. After an introductory slide showing a portrait of Longfellow and a view of his birthplace, we see that "Under a spreading chesnut tree, the village smithy stands," the blacksmith at work, doorway of the smithy, two views of the interior of a village church, with two effect slides, finishing up with the interior of blacksmith's cottage, also with an effect slide.

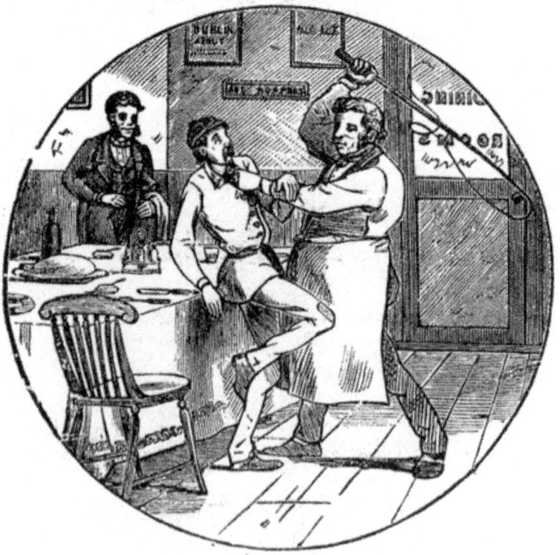

No. 1214.

1214.—Vat You Please (PUMPHREY, 6 slides).

Describing the way one Frenchman gets a dinner, and another a hiding, by saying "Vat you please."

1215.—Well of St. Keyne (YORK, 3 slides; PUMPHREY, 4).

Illustrations for these well-known verses.

1216.—Witches' Frolic (YORK, 12 slides).

From the Ingoldsby Legends. Grandpapa's tale to little Ned of how the witches in the old gray ruin treated Robin Gilpin.

1217.—Weather Prospects (YORK, 6 slides).

Mr. Punch here illustrates "Rather fresh, cloudy and cold, but threatening," "Glass falling; stormy, with much rain," and other prospects.

1218.—Women of Mumbles Head (NEWTON, 5 slides).

Illustrating the poem by Clement Scott. We see portraits of the women, Mumbles Head, women to the rescue, etc.

1219.—Wedding Bells (YORK, 10 slides).

A powerfully dramatic poem, by Charlotte M. Griffiths, of a wedding and the bridegroom's secret, almost revealed by a dead form on a mother's grave.

1220.—Webb's (Captain) Wonderful Feat (HUGHES, 6 slides).

Shows Captain Webb swimming across the English Channel, swimming the Rapids, his death, etc.

1221.—Walrus and the Carpenter (PUMPHREY, 3 slides).

Describing how they wheedled the oysters, and ate them up.

1222.—Well-read Hunter (PUMPHREY, 4 slides).

Story in verse of a little boy and a lion.

1223.—We are Seven (PUMPHREY, 6 slides).

These are photographs from life taken to illustrate Wordsworth's poem.

1224.—Wordsworth (NEWTON, 7 slides).

Here we have a portrait, and views of Rydal Mount, Lake, Park, etc.

1225.—Warranted all Bristles (WOOD, 12 slides).

The history of Tom and his broom.

1226.—You are old, Father William (PUMPHREY, 4 slides).

A parody. Copied from designs by John Tenniel in "Alice in Wonderland."

1227.—You are Littler than I (PUMPHREY, 3 slides).

Verses. A discourse on the cruelty of fishing, the boy telling the fishes, "You'se littler dan me." A crocodile catches the boy, saying the same thing.

1228.—Young Ragamuffins (NEWTON, 24 slides).

Shows a boy careless of his clothes, the boy that robbed an eagle's nest, the boy who was so frightened at soap and water that he got sold for a nigger, and other equally naughty boys.

1229.—Young Ellerby (12 slides).

A story to illustrate that "good deeds are never without their reward."

1230.—Young Crossing Sweepers (WOOD, 27 slides).

The lives and struggles of two boys after the death of their mother.

1231.—Zoological Alphabet (PUMPHREY, 24 slides).

Illustrating the Alphabet by animal pictures, the reading in rhyme commencing, "A was an amiable ape, Who lived on an African cape, He climbed up the trees, On his elbows and knees, And came down by the fire escape."

Temperance.

1232.—An Old Story (YORK, 26 slides; NEWTON, 26).

A set of pictures illustrating the evil effects of drink, contrasts between the sober and drunken man, happy and wretched homes, &c.

1233.—Alcohol (PUMPHREY, 8 slides).

A Band of Hope address, taken from the "Band of Hope Chronicle."

1234.—Boons and Blessings (YORK, 16 slides).

Showing the advantages of temperance, with such slides as the drunkard's Bible, the worn thimble, building a house with a teacup, digging a grave with a wine glass, &c.

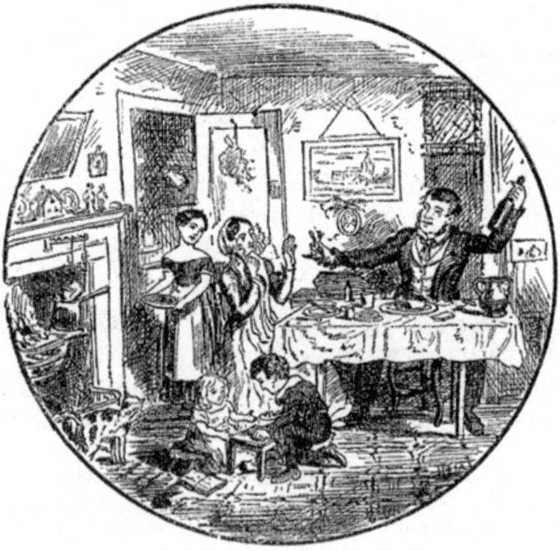

No. 1235.

1235.—Bottle (YORK, 8 slides; PUMPHREY, 8).

(Cruikshank.) The evil work wrought by the bottle to all the family.

1236.—Buy your own Goose (Wood, 6 slides; York, 6).

A story of a workman who, instead of joining a goose club, as usual, at the public-house, determines to give up drink and buy his own goose. The result is happiness all round and a better life.

1237.—Buy your own Cherries (Pumphrey, 10 slides, York, 7 and 10).

Illustrations to this popular temperance story.

1238.—Beast and the Brute (Hughes, 3 slides).

The beast who squanders all in drink, the brute and wife-beater, and the contrast—the happy home of temperance.

1239.—Dan Dabberton's Dream (York, 18 slides).

The drunkard's dream of what was and what might have been, with the ending; a resolve to grasp the sunshine seen in the dream.

1240.—Dear Father Come Home (Hughes, 7 slides; Pumphrey, 3).

Illustrations to the song beginning "Father, dear father, come home with me now," one of the most pathetic ballads ever written.

1241.—Drunk as a Brute (Pumphrey, 14 slides).

A fairy tale. Giles Fallowground, when extremely drunk, is caught by a demon, and, of course, ever after remains sober.

1242.—Dream of the Reveller (Pumphrey, 12 slides).

Chas. Mackay. "Body and soul are mine," said he, "I'll have them both for liquor."

1243.—Drink Facts (Pumphrey, 11 slides).

Startling figures, facts, and statements in favour of temperance and against drink.

1244.—Drunkard's Children (Pumphrey, 7 slides; York, 8).

Neglected by their parents, they are led to the gin shop, and from thence to ruin and death.

1245.—Drunkard's Progress (Pumphrey, 13 slides; York, 14).

Thirsty, muddy, communicative, oblivious, and so on to dead drunk an death.

1246.—Drunkard's Dream (Hughes, 8 slides).

A man is represented asleep in his chair at home, who sees in visions through the cottage window various terrifying scenes of a drunkard's life.

1247.—Factory Chimney, The (Pumphrey, 4 slides).

Or the Little Badge of Blue. A story by Joseph Malins, R.W.G.T., of a man left on the top of a tall chimney, who escapes death by unravelling his stocking, to which a cord is affixed and then rope, with analogy to gospel temperance.

1248.—Gin Fiend (York, 4 slides; Newton, 4).

The gin fiend's triumph. "He hath slain his wife, he hath given his life, and all for the love of me."

1249.—Gin Shop (Pumphrey, 12 slides; York, 12).

Cruikshank. Told after the fashion of "The house that Jack built."

1250.—Give me a Penny (PUMPHREY, 4 slides).

"Please give me a penny, sir, a penny, please, for bread."

1251.—Home of John Hampden (YORK, 6 slides; WOOD, 6; NEWTON, 6).

The elevating influence of the Bible, the only help for the drunkard.

1252.—Human Stomach (PUMPHREY, 7 slides).

Pictures of the stomach in health and under alcoholic excitement and disease.

1253.—I have Drunk My Last Glass (4 slides).

The story of a reformed drunkard, who explains why he will drink no more.

1254.—Jack and his Hard Lump (PUMPHREY, 2 slides).

Jack is recommended to a drink by the landlord as a "cure for a hard ump," which he refuses, and shows the l imp to be a bag of gold.

1255.—John Tregenoweth: His Mark (YORK, 18 slides; NEWTON, 18).

The drunkard who, in a maddened state, strikes his faithful attendant daughter, but afterwards is brought to his senses by the scar on her forehead, and treats it when signing the pledge as "His Mark."

1256.—Little Shoes (6 slides).

Verses by Mrs. Sewell; the slides are from life models.

1257.—Legend of St. Swithin (NEWTON, 12 slides).

This is an orginal set with manuscript reading. "A Rhyme for Rainy Weather."

1258.—Little Tiny (YORK, 14 slides; NEWTON, 14).

Story by the Rev. David Macrae, and slides from life models. The child who saved a man from drinking to death, and was the indirect means of foiling two villains who were plotting mischief

1259.—Mat Stubbs's Dream (13 slides).

A dream of rum-drinkers. "There was every class, from the king to squalid rags."

1260.—Nelly's Dark Days (PUMPHREY, 6 slides).

A drunken father, who even pawns his child's doll for drink, saved by signing the pledge.

1261.—Poke your own Fire (WOOD, 12 slides).

Inside the public-house, signing the pledge, new suit of clothes, poking his own fire.

1262.—Return from the Tavern (YORK, 4 slides).

Illustrating the Departure, the Misconception, the Remonstrance, and the Return to the Tavern.

1263.—Sir John the Giant Slayer (YORK, 24 slides)

An allegorical temperance tale, by Mrs. Noel-Thatcher, somewhat after the style of Pilgrim's Progress. Sir John Christain Abstainer, Barleycorn, Old Tom, Sham Pain, Mr. Moderation, Richard Drinkwine, and other characters.

1264.—Sultan of Ragabaga (YORK, 15 slides).

His travels in Grogolfland. An allegorical legend.

1265.—Sam Bowen's Dream (12 slides).

When driving, his horse bolts and upsets the trap. Whilst in bed a dream effects a reformation.

1266.—Temperance Mottoes (YORK, 17 slides).

The mottoes are such as "Be Temperate in all things," "Taste not—touch not the accursed thing"

1267.—Trial of Sir Jasper (PUMPHREY, 25 slides, YORK, 25).

A renowned temperance story or recitation.

1268.—Temperance Sketch Book (YORK, 32 slides).

The idea is a sort of album of pictures dealing with the temperance question, with examples from the Bible, the Brewery, Bay of Naples, etc.

1269.—Temperance History of Two Lives (PUMPHREY, 16 slides).

One subjected to the influence of the drinking customs of the time, and another guarded from those customs by total abstinence from alcoholic drinks.

1270.—Temperance Pictures (PUMPHREY, 10 slides).

Various pictures.

1271.—Ten Nights in a Bar-room (HUGHES, 14 slides).

A well-known and very popular story, contrasting the lives of Joe Morgan, the teetotaller, and his drunken friends.

1272.—Whiskey Demon; or, the Dream of the Reveller (YORK, 12 slides; NEWTON, 12).

Illustrating the poem, by Charles Mackay, commencing, "Body and soul are mine," said he, "I'll have them both for liquor."

1273.—Worship of Bacchus (YORK, 14 slides; NEWTON, 14).

Shows only 1d worth of nutriment in a gallon of ale, what ale is composed of, two ways of spending 2s., bread or beer, analysis of a bottle of fermented and unfermented wine, etc

1274.—Wine *versus* **Water** (HUGHES, 17 slides).

A story graphically illustrating the rise and fall in life of the children of temperate and intemperate parents, the story of two lives.

1275.—Which Side Wins? (WOOD, 25 slides).

Arguments and pictures for temperance, as told by Jim Fergus.

1276.—Whiskey War (WOOD, 13 slides).

The sentinel, intemperance and misery, Bacchus drowns more than Neptune, modern Juggernaut, blue devil, etc.

Services of Song.

1277.—The following sets are obtainable, with hymns, &c., and so arranged as to be shown during the progress of the Service of Song:—

Babylonish Captivity ..	WOOD.
Christie's Old Organ ..	YORK.
Christiana	,,
Dan Dabberton's Dream	,,
David	WOOD.
Daniel	,,
Eva	YORK.
Elijah	WOOD.
Holy War	,,
Home, Sweet Home ..	,,
Israel in Wilderness ..	,,
Jessica's First Prayer ..	YORK.
John Tregenoweth.. ..	,,
Joseph	,,
John Knox	WOOD.
Little Tiz	YORK.
Luther..	WOOD.
Messiah (children's) ..	YORK.
Monk that Shook the World	WOOD.
Prodigal Son	YORK.
Pilgrim's Progress ..	,,
Paul	,,
Sayings of Jesus ..	,,
Samuel	WOOD
Solomon	,,
Voyage of Life	,,
Which Side Wins? ..	YORK.

Entertainments, Pantomimes,

ETC., ETC.

1278.—Aladdin the Scamp (W. C. HUGHES, Kingsland).

The Wonderful Lamp. The geniis two, and the magician too, and the Emperor's daughter whom Aladdin did woo. An extravaganza on the Arabian Night's story.

1. *The Magician's Cave*—with effects of sudden appearance of the Genii of the Ring, who shows the boy Aladdin to the Magi.

2. *Circular Street, Pekin.*—The Magi finds young Aladdin playing marbles with his idle companions

3. *Pretty Rustic Scene in the Suburbs of Pekin* —The Magi coaxes Aladdin to take a ramble in the country.

4. *A Dark Gully between the Blue Mountains.*—A weird scene, with effects of flying bats; the learned owl; the Magi and Aladdin light a fire over the magic stone; the stone appears, and Aladdin descends into the bowels of the earth to seek the lamp.

5. *The First Cave (or entrance) below the surface of the Earth* —The Magi directs Aladdin where to find the lamp, effects—Magi disappears and Aladdin gathers some of the jewel flowers and fruit.

6. *Grand Panoramic effects*—showing Aladdin's passage through the subterranean caverns leading to the garden of jewels, effects—Aladdin gains the wonderful lamp, the garden of beautiful jewels.

7. *The Exit from the Caves*, with effects —The Magi demands the lamp from Aladdin, who refuses; whereupon the Magi closes up the cavern and leaves Aladdin to perish. But the Genii of the Ring suddenly appears, and, at Aladdin's command, carries him home out of further danger.

8. *Interior of Aladdin's Home* (pretty scene), with effects.—Aladdin returns and embraces his mother, and shows her the lamp, which, being dirty, she cleans, when the Genii of the Lamp suddenly appears before them, soliciting their commands, whereon they ask for a grand banquet, and it appears; also for some new clothes, whereon they are magnificently dressed.

9 *The Hebrew Jeweller's Shop.*—Aladdin sells some of the plate; but is cheated by the crafty Jew.

10. *Grand Entrance Hall to the Royal Baths*—beautiful panoramic effect.—Aladdin watches the Princess and her attendants entering the baths. Panorama.

11. *Audience Chamber in the Emperor's Palace, Pekin*—Aladdin and his mother bring a rich present of jewels to the Emperor, and a wedding present to the Royal Princess.

12. *Courtyard of the Royal Palace.*—Gorgeous panoramic effects of the wedding procession leaving the palace on its way to the temple.

13. *Vestibule of Aladdin's Palace*, effects—The Magi, disguised as a dealer in lamps, offers a new lamp for an old one, and the Princess unwittingly exchanges the wonderful lamp for a new one; whereon the Magi, getting it into his hands, summonses the Genii of the Lamp—who appears—and commands the Genii to carry the palace, himself and the Princess off to Egypt.

14. *Exterior of the Palace, with Terrace, Gardens, &c.*—A splendid scene, the Magi informs the Princess of his power.

15. *Grand Panorama, from the Great City and Wall of China—across the sea—to the Shores of Egypt.*—With distant view of the Sphinx, Pyramids, and desert, and effects of the Genii of the Lamp carrying off the Palace through space; and Aladdin, carried by the Genii of the Ring, giving chase (most exciting effects).

16. *Aladdin's Palace (now) in Egypt* (a beautiful tropical scene).—The palace and Princess in the possession of the Magi

17. *The Princess's Boudoir in Aladdin's Palace* (a delightful scene).—The Princess, by Aladdin's directions, invites the Magi to a special respast, and having drugged his wine, Aladdin soon disposes of his crafty and wicked enemy. And they then command the Genii to take them safely home again.

18 *The Return of Aladdin's Palace to China*—and happy meeting of the Princess and her father, and Aladdin and his mother; on which special event Aladdin commands the Genii to illuminate the palace and grounds with a hundred thousand lamps, lanterns, and other grand devices—a most gorgeous effect is produced here of the grand illuminations, &c, which terminates this, one of the most elaborately illustrated versions of the "Arabian Night's" stories.

1279.—Aladdin (J. H. STEWARD).

With a special libretto written by Ellis Reynolds.

1. Introductory slide of Genii holding Lamp with Fairy Palace in the distance, and appropriate group. 1A. Effect of rays radiating from Lamp. 1B. Effect of title, Aladdin (the wonderful), or the Scamp, the Lamp, and the Tramp

2. Aladdin meets Magician.

3. Incantation scene outside Cave. 3A. Effect to fire burning.

4. In the Cave—two effects for Aladdin to move along pathways.

5. The Shrine of the Lamp.

6. Aladdin returns to the mouth of the Cave. 6A. The Cave closed (flashed on effect).

7. Choreutoscope—effect of Genii of the Ring dancing.

8 Aladdin's Home. 8A. Effect for appearance of Genii of the Lamp.

9. Princess going to the bath.

10. Aladdin's mother with jewels.

11. Vizier's son out in the cold at night

12. The Sultan visits his daughter.

13. Aladdin takes more jewels to court

14 The Fairy Palace of Aladdin seen from the Sultan's window (effect).

15. The Magician selling new lamps for old.

16. Effect of disappearance of Aladdin's palace.

17. Aladdin contemplates suicide.

18. Palace of Aladdin in Africa by moonlight.

19 Aladdin's triumphal return, grand moving panorama of Aladdin and soldiers on horseback. Very effective.

1280.—Cinderella and the Fairy Glass Slipper (J. H. STEWARD).

The story of Cinderella and Iko Joko, the Comical Black Page, the two Ugly Sisters, and the Fairy Godmother.

1. Introduction—Curtain with Fairy Godmother revealing the special emblems and title

2. Forging of the fairy glass slipper in the Stalactitic Cave.

3. Street in Dontnowhere—town crier, escorted by herald and halberdier, gives invitation "to all, both great and small, to the Prince's ball."

4. Special reception by the Baron and the Ugly Sisters of the formal invitation.

5. The sisters dressing for the ball—great exertions of Iko Joko the Black Page in assisting in the Ugly Sisters' toilet.

6. The departure of the Ugly Sisters for the ball.

7. The kitchen—Cinderella at the fireplace. 7A. Effect to blowing bellows

8. The kitchen—Cinderella asleep in the chair, and the appearance of the Fairy Godmother. 8B. Effect panoramic of figures representing 22 fairy tales (which pass in the smoke of chimney), conjured up by the Fairy Godmother for Cinderella's pleasurable dream 8C. Effect of Cinderella in grand coach drawn by six horses.

9 The kitchen—pumpkin, rats, lizard, etc.

10. The kitchen—Cinderella dressed ready for the ball.

11. The fancy dress ball—arrival of Cinderella.

12. The conservatory—a most beautiful scene.

13 The vestibule—midnight—all changed to original forms.

14. The Prince holds a council—all the ministers are present to give advice as to how to find the owner of the slipper. (Clever character sketches of representatives of war, navy, school board, art, law, church, medicine, and agriculture.)

15 Cinderella asleep in her underground bedchamber—Fairy Godmother sppears and shows her in a dream the coach and horses (repeat effect 8C), and also 15A. Effect, marriage of Prince and Princess.

16 Interior of baronial hall—"Trying it on."

17. Cinderella is successful, and produces the slipper to match.

18. The kitchen in the Prince's castle—great preparations for the marriage feast. 18A. Effect of Iko Joko dancing. Choreutoscopic design of six figures, alternating and interchanging so as to produce life-like dancing in unlimited variety.

19. The wedding feast.

20 The fairy grotto of the Prince's home 20A. Effects for appearance of the fairies in the shells and bowers. The Godmother dressed as Queen of the Fairies.

1281.—Cinder Ella (W. C. HUGHES, Kingsland).

The pantomine of the Slippery Slipper that wouldn't go on, and the two Wicked Sisters, the Pretty Prince, and the Fairy Godmother. The first slide shows the Elgin Glass Mine, and the manufacture of the slippers. Thence to No. 16 the old tale proceeds, and at No. 17 we have a grand transformation scene, "Flora's Palace of Golden Palms," the remaining four slides being devoted to Mr. Joey Clown and his operations.

1282.—The Corsican Brothers (J. H. STEWARD).

An old story retold in a new relation, with a specially-written humorous libretto.

1 Curtain. 1A. Title for curtain.

2. Pine forest in Corsica—in winter.
3. Ball room, Paris.
4. Boudoir in Fabian's Villa, Corsica—by day.
5. Ditto, ditto, ditto by night. 5A. Ghost effect for ditto. (Mechanical slide for ghost to move and arm to raise.) 5B. Vision of duel.
6. Forest of Fontainebleau. 6A. Choreutoscope effect of duel. Showing:—Duel with swords—the broken sword—duel with broken swords used as daggers—and death of the Frenchman.
7. Boudoir in Fabian's Villa, Corsica, showing friends and villagers offering congratulations.

1283.—The Enchanter Foiled (J. H. Steward).

How the fairy helped the Princess that was kind to her when in the form of a bird. Libretto by Ellis Reynolds. List of views and effects —

1. Exterior of the King's palace
2. Interior—King weeping at the loss of his wife.
3. Apartment in palace—Princess at window—a prisoner. 3A. Effect to bird fluttering at window
4. The same apartment—the bird turns into a fairy. 4A Effect for ditto. 4B Effect for vision of Prince.
5. Princess's apartment, with the King's second wife. 5A. The Enchanter or Magician appears and claims his reward for helping her to marry the King 5B. Effect of Page appearing to announce the arrival of the Prince.
6 Audience chamber—Queen trying to pass herself off as the Princess.
7 Princess having escaped from her room enters the audience chamber, and finds the Prince she has seen in her vision
8 Prince claiming Princess as his Queen, and Magician summoned by the Queen to her assistance.
9 Magician takes off Princess in the dragon chariot
10. View of forest—Prince asleep—vision of Princess.
11. Same with fairy—Prince aroused.
12. The voyage—panoramic slide (9 inches long) of sea shore—at sea and arrival at Eastern City—the Prince on dolphin.
13 The Castle—Magician intimidating Princess.
14 View of ramparts and clock tower—arrival of the Prince to the rescue.
15. Prince consoling Princess. 15A Effect of fairy.
16 Clock illuminated. The hour approaching when the Enchanter shall lose his power—Prince puts back clock to foil the Enchanter.
17. Magician returning at expiration of half-hour's grace.
18 Delays are dangerous—the Magician is about to kill the Princess when the clock strikes twelve, and he becomes almost helpless, but by a superhuman effort he manages to discharge an infernal machine. 8A. Effect to explosion 18B. Fortunately, the damage is only local, and no one suffers but the Enchanter, who is blown to atoms, the Prince and Princess being left free to marry. Good fairy appears to give blessing and wish joy and happiness to the young couple

As some novel and pretty effects are introduced in this, and the designs generally are of a high-class character, it will be found an attractive and interesting subject for both adult and juvenile audiences

1284.—The Fairy Wand (J. H. Steward).

A set of photographic slides, from special designs, to illustrate G. R. Sims's beautiful and pathetic poem of "The Fairy Wand."

1. Title picture. 1A. Effect—lace curtain (to roll up by curtain shutter). 1B. Effect—title

2. "Horrible dens, sir."
3. "Knock at the door!" "Pooh, nonsense!"
4. "That's lucky, the place is empty."
5 "The father gets drunk."
6. "The land of shadows," in frame; and effect—moving eyes.
6A Effect—Rack slide, smoke from torch.
7. The land of shining day.
8. " night. 8A. Effect—rising moon and ripple on water. 8B. Effect—panorama for sky. 8C. Effect—panorama for water. 8D. Fairy queen rises out of the water, in basket of roses, and drops of flowers.
9. She knelt by the wretched pallet.
10. With her shawl wrapped over something.
11. Then raising her wand, she waved it. 11A. Effect of raising wand. Lever.
12. While a man from the playhouse watched her. 12A. Effect—angel and mother.
13. He told how he watched her waving.
14. The manager takes charge of Sally. 14A. Effect—angel and mother.

1285.—Macbeth (W. C. HUGHES, Kingsland).

A set of 18 slides with effects, a burlesque on the tragedy, with music and words to make an entertainment. Very amusing.

1286.—The Miller and his Pretty Daughter (J. H. STEWARD).

How the straw was spun into gold by the aid of Rumple Stiltskins, a new fairy tale from original designs, with libretto.

1. The miller and his daughter in their cottage.
2. The castle—the miller and his daughter seeking an interview with the king.
3. Introduction to the king.
4. The king commands the miller's daughter to spin straw into gold.
5. The miller's daughter in prison, disconsolate.
6. Effect of dwarf appearing.
7. The dwarf showing the maiden how to spin straw into gold (effect to moving wheel, etc.)
8. The surprise of the king at the result.
9. The dwarf demands his bond from the miller's daughter (now the queen).
10. Interior of dwarf's hut—effect (choreutoscope) of dwarf dancing.
11. The queen, by guessing the dwarf's name, frees herself from the bond contracted at time of spinning straw into gold.
12. The dwarf leaving the court discomfited.

1287.—Ramsinilus (W. C. HUGHES, Kingsland).

A refined, laughable, irresistible, and facetious story, supposed to be told by a mummy 2,000 years old.

1. The Story-teller—oriental garden, tourists' sojourn
2. The Treasury with title.
3. The Legacy, death of the Architect.
4. Landscape and funeral procession, viz. (equal to six slides). 1—Priest. 2.—Slave with emblems of deceased's profession. 3.—Band of singers. 4.—Slaves with a god, Osiris. 5 and 6.—Priests. 7 and 8.—Men and women with wine and bread for the dead. 9 and 11.—Mutes. 10.—Slaves with coffin. 12.—The wife. 13.—The sons. 14, 15, 16, and 17.—Relatives. 18.—Priest. 19.—Doctor. 20.—Beadle.
5. The Midnight Robbers.
6. Courage and fear (song).
7. Caught in a trap, with two effects.
8. The Decapitation, a brotherly caution.

9. The Headless Trunk, awful discovery.
10. The Ruse, "between two stools."
11. The Dead Body Guard, Egyptian police.
12. The King and his daughter in the palace gardens. Delightful scene.
13. The Dead Body Guard get drunk—the escape with the body (song).
14. The Trick on the Princess, with effect (song).
15. The Herald reading the King's proclamation.
16. The Thief presents himself at court.
17. Landscape and Grand Wedding Procession, viz. (equal to 12 slides): 1—Populace. 2—Officers. 3.—Mask and children. 4.—Hobbyhorse. 5.—Performer with sacred crocodile. 6.—Band. 7.—Troupe of dancers. 8.—Boats. 9.—Sacred crocodile and ibis. 10.—Barge with band. 11.—Obstructionists. 12.—The King's state barge. 13.—Boat with band. 14.—State barge with bride and bridegroom, drawn by swimming slaves. 15.—A Noble's state barge. 16.—Gentlefolks. "Now comes the national presents for the bride." 17.—John Bull, England. 18.—Scotland. 19—Ireland. 20.—Africa. 21.—America. 22.—Germany. 23—France. 24.—Russia. 25.—China.
18. Grand Wedding Banquet in the palace of a thousand pillars, Thebes.
19. Good night (Egyptian).

1288.—Little Red Riding Hood (W. C. Hughes, Kingsland).

The Maiden Good; or, the little Woodman Prince and his Silver Axe, the Fairy Dove, and the Wicked Wolf, who got the whacks An original extravaganza fairy tale. The usual fairy story mixture, with a grand transformation scene as the final. Twenty slides.

1289.—Robinson Crusoe (J. H. Steward).

A very fine set, with dissolving effects and mechanical movements *ad lib* Designed for an entertainment of about one hour. Mr. Steward has had a special libretto written for the set. Some of the mechanical slides are worthy of mention, the principal being—Panorama of Ship Leaving Hull; the Wreck (lightning effect); Moonlight and Ripple on the Water, Dancing Cat, Panorama, Raft Leaving Ship and the Voyage, Day to Moonlight; Levers—Crusoe and Dancing Kid, Building the Boat, Voyage Round the Island, Cave Scene; Effects (choreutoscope)—Dancing, Friday and Bear.

1290.—The Rose and the Beauty (W. C. Hughes, Kingsland).

A grand juvenile comic Christmas pantomime of the Rose and the Beauty, or Harlequin Beautiful Beast and the Merchant with the Daughters three. A total of 24 slides, numbers 1 to 17 dealing with the fairy tale, the last being the change of the beast to his proper form and defeat of the wicked enchanter. No 18 is a grand transformation scene, the clown and his adventures filling up the remainder.

PART III.

Coloured, Mechanical, Effect and Movable Slides.

THE colouring of slides, especially those produced by photography, is rather a debatable point with exhibitors. Some not only prefer plain photographs, but will not have coloured at any price. Others again like the coloured ones, so long as the tinting is well and artistically executed. We will quote an American opinion on Colour *v.* Monochrome:—

"Colour alone seems to have greater charms than form alone. A sunset is entrancing when the sky glows with radiant hues; the blue is almost lost in red, the yellow is as a sea of transparent gold, and the whole presents a variety and blending of tints which charm, and soothe, and lull to reverie; and yet all form is indistinct and obscure. If so charming when separate from form, what is colour when properly combined with beautiful shape? There are few objects to which colour may not be applied, and many articles which are now colourless might be coloured with advantage. Our reasons for applying colour to objects are twofold, and here, in fact, we see its true use:—

"(1.) Colour lends to objects a new charm—a charm which they would not possess without it.

"(2.) Colour assists in the separation of objects and part of objects, and thus gives assistance to form.

"These, then, are the two objects of colour. Mark, first it is to bestow on objects a charm such as they could not have in its absence. In the hands of a man of knowledge, it will do so—it will make an object lovely or lovable, but the mere application of colour will not do this. Colour may be so applied to objects as to render them infinitely more ugly than they were without it. Here, again, it is knowledge that we want. Knowledge will enable us to transmute base materials into works of marvellous beauty, worth their weight in gold.

"The second object of colour is that of assisting in the separation of form. If objects are placed near to one another, and these objects are all of the same colour, the beholder will have much more difficulty in seeing the boundaries or terminations of each, were all coloured in the same manner, than he would were they variously coloured; thus colour assists in the separation of form. This quality which colour has of separating forms is often lost sight of, and much confusion thereby results. If it is worth while to produce a decorative form, it is worth while to render it visible; and yet how much ornament, and even good ornament, is lost to the eye through not being rendered manifest by colour! Colour is the means by which we render form apparent.

Colours, when placed together, can only please and satisfy the educated when combined harmoniously, or according to the laws of harmony."

The foregoing very clearly defines what colouring ought to be. We can add nothing to the definition, and so will proceed with the notices of work.

Slide Makers & Slide Painters.

We have dealt with the great majority of slides issued in sets or series, crediting them, as far as we were able, to their respective makers or proprietors. But there are still several general points, such as the colouring, or the process by which the slides are produced, which cannot be referred to in any one special set, considering that all the sets by that maker no doubt possess the same feature. To obviate this difficulty, and at the same time to give details which must prove of interest, the following notes are written, dealing with the firms alphabetically. The reader will therefore please understand this chapter to be a *general* notice of the slides issued by a firm as distinct from the particularising of sets, stories and series —

Appleton & Co.—This firm's speciality is a very fine series of slides representing views and life in India. They are produced from a series of negatives purchased by them, and of which they, of course, hold the copyright The views, according to their list, number 464 in all, and embrace the following.—Agra 23, Ahmebadad 13, Delhi 27, Secundra 10, Madras 70, Conjeveram 14, Vellore 5, Tirukoilur 4, Tadpatri 18. Shevaroy Hills 56, Karli Caves 4, Kaichore 4, Beypore 28, Tierns 28, Behari 4, Namburi Brahmins 8, Nairs 13, Chermas 5, Futtehpur Sikri 7, Gooty 4, Mount Abu 14, Rookee 8, Madura 7, Lucknow 3, Ganges 5, Calicut 42, and others The tribes, natives, customs, religions and social life are all well represented, in addition to the ordinary views of special places Altogether, this must be pronounced a very fine series, and judging from the samples seen and tried by us, the slides are carefully produced and turned out.

Cubley and Preston.—A set entitled "The Story of Eyam" is their registered property. A notice of the story has already been given, and we can but add here that the slides are produced carefully by, as far as we can judge, a dry plate process They are quite clear, and give excellent results on the screen. We mention this to overcome the scruples of some who deem all dry plate slides to be mere "amateur rubbish."

W. C. Hughes.—The majority of this maker's slides are made by his own special enamelled albumen process—that is, those in which photography plays any part; and it does in very many more than the general reader may imagine. Not only the ordinary views are photographic, but nearly all those copied from engravings and sketches are first photographed, and then coloured. Space will not permit, of course, our entering into much detail; his catalogue is bewildering in its variety of styles. It reminds the compiler of his visit to Mr. Hughes's slide exhibition room, which gave him a headache for a week after, and he'd no more attempt to grasp the slides - coloured and plain, effect and dissolving, mechanical moving, the million and one described in Mr. Hughes's lists—than he would—well, enter that chamber again. We can, however, refer to some we have had the opportunity of inspecting at our leisure. The coloured scriptural at 2s. each are good value for the money, as are those from foreign transparencies at 3s 6d. each; but, as is implied in the catalogue, the better the price the better the work, and we have a good example of that in one of the Pyramids at 7s. 6d, which is beautifully done. In fact, any price almost can be paid, and from what we have seen we are sure the purchaser will receive his money's worth. Mr Hughes goes in very largely for the mechanical and best effect slides, his "Ship on Fire" set being equally clever in mechanical movement as in artistic colouring. The "Pandiscope," or lantern sketcher, is a clever mechanical slide, an account of which, however, will be found further on, as also the "Giant Choreutoscope." Altogether, this maker's slides are apt to make a lanternist's mouth water.

Mason & Co.—This firm have only recently introduced their slides into the market, and are themselves quite satisfied with their success. Starting last season with scriptural sets, they have this year gone a little deeper into the matter, and are issuing other sets, such as Scotch views, tour in Norway, &c. The coloured scriptural at 36s. per dozen are wonderfully cheap, the colouring being carefully done, and without losing any of their transparency. The price of these plain is 18s. per dozen. They have also a second and lower price of 6s. per dozen, for plain slides only, which are produced from the same negatives, but in a cheaper manner. We believe the best coloured and best plain would be hard to beat at the price.

Newton & Co.—A firm dating back nearly 200 years (Sir Isaac Newton was knighted in 1704), and who gained honours and medals in 1851, command attention at once. The slides issued by this firm are all painted, and to a great extent made by photography, on their own premises. We can only say that they are in every respect of first rank. Taking their catalogue as a guide, the series of countries and people, such as "China and the Chinese," "Russia and the Russians," &c., are most complete. But we must give special praise to their elaborate series of "English Church History," "Missionary Scenes," "Evenings with the Poets," and one or two others which lie quite away from the usual run. In conclusion, their catalogue is the best arranged one published, a feature quite worthy of notice to those who do much catalogue rummaging.

Noakes & Son.—We must call special attention to the Messrs. Noakes's great feature, viz., all plain slides at 15s and coloured at 24s per dozen These are very low prices indeed, but in addition to this they offer to supply any photographic slides that are published in England at the same prices. Had we not seen and tried their coloured ones at 2s., we should have been, and with some measure of right, a little sceptical as to the quality of the colouring. But having done this, both by oil and limelight, we found them equal to many sold at much higher prices. What difference there exists between the two lights when used for this purpose we know not, but some difference has often been, and is still stated to exist, and therefore we tried them both ways, and were perfectly satisfied that at 2s. each the coloured slides are good enough for any ordinary use. How it is done for the money we don't know, and, of course, it is not our business to care, nor will the public therein concern themselves. But they are there—try them.

Chatham Pexton.—A well-known slide painter and producer. His great speciality, perhaps, is colouring, and he works for the trade so largely that one is never sure that he has not had some hand in the production. We, of course, do not mean to infer that no one else but Mr. Pexton colours slides, but we do know that a glance at his ledger would show the great number sent out in trade circles. Turning now to the actual painting, we cannot do better than quote from an article in the *British Journal of Photography*, which all the same will be new matter to the great majority of our readers:—

"Having been ushered into a studio in which a number of young ladies were seen seated at easels, and engaged in one or other of the multifarious branches of transparency painting, we first of all inquired of Mr. Pexton if there were any secrets in his profession that we would witness, yet which he might not consider desirable to be described to the public. The reply of Mr. Pexton was—'Write and describe whatever you see, we have no secrets here; we trust to skill alone, and not to any fanciful methods of mixing or applying colours.'

"While watching the painting of a lantern transparency on a photographic basis, we could not but admire the skill displayed in making a sky possessing uniformity, and which is considered a difficult feat. This was effected in the following way:—The transparency having been placed on the easel, and the palette set with a range of colours from oil tubes, the blue pigment, previously modified with an admixture of varnish, was applied at the top with broad sweeps from left to right in such a manner as to become lighter and lighter as the horizon was approached. But just previous to this stage being reached, other colours of the rose-madder class were blended so as to bring the sky down to the horizon in a warm and pleasant tone. The sky was now ready for having the colours blended together and made to run into one another with such imperceptible grading as to present the appearance of one harmonious and continuous whole, rather than a succession of tints. The deft application of the point of the finger effected this, and although it appeared to be accomplished with much ease, yet we could plainly see that it required great care and skill to ensure a uniform result. In reply to our question concerning the advantages of other methods of equalising skies or broad tints—such as dabbing brushes, the fingers of leather gloves, and expedients of a similar kind—Mr. Pexton informed us that he found the finger to be the best means for obtaining this effect. Smoothness of texture of the skin and the most scrupulous cleanliness were essentials towards producing the highest class of results. To show the perfection that could be attained, he requested one of the lady assistants to lay a uniform tint upon a piece of plain glass. This was done, and such was the smoothness and uniformity secured that, had we not witnessed the operation, we would have thought it had been produced by pouring a clear-coloured varnish over the surface of the glass.

"The sky having been laid in, the production of clouds is the next step. This involves a high artistic perception. Everyone knows how exceedingly beautiful are some of the clouds in a high class lantern slide. Nay, so beautiful and fascinating may they be rendered as sometimes to prevent the eye from readily resting upon the scene below; and the character of the scene is greatly modified by the clouds. In no department of slide painting is there greater scope for artistic revelling than in the formation of clouds and skies. Taking up a photographic transparency of a Scottish coast scene, in which were several fishing boats careering upon a glassy sea, Mr. Pexton said to paint such a scene in the most effective manner, so as to 'bring down the house,' it ought to be made a moonlight study. At our request the transparency was treated in this manner, and never was a more magical change presented. The moon, which was 'picked out after the sky and sea had been draped in the solemn blue and green characteristic of moonlight, shone brightly and tipped the edges of the sails and rigging of the boats, and by the skilful use of what we may designate 'delicate little artistic scalpels' its beams danced upon the scarcely rippling water.

"We inquired concerning the production of what is known in the lantern trade as 'tales,' and were told that these subjects are first drawn upon cardboard on a somewhat large scale, and finished in light and shade with the most scrupulous care. The test for good finishing in this drawing consisted in enlarging a 3in. photograph obtained from it up to forty feet. The production of these originals was attended with great cost, as a high degree of artistic skill was involved; and, although many figures were frequently introduced, it was necessary that the *likeness* of each characteristic figure should be maintained throughout the whole series. Mr. Pexton showed us the original drawings of some sets which are now being brought out commercially, and so fine were they that we have conceived a respect for pictures of the fairy tale and kindred classes we had not previously experienced."

Since the change, however, of the official premises from Holloway to Gray's Inn Road, and partly because of increased business, much of the work is done at the respective homes of the artists The samples tried by us were in every way first class, the animated statuary effects being something lovely His general prices are—Plain, 15/- per dozen; coloured, 24/-, 36/-, and 60/- per dozen. Some of the original sets will be found in our previous pages.

Alf. Pumphrey.—Having already spoken about cheap slides, it is rather difficult to further enlarge on the same subject. But here we have a manufacturer, with many thousands of slides, all made and coloured on the premises, at the price of 6/- per dozen plain, 12/- per dozen coloured At least this season, we are informed, these are to be the universal prices, with but one or two variations The plain slides do not alter, but some of the coloured landscapes will be from 15/- to 18/- per dozen However, the popular stories and sets will be coloured at the 12/-, and Mr Pumphrey has an enormous number of these. Surely no one can want the prices reduced below that He says he can produce good slides at those prices and—well, no one will quarrel with him for it The slides are made, as is usual with big manufacturers, by the wet collodion process, and are very clear and transparent If any reader is not satisfied, after a purchase from Mr Pumphrey, he can send them on to the compiler, who can do with any number of them

Riley Brothers.—A firm who do an immense hiring out trade, to which we refer later In addition, they do their own colouring, and manufacture several original sets of slides We will refer to one that has especially taken our fancy, viz., "People we Meet." These are not made up from life models in the ordinary understanding of that phrase, that is, live people suitably attired photographed in a studio in front of a consistent background. But they are taken directly in the street, and show the natural surroundings Some perhaps are "People we would rather not Meet," such as the gas inspector. The slides we have seen are extremely good and realistic Then there are the political series, representing Ireland from the Gladstonian Liberal point of view, which are quite unique in their effect and completeness Also one upon a new country, or, rather, we should say the set is new—"Fiji and the Fijians." The firm have not quite a universal price, but the majority are 1/- each plain, and 1/9 coloured, the colouring being very well done for so low a price.

Sciopticon Company.—The slides issued by this firm, under the able supervision of Mr. Geo. Smith, are made by the Woodbury process—a photomechanical one. A mould is first made from the negative, and then a metal plate, the slides being reproduced by inking We had the pleasure of a personal inspection of the process, the hydraulic press, the plates so systematically arranged and numbered, the little levers for producing the print—for it may be termed a print—and, in fact, the complete arrangements. But readers will perhaps be more interested in details of the slides than in the means of production. Our opinion is that they are unexcelled for delicate detail and beauty. Take the tracery of a lattice-work window, or the fine details of architecture as in a Moorish building, no other process renders it so delicately and so full of minute detail The tone adopted—a rich warm brown—is most pleasing to the eye. Altogether there are no more beautiful slides made The price starts at 18/- per dozen plain, with reductions according to the number of dozens purchased.

A. W. Scott.—This gentleman is an expert lanternist, and also a painter of slides. Some ordinary dry plate slides (made by the compiler) coloured by him are extremely well done. The commercial dry plate slide is rarely so clear as one produced by the wet process, and therefore the colouring is attended with greater difficulty in avoiding loss of transparency. Mr. Scott, however, does not appear to have made any trouble about it, and the results on the screen are decidedly good.

The Solgraph Co.—We have inspected samples of their slides at 6d. and 1/- each plain, and 2/- coloured. Finding such a ready sale for the latter they are now issuing all their coloured slides at that price. They say they find they give satisfaction, and judging from those we have tried they ought to. The firm also make a speciality of micro-photographic slides, which are well executed.

J. H. Steward.—The hand-painted or coloured photographic slides issued by this well-known maker are in keeping with his reputation -first-class. The dissolving effect series, such as the Riallo Venice by day, night, moonlight, illuminated, &c , and others, are marvels of painstaking colouring work Although Mr. Steward sells all the other slides, plain photo, lecture sets, tales, &c., yet we consider his speciality to lie in the very best hand-painted work. Under the heading of entertainments (see No 1283, for example) will be found some most elaborate series, entirely original both in conception and treatment. Some of them have specially written librettos and cost as much as £22 odd. But then the colouring is magnificent, and the effects introduced most ingenious.

Theobald & Co.—A firm perhaps better known for their children's series of slides. Of course, the cheap and small hand-painted series we are most of us well acquainted with; but when it comes to lever and slipping effects, to get those in the small and cheap way is rather new. The methods, too, are most ingenious. However, as this is rather beyond the scope of our work, we merely mention them. Messrs Theobald go in very largely for the lithographic transfer slides. We quote from their letter —

"These are our own special manufactures. The litho stones have the illustrations photographed upon them, and then coloured by hand. These are then printed on sheets, like an ordinary coloured picture would be, with the exception that by varnishing the designs, and then placing the sheets in water, the pictures can be transferred from the paper on to glass, and being transparent (there being the colours only transferred) make capital slides. Although the expense in the first instance is enormous—every sheet of twelve pictures costing us £25 for the stones alone—yet the large number of them sold enables us to retail the sheet of twelve designs for 4/-, with a liberal discount to the trade, or to sell the pictures transferred on glass, complete as slides, at 7/6 per dozen retail. By this process, people get all the delicacy of outline and fineness of work for 7½d., which, in a hand-painted slide, they would have to pay from 5/- to 7/6 for. We are also producing now, for next season, the same process for comic slipping slides, which will be in frames the same as usual, but the quality will be far superior to the ordinary hand-painted work, whilst the price will be considerably lower. We are also making for the coming season all the rackwork pictures, such as windmill, watermill, magician's cave, waterfall, beehive, knife-grinder, Vesuvius, changing heads, house on fire, etc , in these same lithographic pictures, so that anyone will be able to purchase a sheet of twelve rackwork designs for 4/-; and in conjunction with these, we have made a special machine for making the rackwork frames, turning them out with the very greatest exactness, and to retail at 3/6 per frame. Consequently anyone can possess frame and subject complete for 4/-, which formerly cost 7/6, and was even then of an inferior quality to what these will be. We likewise produce all the well-known nursery tales by similar process for the small children's lanterns 1, 2, 3 size, and so on, thus enabling them to have slides with all the exactness of work and detail of the larger and more expensive ones."

Alfred Underhill.—A slide painter and producer whose work we have carefully and critically tried, both with oil and limelight, to our complete satisfaction. Taking the heavier and more elaborately painted slides, such as curtains, mottoes, etc., where there is hardly any clear glass left uncoloured, we were surprised to find such a degree of transparency, the light getting through in a marvellous manner. This maker has furnished us with some details of his methods, which, though of course more especially applicable to him, are still sufficiently interesting to merit insertion. Speaking of drawings specially prepared for reproduction as lantern slides by photography, he says :—

"The whole of the subjects (except five of the 3¼in. slides, and the set of 18 skipping slides) described in my catalogue are photographed from my own drawings (original, in majority of cases), prepared specially for the purpose. There are several advantages to be gained by producing slides in this manner. 1st—The slides (when exhibited on the screen) appear new to the audience. This is not the case when the slide has been copied from some well-known illustrated paper, and has been seen before as a book illustration. 2nd—The subject appears so much finer and more artistic, and is better adapted for colouring purposes, than slides copied from engravings, etc. In the latter it sometimes happens that very coarse lines appear, which look bad on the screen. 3rd—It is often found that the engraving to be copied is actually smaller than the standard sized slide. If so, the coarse appearance before mentioned looks still worse, as it cannot undergo a fining down, and a consequently better appearance, as it would do if a reduction were able to be made. By redrawing and enlarging a design, this bad effect is entirely got rid of, and additions or alterations can easily be made The price for making a design, or redrawing any subject, varies considerably, according to the amount of work required I think 3/6 is about the lowest amount I could quote for making a drawing for reproduction. I need hardly point out that, for commercial purposes, the idea of photographing from drawings is a very considerable advantage to either a private customer or dealer, as, after the first slide is made, any further copies can be had at the same price as an ordinary stock subject. This is not the case if the slide is hand-painted Should the latter get broken or lost, the same amount would have to be paid for a slide to replace it as was paid in the first place for the production of the original slide. During the last four or five years I have made between 600 and 700 drawings. A large number of these have been executed to my customers' orders, and the remainder for my own subjects. I have a large number of books, pictures, engravings, photographs, cards, scraps, and illustrations of all kinds. I use these for reference for carved designing and colouring, also for my customers' convenience when it is desired to prepare anything for special lecture purposes. I cheerfully give my advice and assistance at all times. All my drawings are made on white cards. I use Indian ink chiefly, and sometimes pencil."

Then, as to the production, first of negatives and then of the transparencies, he says :—

"After the drawing is finished, great care is taken in making a suitable negative, so that none of the fine details are lost. The slides are then produced from the negative by the wet process. (This process holds its own for clearness in the high lights, so essential to the production of a good transparency.) I prefer this process to any other, all my slides are produced by it. Nearly all my slides are photographed right up to the edge of the glass, so that either a square or circular mount may be used. Personally, I prefer the slides with circle mounts."

And lastly, as to the colouring, both of photo and entirely hand-painted slides, he adds:—

"Colours are used, made by the best makers, and of fine quality and well ground. I use a varnish which I consider the most transparent medium and best suited for the purpose. The same colours, etc, are used for colouring transparencies as for hand-painted slides The greatest care is taken with the colouring, and the reading, if lecture or set, is generally read before commencing, in order to obtain any hints or peculiarities of the various colours of buildings or figures which may be described."

Valentine & Sons.—One of our largest photographic publishers—that is, publishers of views of places all over the kingdom. They have something over 20,000 views in England, Scotland, Wales and Norway. As regards lantern slides, the firm have for a long time been noted for their excellent productions, and, in fact, are so well known that further comment is unnecessary. They have many sets, as will be seen in our previous pages, of views and landscapes, as distinct from tales and stories. In addition to these they issue the following.—

Yachts and Yachting	24 slides.
Instantaneous Shipping	24 "
Flowers	12 "
Cloud Effects	8 "
Loch Scenery	12 "
Trees	8 "

These are in all respects most beautiful slides.

West & Sons.—Though well enough known as the great yacht and instantaneous shipping photographers, they have not done much as yet in the slide line, but intend dipping into it more fully in the future. The samples we have seen are produced by a process giving the slides a warm brown tone, and are in every respect good productions. Of course, with such splendid negatives to produce from, the slides are bound to be most interesting. Mr. West says:—

"A few years ago it was thought a matter of impossibility to photograph any object whilst the camera was unsteady and moving, but it has been proved not only possible, but can be done with great success, as will be easily understood when we state that all our studies of yachts in full sail are photographed from a small sailing boat. The camera is fixed so that no matter how rough the sea is it can always be levelled at the object to be photographed, and if the shutter is fired at the precise moment, seldom fails in obtaining a good shot (or picture). I have had great experience in this kind of work, and although oftentimes attended with considerable amount of danger, it is a source of great pleasure to find oneself sailing along in quest of any vessel which will make an artistic photograph.

"I have often compared myself to a harmless pirate, lying in wait for a vessel, and choosing the best position (which can only be done by being in a boat, whereas on shore, or on a vessel at anchor, the vessel can only be taken in a certain position) to capture her, and, having fired, look around for more plunder. There is enjoyable excitement when I find myself in a cluster of yachts which are tearing along at a fearful pace, plunging through the water, dashing the white foam on one side and heeling over, so that the lee side of the hull cannot be seen owing to the water rushing over it."

Wilkinson & Co.—A firm of old-standing reputation in the painting and colouring of slides. Many of their productions were for the now-defunct Polytechnic, which may well be said at one time to have been the home of the optical lantern. The exhibitions at that institution, by the aid of limelight, were remarkable for ingenuity, effect, and elaboration. In no other place were exhibitions worked to such a high state of perfection. Messrs. Wilkinson are firm believers in hand-painted slides. To give examples of their work, they coloured for us a dozen photographic slides, one or two by the compiler, the others by York, Wilson, and Valentine.

This is what they say about photo slides —" They are fairly good specimens of photos, but are not pictures. We do not admire photos at all, and consider them an uneducating class of picture or thing, calculated to breed contempt for the beauty and colour with which nature everywhere abounds. Photographers substituting cold, matter-of-fact, common-place, and unnatural sombreness in lieu of sunlight, brightness, and cheerfulness. But enough, we only get vexed over these chemical and optical abominations (photos). You will find an amount of specks and flaws in the glass, and in the photos these ought to have been rejected by the makers, but again, unhealthy competition makes them sell what ought never to leave the studio. You will see the most promising slide of all, Swan and Castle, moonlight, is flawed in process of painting, through a dirty glass being used to take the photo upon in the first instance (competition and carelessness)." We have no desire to discuss the relative merits of coloured photos as against entirely hand-painted slides, but those coloured for us by Messrs. Wilkinson are remarkably well done, and have given us pleasure to show on the screen. When sending these slides Messrs. Wilkinson also submitted some of their hand-painted work, which, after trial, we feel satisfied to say is in every respect of the finest description. Detail is so finely drawn as to be in that one respect equal to a photo outline, the colours are artistically used, and in the Scriptural slides especially, with fitness and propriety.

Wilson & Co.—This is another firm in Scotland whose speciality consists of what might be termed landscape photographs for albums. But what dumfoundered us, on arrival at their catalogue in the compilation of the epitomies of sets and series, was the fact of their advertising that any picture in their catalogue could be obtained as a lantern slide When we got this fairly in our grasp as to what it meant for us, we were quite willing, nay anxious, to back out of the battle and declare it a draw. Fancy our dismay at seeing thousands of views described which were obtainable as slides. But a serious catastrophy was avoided by the discovery that they issued several series of pictures specially as lantern lectures. Nature once more resumed its brightness, the world seemed once more gay. These sets will be found included under their respective divisions. We have, however, a few words to say upon a series we ourselves have christened "Cloudland" —more lovely, more marvellous cloud effect slides have not been produced. We will only mention two, one of a sunset on the sea, the other of the cloud with the silver lining Every cloud is credited with a silver lining, according to the good books, but we don't often see it—we mean in a lantern slide. In this one it is shown to perfection.

York & Sons.—Now what can we say about a firm of such reputation as this one, whose manufactures are known the world wide. They have slides on almost every conceivable subject, however distant the place referred to, however abstruse the subject. And no one has a greater variety of short stories, and the well-known tales in prose or verse. The firm, too, keep well up to date, and are constantly bringing out new sets. Now as to the slides themselves, they are exclusively photographic (with the exception of just a few hand-painted effect slides), and, of course, they are of the very best production. In fact, we need say very little about Messrs York & Sons, for everybody knows their slides A universal price is adopted of 18/- per dozen plain; 36/- per dozen coloured.

Dissolving and Effect Slides.

By these we mean such as are especially adapted for the exhibition of effects, as apart from those which, though possessing mechanical movements, are not used, in conjunction with others, to produce dissolving and other effects, but stand on their own merits. Space will not permit our mentioning all the series published, but we describe a few to give some idea as to what can be got and the cost.

	£	s.	d.
Emigrant Ship. Departure—Interior of steerage—Last sight of land—Calm—Double effect of lightning, ship struck, and set on fire—Ship on fire—Saved. Eight slides	3	16	0
Lever Slide and Panorama of Ships. The ships are seen through the porthole in cabin, which is all in motion—Man seen in berth, sick. Two slides	1	3	0
Deserted Mill. Summer and winter—Night effect, with encampment of gipsies Four slides	1	13	0
Bear Hunt. View in Arctic regions—Moon plays on the water and disappears—Bear comes on rock—Boat sails up—Men fire—Bear falls on ice. Two slides. Five effects	0	17	0
Vision of the Golden Candlestick and Angel Two slides.. ..	0	15	0
Mount Etna With rackwork smoke effect. Two slides	1	1	0
Crossing the Desert. With effect of sand-storm. Two slides ..	0	19	0
Adam and Eve. Expulsion from Eden. With angel effect. Two slides	0	18	0
Hour of Prayer. Crossing desert by moonlight Two slides ..	0	19	0
Cairo. Day and night—With torchlight. Two slides	0	19	0
Refraction of Light in the Polar Seas. Spectral ships	0	15	0
Will o' the Wisp. Two slides	0	15	0
Christmas Waits Magnificent view by night of the old hall—Effect of moon and musicians appearing Two slides	0	17	0
Peace and War. View in the corn-fields at home—View on the battle-field, finding the killed and wounded. Two slides	0	19	0
Child's Dream at Christmas. Child sleeping—Dreams about the pantomimes, etc.—Rackwork effect of figures Two slides ..	1	5	0
Spider's Web. Fashionably dressed young lady in the centre, around whom a number of admirers revolve. Two rackwork slides ..	1	7	0
View on the Grand Canal, Venice. Moon rising—Beacon fire—Smoke effects—Gondolier playing in boat—Lady appearing on balcony. Two slides	1	5	0
H M S. Agincourt. Displaying electric light—Illumination—Setting off torpedoes—Effects. Three slides	2	2	0
Interior of Ironclad. Panorama of forts, ships, &c., seen passing by port-hole Two slides	1	3	0
Harpooning the Whale. Effect, smashing the boat. Two slides ..	0	19	0
The Mail Steamer Arizona. Striking the iceberg (movable). Two slides	1	1	0
View. A woodland stream, with effect of fairies out of the water. Two slides	0	15	0
Smuggler's Cave. View of the cave, with moon effect on the water—Interior, with smoke effect. Four slides	1	10	0
Castle of St. Angelo, Rome. By day—Night—Carnival—Clock lit up, and fireworks. Three slides	1	5	0
Traveller attacked by Wolves. With effect—Shooting the wolves. Two slides	0	15	0
Cave Scene. Colleen Bawn Cave—Moon effect—Miles crossing over with rope—Danny Mann, with boat coming on—Danny shot—Boat drifting—Miles diving for Colleen Bawn - Saved. Five slides ..	2	2	0
The Cathedral door at Nuremberg, with door to open, showing interior. Two slides	0	15	0

	£	s.	d.
Children's Fancy Dress Ball at the Mansion House, London. With curtain to roll up, and procession of children passing. Three slides	1	10	0
The Earthquake at Ischia. Day scene—Earth opening—Ruins. Three slides	1	2	0
Bombardment of Alexandria. Effects of shells bursting—Guns firing—Forts blown up. Two slides	0	16	0
The 40-ton Gun on the Railway (Egypt)—Train moves along—Gun fires. Two slides	0	16	0
Torpedo boat sails up to Ironclad, leaves torpedo, and moves away—Effect of explosion. Two slides	1	5	0
Extra slide, showing men and wreckage in the water	0	10	0
Lever Slide Panorama of Ships. Interior Saloon—All in motion—View through port-hole—Ships passing. Two slides	1	3	0
Ship in full sail—Moves away and moon appears—Ship on fire, with little boat moving away. Two slides	1	1	0
H.M.S. Lively running on the Hen and Chicken rocks, Stornaway—Steaming for the rocks—Destruction. Two slides	1	0	0
Bombardment at Tonquin. Two slides	0	15	0
Departure and return—Lifeboat setting out—Return into Harbour. Two slides	0	19	0
Light-ship at the Nore—Calm and Storm. Two slides	0	19	0
Last appearance of the sun—Arctic regions. Two slides	0	15	0
Epsom races—View of the grand stand—The start and finish (long slide). Two slides	1	5	0
Old Ferryboat—Summer—Winter—Skating—Night with boat lit up—Moon effect. Four slides	1	15	0
Rialto Bridge, Venice—Day changing to night—With boats passing, lit up with lanterns. Two slides	1	0	0
Mother's grave—With effect of angel. Two slides	0	16	0
Yacht race—View of the race, and interior of cabin. Two slides ..	1	1	0
Macbeth and the Witches—Cauldron—Imps rising up. Two slides ..	1	0	0
Niagara Falls and Suspension Bridge by day—Night, with effect—Train passing, lit up. Two slides	1	0	0
Ruins of Sprites' Hall, with sprites at their gambols—Moon effect. Three slides	1	5	0
Hamlet and Ghost. Effect of Ghost receding out of sight. Two slides	0	15	0
New Eddystone Lighthouse. Day—Night—Storm—And effect of Lightning. Four slides	1	13	0
Water Wheel. Summer view—Wheel turning—Winter—Effects of Moon playing on the water, and swan with moving head. Four slides	1	18	0
Mosque of Omar, with illuminations. Two slides	0	15	0
St. Peter's, Rome. Day—Night, with illumination and effect of fireworks. Three slides	1	4	0
Windmill, sails in motion. Summer and winter. Two slides ..	1	1	0
Ascent of Mont Blanc, arrival at the Grand Mulets, effect of camp fire at night. Three slides	1	5	0
Spectre of the Brocken. Two slides	0	19	0
Lurline. Fine view of the Danube, with boat moving, moon rising behind rocks, and spirit of Lurline playing on the harp rises from the water and floats along. Two slides with lever	0	19	0
Wave slide with eccentric motion, boats and ships moving to and fro, birds flying about, &c., &c.	1	1	0
The Village Smithy. Winter and summer, moon rises and plays on the water, smith at work inside. Lever movement. Four slides ..	1	7	0
Dear Father, come home. Three slides (the set)	0	10	0
Soldier's Dream. Two slides	0	15	0
Mount Vesuvius by day and night, with eruption rackwork, slide of fire coming on Three slides	1	10	0
Magician and Cauldron. Rackwork effect of spirits rising while the the magician waves his wand. Two slides	1	1	0

	£	s.	d.
The Magic Fountain. Rack effect of water playing. Two slides ..	1	1	0
Rembrant's Mill. Summer, mill in motion, and winter. Two slides ..	1	3	0
Signal Gun on the Bosphorus. Moon plays on the water, and camp fire burning—Effect of gun fired off, lever movement. Two slides	0	17	0
Children in the Wood. Abandoned in the wood, with beautiful view of lake—Children dead, with effect of birds flying down and covering the children with leaves. Three slides, or with moon rippling on the water. Four slides	1	9	0
Ringing the old year out and the new year in. Old church in winter, procession of bell-ringers, who enter the church door, and as they go in the windows are lighted up—In the steeple ringing the bells—As an old man, representing the old year goes away, a young maiden, the new year comes. Three slides	1	5	0
Haunts of the Fishes. Sea fish, lobsters, crabs, &c., crawling about. One rackwork slide	0	16	0
Fairy Glen. Bettws-y-Coed, by day—Night, moon effect, and fairies appearing on lilies. Two slides	0	15	0
Lisbon. By day, the great earthquake, effect of the city being destroyed. Three slides	1	5	0
Fog Bow on the Matterhorn. Natural phenomena view and effect. Two slides	0	15	0
Figure of a man whose head changes to that of all nations. One rackwork slide	0	17	0
Simplon, a village in the Alps, effect of the destruction by an avalanche. Two slides	1	0	0
Sailor's dream in the Arctic regions. Sailor asleep in his bulk dreams of being on a whales back, effect. Two slides	0	15	0
Christmas Carol. Children singing opposite a house, effect of door opening and windows lighted up. Two slides	0	17	0
Stag Hunt. View of stags drinking at a river, boat comes up, from which a man fires a gun, and stag falls in the water—Moonlight, double effect. Two slides	0	17	0
Mont Cenis. Past and present times, old stage coach coming down the mountain, and railway coming out of the new tunnel. Two slides	0	19	0
Past and present on water, old wooden walls and modern ironclad. Two slides	0	19	0
Arctic view of icebergs and ship, effect of iceberg falling on the ship. Two slides	0	19	0
Arctic view of icebergs, effect of Aurora Borealis in the sky and on icebergs. New effect. Two slides	0	15	0
Tobogganing in Canada. View showing effect of slide. Two slides ..	1	5	0
"Coming events cast their shadows before." View of old hall—Shadows on walk, and moon effect. Two slides	0	17	6
Enchanted Wood by Night—Changing to Fairies dancing round tree. Two slides	0	16	0
Mount Etna. From the village of Catania—Day changing to night—With effect of eruption—Rackwork. Three slides	1	7	
Ascent of the Rigi. Distant view—Ascent by moonlight—With train lit up, and view from the summit. Three slides	2	2	0
Return of the Prodigal Son. View of the old home, with son moving to the window—Effect of window showing company within. Two slides	1	1	0
Shadows in the Fire. Young girl reclining before the fire, in which her admirers appear in succession—Rackwork. Two slides ..	1	8	0
Spanish Bull Fight. View of the arena on which appears the Bull—Effects of Bull pierced and falling. Two slides	0	17	0
The Ghost of the Old Hall. Four tableaux. Two slides	2	0	0
Old Man in Churchyard. Skeleton appears. Two slides ..	0	13	0
Camp at Aldershot. Shooting season. Two slides	0	13	0

Movable & Mechanical Slides.

The mechanical movements of slides may be divided in three principal methods—

SLIPPING.—A piece of glass moving in a straight line.

LEVER —An up-and-down or see-saw motion, produced by a brass lever attached to the rim holding the movable glass.

RACKWORK.—A revolving motion imparted by a rack actuated by a handle. The above are certainly the principal movements in general use. There are others, however, which will be described in due order.

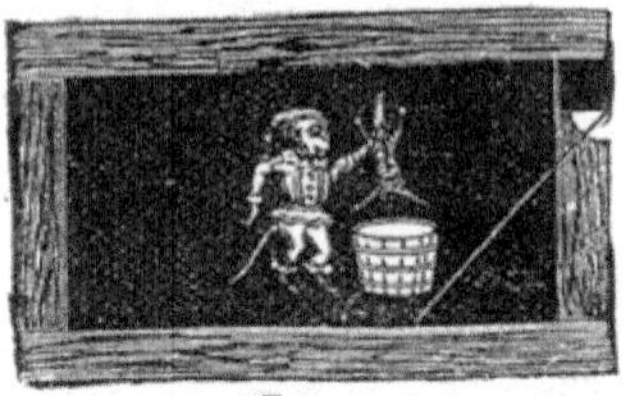

FIG. 1.

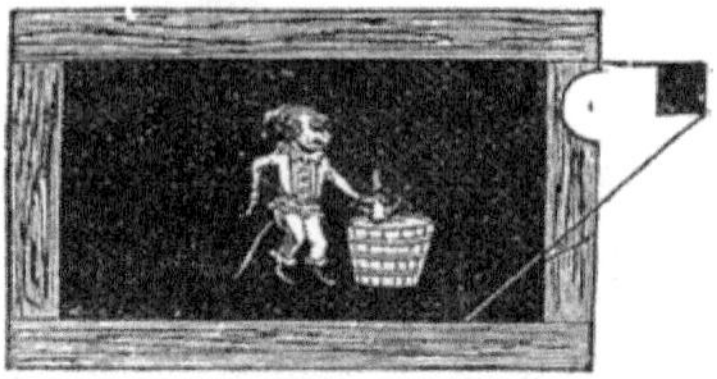

FIG. 2.

No. 1291.

1291.—Slipping Slides.

In these a movement, changing the picture, is effected by means of a second piece of glass pulled quickly over the first to a certain point. In Fig. 1 we have the slide as first shown, and in Fig. 2 the slipping glass pulled out, altering the picture. The usual price for these is about 13/6 per dozen for the common varieties, and for the better class coloured photographic ones about 3/- each.

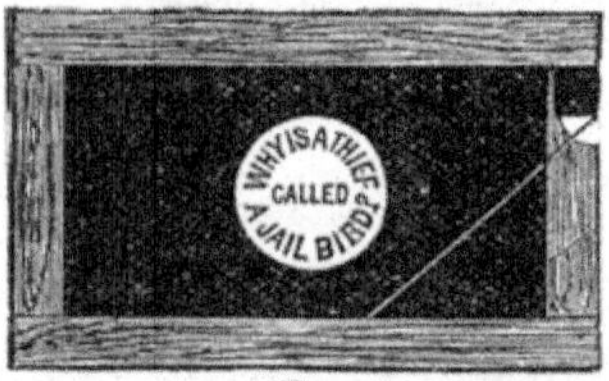

FIG. 1.

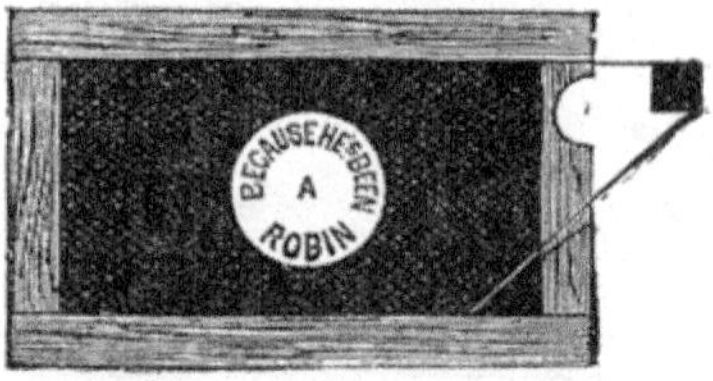

FIG. 2.

No. 1292.

1292.—Conundrum Slides.

A slipping slide, the question being painted on one part of the glass, as in Fig. 1, and the answer on another, as in Fig. 2, being pulled out to effect the change. Prices average .. .. each £0 1 0.

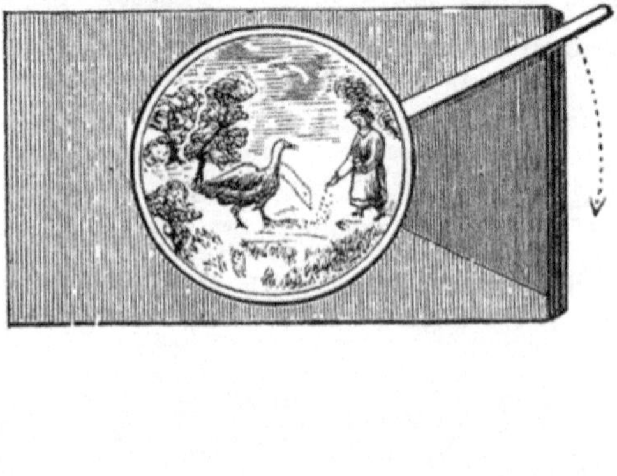

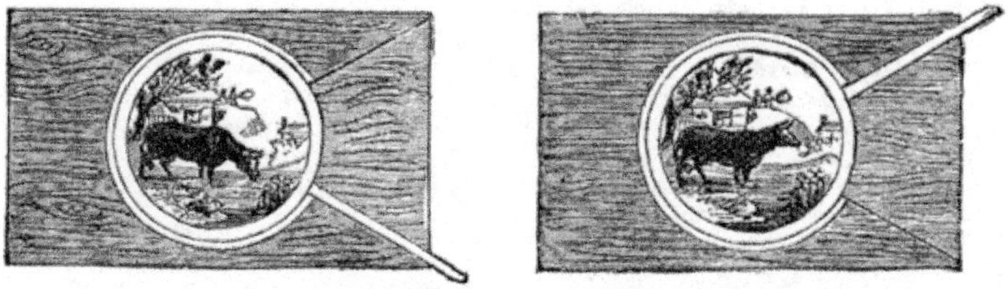

No. 1293.

1293.—Lever Slides.

In these, part of the picture or body is painted on a glass enclosed, as will be seen by the illustration, in a brass circle. This when moved up and down imparts motion to that particular part. Or, in the case of ship at sea, the ship moves up and down in the water. And so on with the different slides. Prices are about 4/-, 5/-, and 6/- each, according to quality.

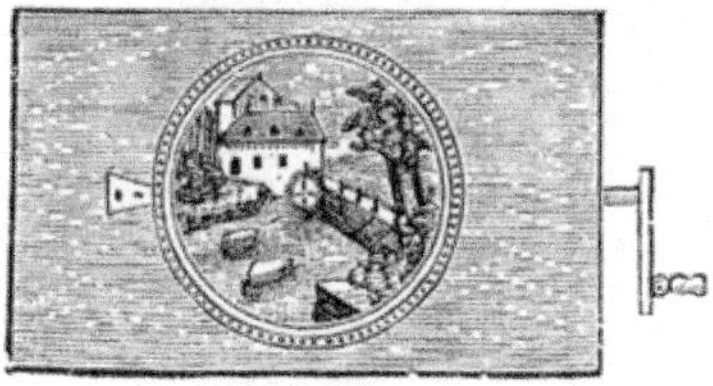

Fig. 1.

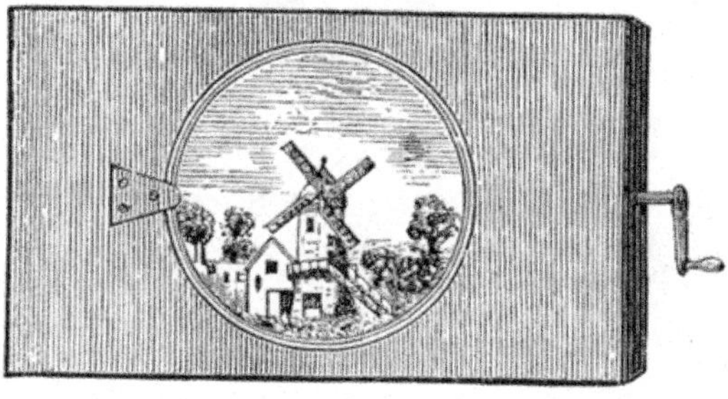

Fig. 2.

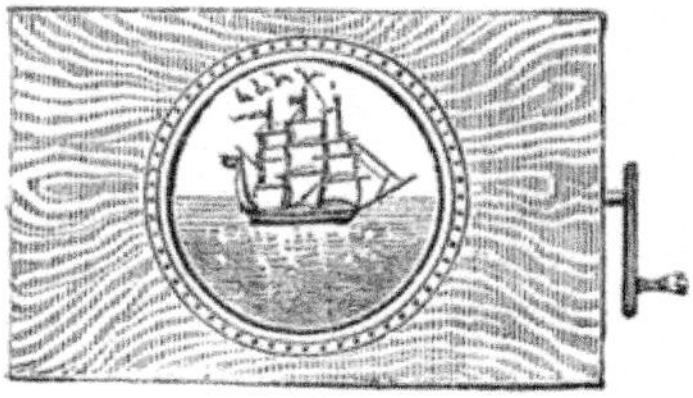

Fig. 3.

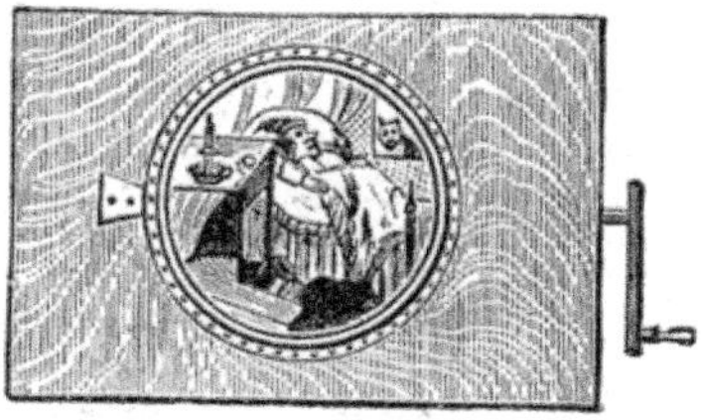

Fig. 4.

No. 1294.

1294.—Rackwork Slides.

These in the ordinary form have the second circle of glass to revolve completely by means of a rack. The sails of the windmill, for instance, are made to turn most naturally. The well-known "man swallowing rats" slide is actuated by rackwork. Fig. 1, the water-wheel revolves; Fig. 2, mill sails revolve; Fig. 3, ship has eccentric motion, birds flying; Fig. 4, the rats pass down the sleeper's mouth. The prices range from 8/6 each to as high as 21/-, or even more.

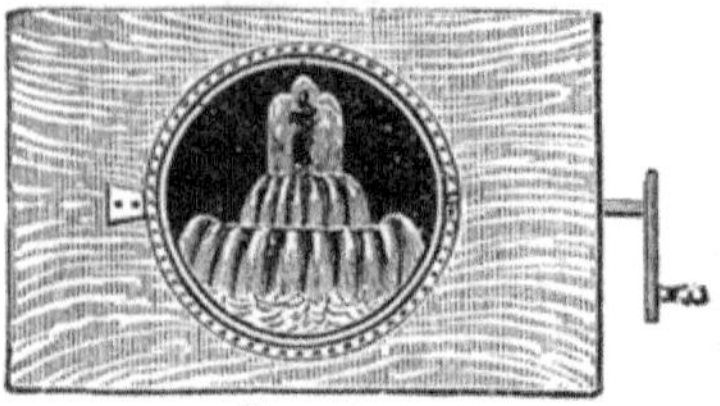

FIG. 1.

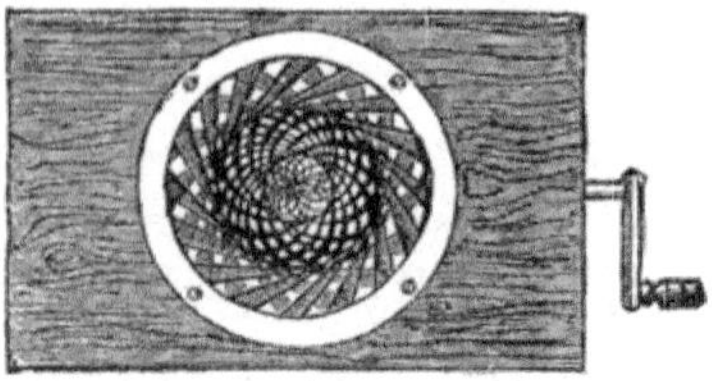

FIG. 2.

No. 1295.

1295.—Chromotropes.

Briefly stated, these slides are rackwork ones, in which a glass of coloured design passes over, in revolving, a similar but fixed one, the effect gained by the constant changing and flashing of colours being remarkably pretty. Fig. 2 shows this exactly. But they are sometimes made with pictures in the centre, such as a fountain playing (Fig. 1), portraits, views, &c. Prices about £0 12 6

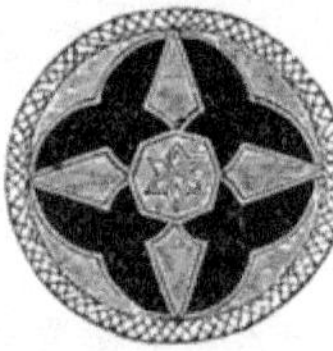

No. 1296.

1296.—Interchangeable Chromotrope.

In these the designs can be removed to form, with new ones, fresh combinations. It is a wooden frame containing two brass recesses moved different ways by a pinion. Into these recesses, and held by a spring, the painted designs are placed. The illustration shows the removable design. Price—frame, 5/-; designs, 2/- each.

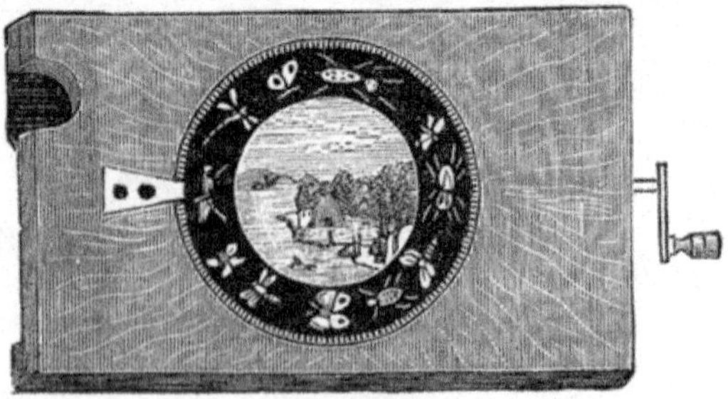

No. 1297.

1297.—Combination Rackwork.

A slide compounded of the rackwork and slipping motion. In the example illustrated, the sails of the mill revolve, the butterflies and insects pass round, continually changing colour, while, by pulling the slipping glass, two boats are seen to move across the river. Price about £1 1 0

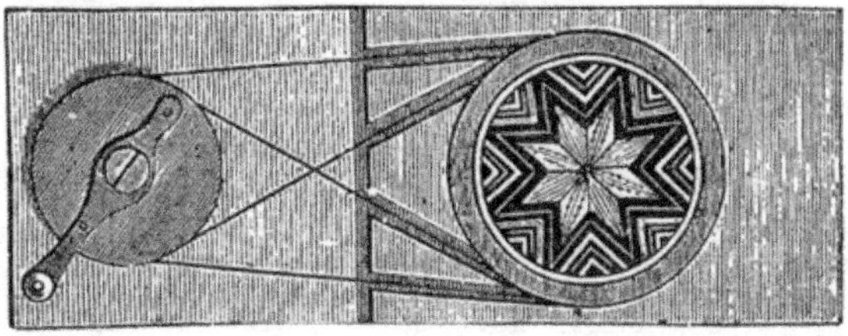

No. 1298.

1298.—Band Chromotrope.

The motion in these slides is produced by a band instead of a rack. The result, however, is the same on the screen, and the price also about the same.

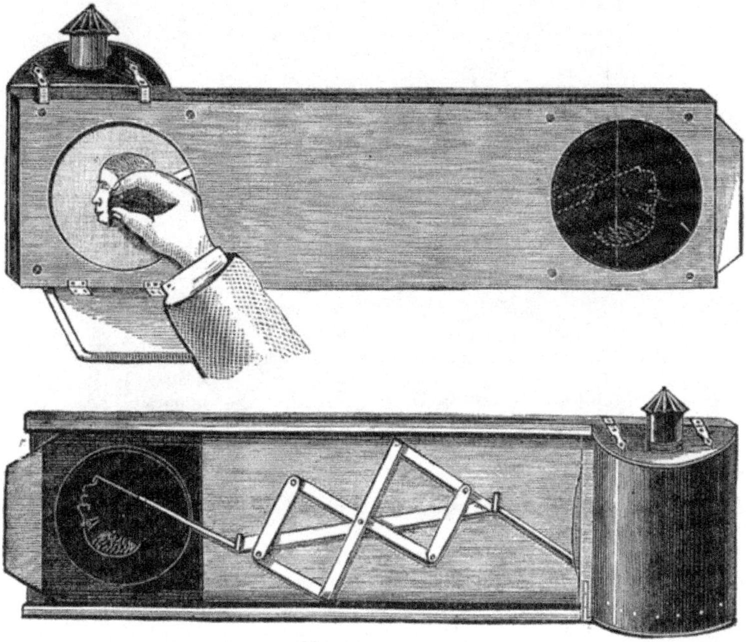

No. 1299.

1299.—The Pandiscope (W. C. Hughes, Kingsland).

An apparatus for drawing or writing on the screen in full view of the audience. A very ingenious piece of mechanism, to which there is no limit of usefulness and instructive power. Beyond requiring a little practice, there is no difficulty in sketching or making mysterious writing to appear on the sheet. Smoked or blackened glasses are used, and therefore the design is traced out of darkness The compiler has often illustrated a humorous lecture on "Noseology—the Science of Noses," by this instrument, the face and nasal organ in each case being sketched in view of the audience. The illustration is almost sufficient to give a clear idea of the method of use, but we quote from the maker's description in order to make the more certain —

"A is the copy-holder, where copies of prints, letters, portraits, hierographs, are placed, which are illuminated from behind by a small lamp screened and made transparent so as to be seen in the darkness. C is the tracing-point, which can be guided by the hand to any angle or degree over the copy with great ease and comfort, the action of which is simultaneously conveyed by an ingenious contrivance to the other tracing point or sketcher, which travels over the glass D, which also occupies the frame B, and when placed in the slide-holder of the lantern is thence transmitted by the rays of light passed through it to the screen."

The price is inclusive of case and supply of extra glasses .. £2 10 0

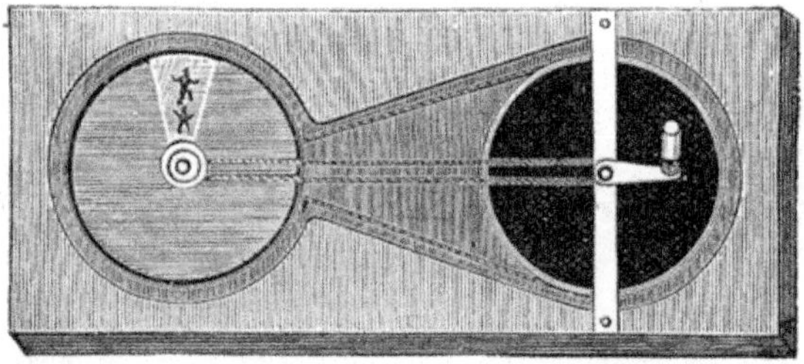

No. 1300.

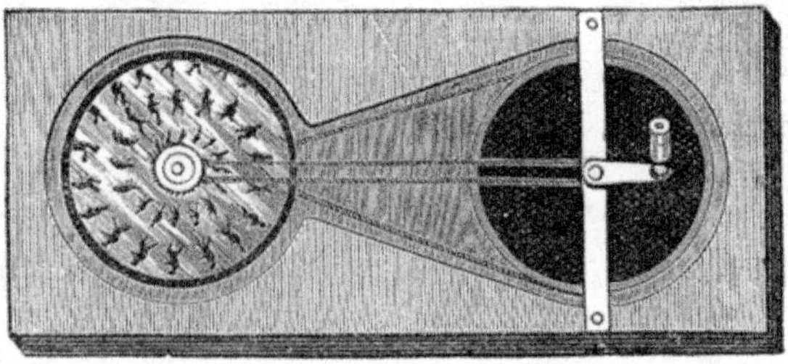

No. 1300.

1300.—The Wheel of Life.

A slide which is an imitation of the effects gained by the old optical toy of the same name, or as sometimes termed, the Zoetrope. Price about £0 14 0

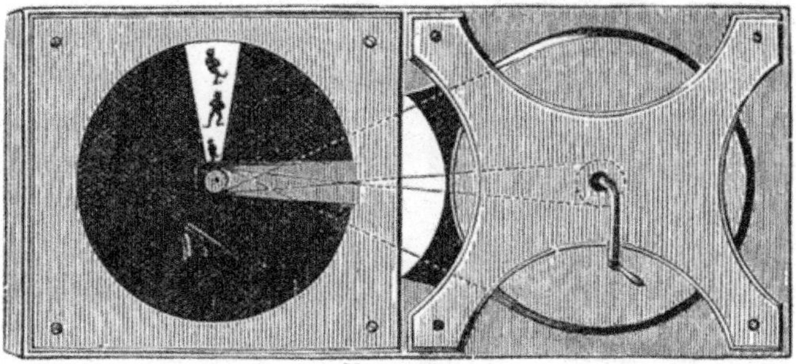

No. 1300.

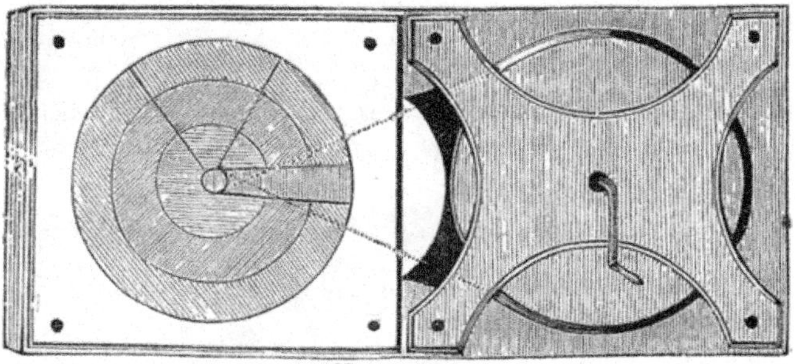

No. 1301.

1301.—The Chromatic Wheel.

An adaptation of the last slide to another purpose. By a simple arrangement, a disc coloured with three primary colours being substituted for the figure subjects. As the handle is turned, so the various colours dissolve into one another, producing interesting and pretty effects. Price about £0 14 6

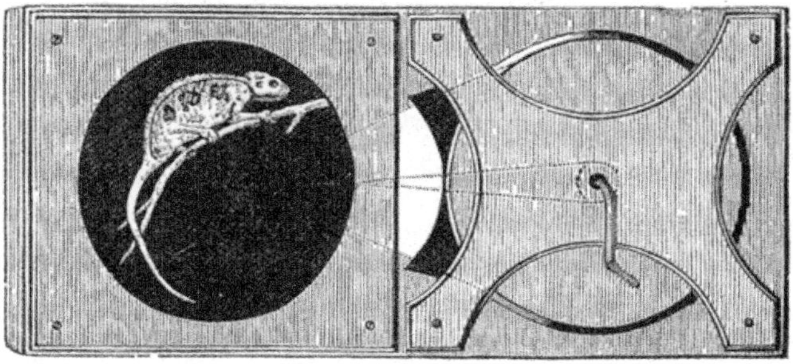

No. 1302.

1302.—The Chameleon Slide.

An adaptation of the foregoing slide, in front of which is placed a drawing of a chameleon. As the instrument is worked, the creature changes to all the colours of the rainbow, the uninitiated being, of course, as much pleased as surprised. Price about £0 14 6

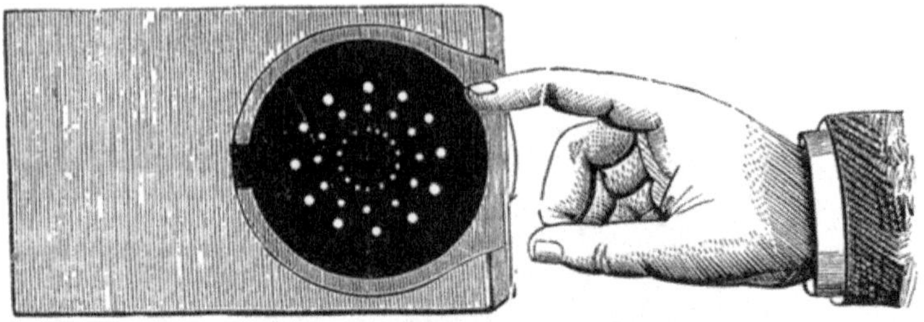

No. 1303.

1303.—The Kaleidotrope.

Like the Wheel of Life, it is dependent upon the persistence of vision, as a lighted stick waved about produces lines of light. It consists of a perforated metal plate, rotatable on the end of a metal bar capable of vibration in all directions in the line of the slide. The perforations are seen as so many dots of light. If the plate be gently tapped, each spot is seen as a line or circle of light. By moving the finger in the direction of the arrow line, a rotary motion may at the same time be communicated, which produces a series of most beautiful geometrical figures, which change with every vibration or difference of speed of rotation. Price about .. £0 5 0

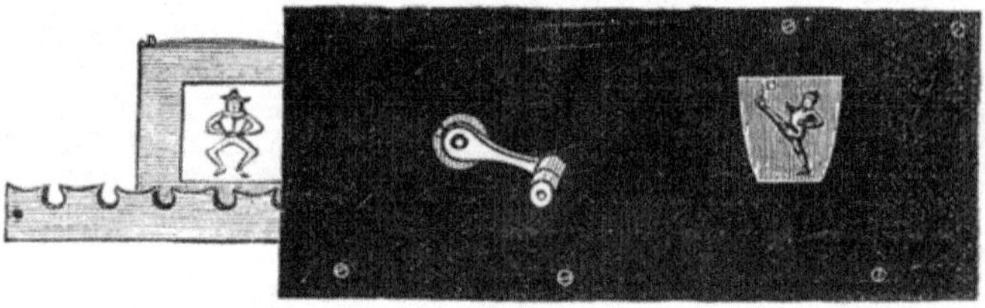

No. 1304.

1304.—The Choreutoscope (W. C. Hughes, Kingsland).

A long slide upon which is painted a figure in eight different positions is contained in a frame. An aperture in the frame allows only one of the positions to be seen at a time. A handle at the back, when turned, brings on each position alternately, and by a cam movement the change is made almost instantaneously. A shutter closes the aperture at actual time of transit, so that the illusion is not spoilt by the appearance of part of two positions at once. The effect produced is grotesque and amusing. Price £1 4 6

1305.—The Cycloidotrope (A. Pumphrey).

The invisible drawing master, it is also termed. Slides in which some movement or action takes place are, of course, always attractive. This slide is especially so, the geometrical designs traced by it on the screen are wonderfully pretty to look at, and the gradual development of a design from the starting point is most interesting. The slide is made to carry a circular piece of smoked glass, and upon the handle being turned the design is traced on the glass, showing a white line upon the screen. The pattern is easily varied by a slight alteration of the different adjustments, six in number; as soon as one design is finished it can be removed, and another glass introduced. Glasses can be carried ready smoked in a suitable box supplied with the instrument. The two following illustrations are designs worked by the instrument, and the price is £1 10 0

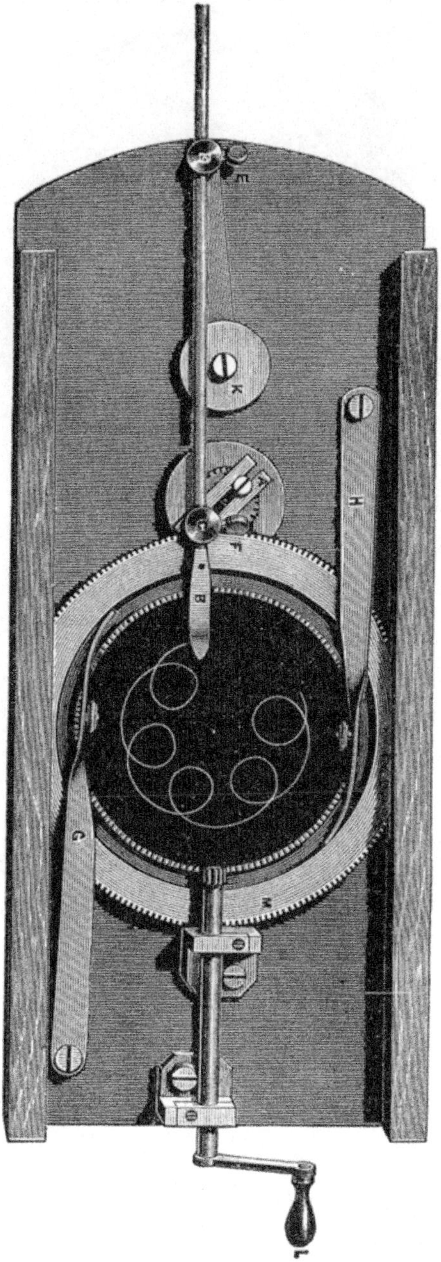

No. 1305.

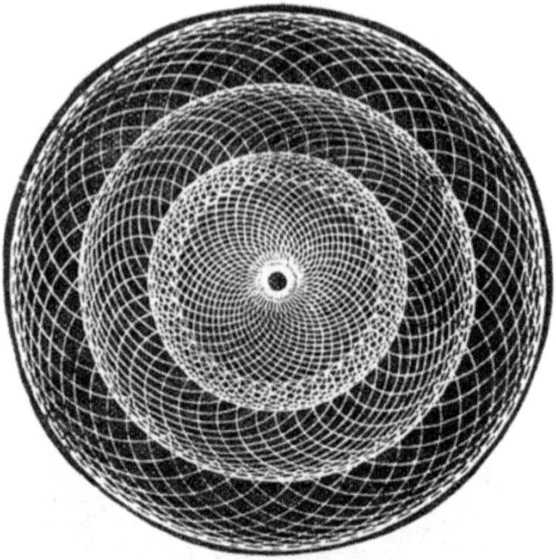

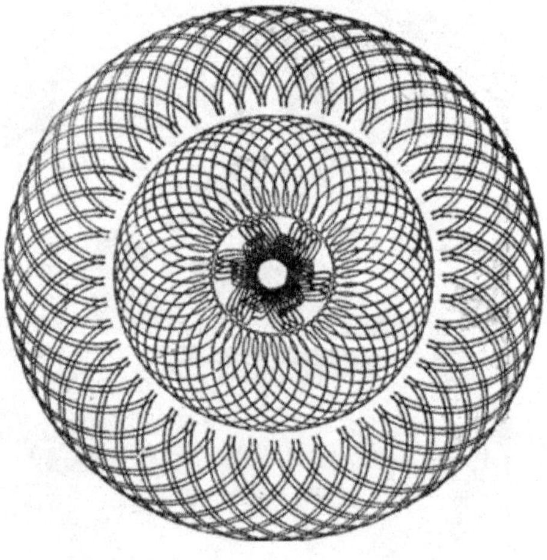

1306.—Snow Storm Effect.

A slide to illustrate the falling of snow, which is attained by the winding up of a piece of leather perforated with small holes. This, passing in front of the light through the condenser, gives a very natural appearance of a snow fall. Price about £0 7 0

1307.—Moon Rising Effect.

A lever slide, representing the rise of the moon in the sky. Also very natural. Price about £0 4 6

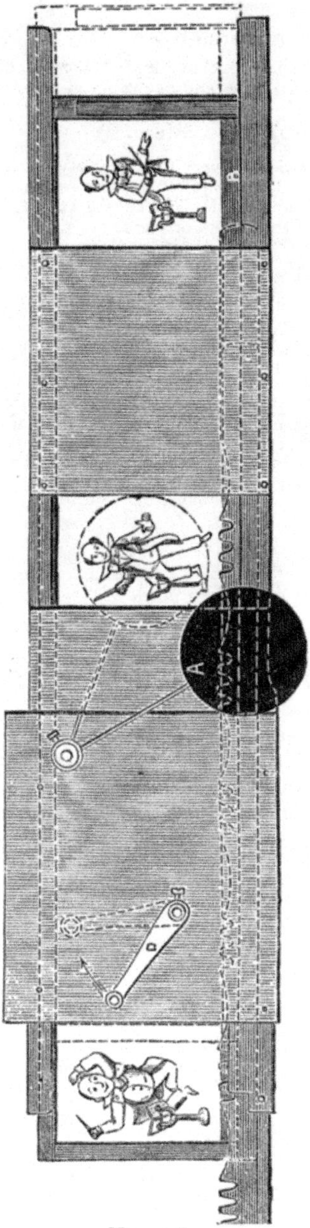

No. 1308.

1308.—The Giant Choreutoscope (W. C. Hughes, Kingsland).

Mr. Hughes's improvement upon No. 1304 (Beale's patent), by which the figures are given full size, and not reduced as in the other. Price £4 4 0

1309.—The Panoramic Slide.

These are painted in very great variety and at every possible price, both as single slides and in conjunction with others as effects. The principle, however, is the same. A long piece of glass upon which figures are painted passes in front of the design or view.

1310.—Lightning Effect.

A blackened slide, with the presumed shape of a flash of lightning scratched on it, and coloured of a fiery hue. Lightning effect is shown in conjunction with a picture representing a storm. The picture is put into one lantern and thrown on the screen. The lightning slide is put into the other, and the hand held in front of the objective. The handle of the "dissolver" moved to the central position, so that both lanterns may be lighted up. The hand which covers up the lightning slide can now be suddenly taken away and immediately replaced, and the flash of light is seen on the screen, but, like real lightning, existing for a moment only. This can be repeated as often as required in the same manner.

Hiring out of Lanterns & Slides.

A considerable business is done in the hiring out, not only of lanterns and accessories, but a specially large trade is done in slides; in fact, it forms quite a business in itself. Mr. Hughes, for instance, has distinct premises for his trade in that direction. Our work would not, therefore, be complete without reference to it.

1311.—Hire of Lanterns.

The following prices are about the cheapest for one evening;—

	£	s.	d.
Single oil	0	5	0
Dissolving oil	0	10	0
Single limelight	0	15	0
Biunial	1	5	0
Triple	2	2	0

1312.—Hire of Accessories.

	s.	d.		s.	d
Screen, 9ft.	1	0	Cylinder	1	0
,, 12ft.	2	0	Pressure board	1	0
If wetted, extra	1	0	Jet .. .. 1/6 and	2	6
Reading desk	1	0	Retort	1	0
Gas bag	2	6	Screen stand	1	6

1313.—Hire of Slides.

We have, in this particular, more variation, and we had better specify a few of the charges.

Archer & Sons	1/-	per dozen.	
Catlin, W.	1/6	,,	
Hughes, W. C.	2/-	,,	4 dozen, 7/6
Medland, J. B.	2/-	,,	2 dozen or more at 1/6
Newton & Co.	2/6	,,	4 dozen or more at 2/-
Noakes & Son	1/6	,,	
Pumphrey, A.	1/-	,,	
Riley, Bros	3/-	for 50, then 6d. per dozen.	
Steward, J. H.	2/6	per dozen;	beyond 4 dozen, 2/-
Theobald & Co.	1/6	,,	not less than 3 dozen.
Tyler, W...	1/6	,,	
Watson & Sons	2/-	,,	
Wood	2/-	,,	5 dozen at 1/6

1314.—Hiring Terms.

They are much about the same in all cases. We, however, quote those of Riley Brothers, a firm who do an extensive hiring-out trade, as a sample.—

If lent out in town or suburbs, to be returned the following day not later than 4 o'clock, or extra hiring to be paid for each day beyond.

When ordered to be sent by rail or post, the time required for reaching their destination must be stated, and will be allowed for, but if it exceeds four days (*i.e.*, slides sent out on a Monday must be back latest on Thursday) extra hiring will be charged, but slides must only be used on the day or days for which they are hired, and must always be returned by the first train or first post on the morning following the night or nights for which they are hired.

The charge for a single hiring for a set of 50 slides, general subjects, is 3/-, and for any number over in the same set, 6d. per dozen, but the minimum charge will be 3/-. N B.—This charge is for uncoloured slides, but where we have them coloured we send at the same price. To anyone binding us, however, to send coloured slides only, the charge will be 4/- for 50, and 8d. per dozen if more than 50 in the set.

Extra hiring will be charged as follows, viz.—For one day half the amount of the first day's hiring, and for every following day one-fourth the first day's amount, *i.e.*, if the first day's hiring be 3/-, the second would be 1/6, and the third 9d., and so forward. If hired for one week (*i.e.*, to leave here on Monday and back the following Monday, and so on with other days of the week), two whole hirings are charged, and if engaged for a month six whole hirings, or for three months 12 hirings only are charged, and 18 hirings for six months; but slides must in no case be kept beyond the time for which hired without our consent in writing or by wire.

Slides are sent out entirely at the hirer's cost and risk. The hiring fees being due with the order, and if to be forwarded by post an additional 1/6 must be included to cover postage of 50, but should it cost less the 1/6 to be refunded, or if more the amount to be charged. Where postage has not been sent the parcel will be forwarded by train, and must be returned to us carriage paid, or it will be refused, and extra hiring charged so long as it remains out. All parcels sent by post are insured, and must be insured on returning.

The value of the parcel and the amount for which insured will be stated on each advice note. The slides sent are all perfect, but should there be any defective, it will be stated on the note. Should they arrive damaged the postal or railway authorities must be called in at once to see them, and the claim for damages made by the receiver. If this be neglected the claim will be jeopardised.

Lantern Readings.

For nearly every series or set, a reading or lecture in print may be obtained. Most of these are prepared specially for use with the slides, a few only, tales from the Religious Tract Societies' Works, Temperance Tales, and such as these by G. R. Sims, are obtainable in the ordinary way as books. The latter of course are not marked with the place for each slide, etc , as are those specially written. The price is generally about 6d. each.

There are, however, collections printed as "Lantern Readings," which consist of good size volumes containing not one, but many lectures, poems, tales, etc. Of these, probably, the most pretentious are those issued by—

1315.—A. Pumphrey.

Three volumes. Vol. I. contains 80 sets, mostly of the short stories and popular tales. Vol. II. contains 35 sets, a few tales and some long lectures. Vol. III.,.19 sets, principally lectures. Each volume runs up to about 300 pages, and the price is 2/- each.

1316.—Theobald & Co.

Publish one at 1/-, containing readings for about 20 sets, and one of Juvenile sets at 6d.

1317.—York & Sons.

This firm print a great many, mostly, however, in single form. In some cases they have a few together; for instance, No. 1 contains 10 sets.

Colours for Painting Slides.

1318.—Colours for Home Painting.

Messrs. Theobald issue a useful box of water colours for this purpose. Price 1/-

1319.—Liquid Colours (A. W. Scott).

We have only at the last moment received samples of the "Rainbow" water colours, and have given them but a rough trial. They are used with a brush in the ordinary way, and diluted with water if too thick. As far as we have found time to try them, we must say they work well. Price 1/- per bottle.

Section IV.

GENERAL

Conduct on the Platform.

IT is of great importance for a person about to deliver a lecture to any audience, great or small, intellectual or half-educated, that he should mount the platform with a mind at ease as to all arrangements made. The writer has had vicissitudes in this matter; he has faced an audience when a mere tyro has been the only available lantern operator; he has had to appear unconcerned when his mind was racked with doubts as to the competence of the party who was to accompany in the songs; he has even, through no fault of his own, felt very dubious as to the sufficiency of the gas supply. Therefore he is in a position to warn others against such contingencies. If a lecture is to be successfully delievered, with confidence by the lecturer, with comfort to the audience, there must be no feeling of doubt as to the perfection of the arrangements made to secure a hitchless performance. In particular, the light must be tried before the audience enters the hall; the amount of gases must be ample to meet even unforeseen accidents, a little wasted gas is amply repaid by the comfort that a large supply produces to the lecturer. The lenses must be allowed to do all their "sweating" before the first picture is shown, and several baked limes should be at hand, kept in an air-tight cylinder of brass, having a lid screwing on firmly. Limes should always be baked in an oven, or roasted on a hob, immediately before the lecture.

A little "swagger" in the appointments of screen and platform is by no means wasted. No doubt bare poles for the screen-frame look very business-like, but we may be too Spartan in our disregard of appearances. There is no doubt that a simple drapery ought to adorn the screen and cover the unsightly poles, especially when the very slender poles are used, which, when the screen is stretched as taut as it ought to be, look very insecure—not to say inebriated. Two plain curtains of light material and warm colour may be hoisted along with the screen, and by an arrangement of cords too obvious to require description here, these curtains may be caused to form a graceful *proscenium*. Great successes are made of little details.

The contrivance for communication between lecturer and lanternist is another detail of more importance than might appear. The exclamation, "Next slide, please," is out of the question, and a bell is very little better. On one occasion, a red-letter evening in the writer's experience in many ways, he communicated with the lanternist by means of an electric wire, operating a muffled bell in the lantern room. The lanternist alone heard the bell, the lecturer kept the push concealed from the audience, and the views appeared on the screen at the proper juncture as if by magic. Even a bell is unnecessary, for an index might be used to warn the lanternist to change the picture. The writer proposes to purchase a length of insulated wire for all future occasions where it can be used; the effect well repays the outlay.

Perhaps the greatest success of all the writer's experience has lain in the use of his cut-off jet, which is doubtless figured and described in another part of this book. On one occasion he gave a lecture, having as lanternist a person who till that evening had never seen the limelight. Before the audience was permitted to enter the hall, the lecturer arranged the jet-taps so as to produce the best light of which his jet and lime were capable. He then turned down the cut-off taps forming his own design, and the inexperienced lanternist had only to turn up the cut-off taps at the proper moment in order to produce the same light as the lecturer had produced before. If, however, this jet is to be used with gas cylinders, the tubes from cylinders to jet must be very firmly fixed to both cylinder and jet, otherwise the tubes may be blown off.

This leads us to another consideration. The oxyhydrogen light has by past accidents got itself the name of a dangerous system. It does not concern us here to enquire into the question of *how* the bad name arose, but with the partly educated public there is a certain feeling of insecurity in the vicinity of gas-bags or bottles. It is therefore absolutely imperative that nothing like a hitch shall take place with the light. The crack of a jet seems to the timorous the "crack of doom," a hiss may be that of something as dangerous as a rattlesnake—if, indeed, that domestic pet does hiss. If bags are used, a clear space should be reserved all round them; the writer's minimum is 6 feet, and a responsible person is usually put in charge over the bags. But cylinders are infinitely preferable to bags in ordinary cases.

The lecturer's personal deportment is a very important factor in success or failure. If he is very nervous, he usually makes such a poor start that he gets into a state of muddle and dejection from which he never recovers till the mischief is past repair. If he has satisfied himself that no hitch can occur, if he has himself superintended everything, arranged the slides, duly instructed the lanternist how the slides go into the carrier, carefully studied the lecture-text and the slides, if any hitch occurs the lecturer is not to blame.

The "cocky" lecturer may feel highly satisfied himself, but the audience may have a different opinion as to the merits of the lecturer and the success of the performance. The writer has been an auditor as well as a lecturer, and if anything in this line riles him it is a conceited, priggish lecturer.

Timely jokes are of great utility in a lecture. Stale puns are very mean in their effect, but a good "conceit" at a suitable juncture is very telling. The writer thinks that no lecture is really complete or perfect without a few sallies of fun, but he admits he has no experience of teetotal lecturing or Bible stories, which do not always lend themselves to *facetiæ*. It is a very good plan—not seldom followed by public speakers—to introduce a joke near the beginning of the speech or lecture; this puts the audience into good humour, and if the joke is not a "miss-fire" it is very encouraging to the lecturer. Unfortunately, jokes near the beginning very often do miss fire; in that case we must simply load again with a different kind of cartridge and fire another shot.

A few remarks should always be made before the light is turned down, and here is the reason. After the light is lowered the proper pitch of voice cannot be gauged. The writer, in his preliminary remarks, always scrutinises carefully the faces of those of the audience near him, furthest from him, and in the middle of the hall. It is easy to tell when all can hear, and when some can only hear with an effort. General views of back teeth are sure signs that the speaker is not well heard. In nine cases out of ten, a rather loud, high-pitched conversational tone is better heard than a shout; but in every case clear deliberate enunciation is better heard than the ordinary hurried speech of conversation, however loud. The pitch of voice depends chiefly on the formation of the hall, and, of course, on the vocal organs of the speaker, but the most suitable voice will be lost if the enunciation be not distinct.

Before lecturing, where there is likely to be any strain on the voice, no liquids should be taken, least of all hot liquids, as tea or coffee. The meal last taken should be thoroughly digested before the lecture begins, that is to say, at least three hours should elapse between eating and lecturing. Pellets of potassic chlorate suit the writer best for clearing the throat previous to public speaking or singing. If there is a dryness in the throat—proceeding probably from nervousness—a glycerine jujube is perhaps the best thing to take.

If there is one maxim better than another for a lecturer it is: "Forget yourself." But this is to be taken in one sense only, and nothing but "self" must be forgotten.

ANDREW PRINGLE.

The Kodak Camera.

FOR TAKING PICTURES SUITABLE FOR LANTERN SLIDES.

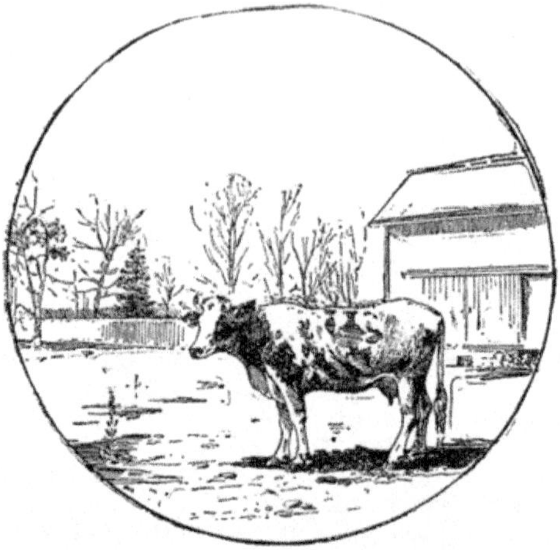

It is not given to every lanternist to be also a photographer, but the new Kodak detective camera is so constructed, and arrangements so made, that with it instantaneous pictures of street life and similar subjects, as also general views, interiors, and even small portraits, may be taken without any photographic knowledge. The principle of the Kodak system is the separation of the work that any person can do, in making a photograph, from the work that only an expert can do. With the Kodak, anybody, man, woman or child, who has sufficient intelligence to point a box straight and press a button, possesses an instrument which altogether removes from the practice of photography the necessity for exceptional facilities, or, in fact, any special knowledge of the art. It can be employed without preliminary study, without a dark room, and without chemicals. This is accomplished – First, by the exceeding simplicity of working the instrument; and secondly, by a system which the makers have arranged to do all the photographic details required. It is for this reason that we devote space to a description, and because it is the latest put into the market for sale.

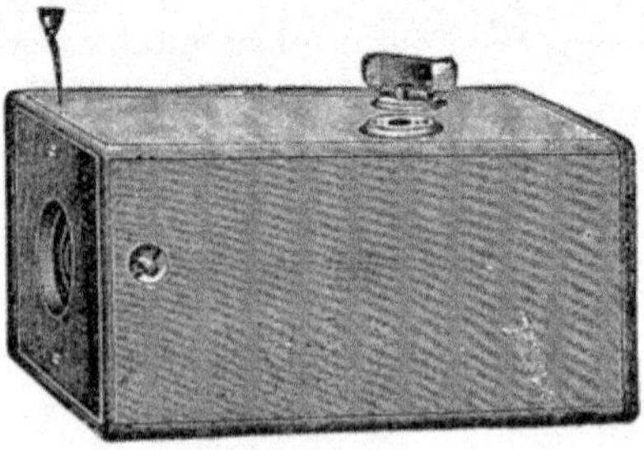

THE CAMERA.

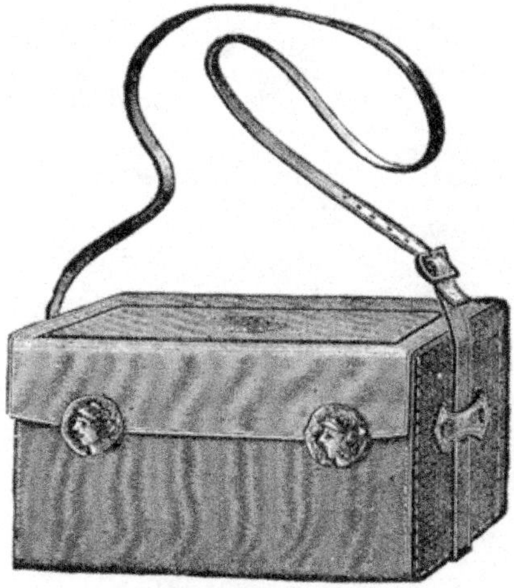

CARRYING CASE.

The illustrations above show the camera ready for use (weight 35 ounces, size $6\frac{1}{2} \times 3\frac{1}{4} \times 3\frac{3}{4}$), and the case, which is hung over the shoulder like a field glass The camera has a roll of material sufficient for 100 exposures. This is wound off as required by turning a key. The roll-holder receives the film and moves it in

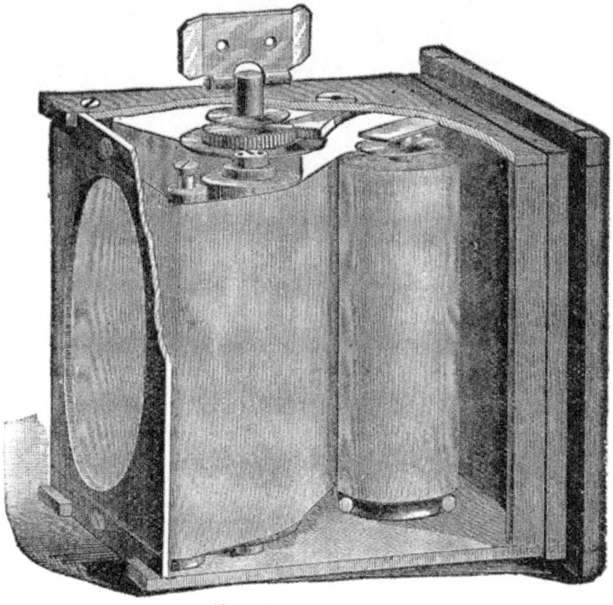

THE ROLL-HOLDER.

a continuous band behind the lens. This part of the camera is fitted with automatic devices which mark the paper so as to show the spaces between the several negatives, and provide for any expansion or contraction of the film due to changes in temperature or humidity. The lens is of special construction, never requiring adjustment and always in focus. The revolving instantaneous shutter reduces the exposure to a fraction of a second, and wholly obviates the necessity of covering the lens during the act of setting the shutter. Now, presuming the happy possessor of one to be ready for work, he pulls the string on top, which winds the spring of the shutter, then holds the camera as shown, the right hand

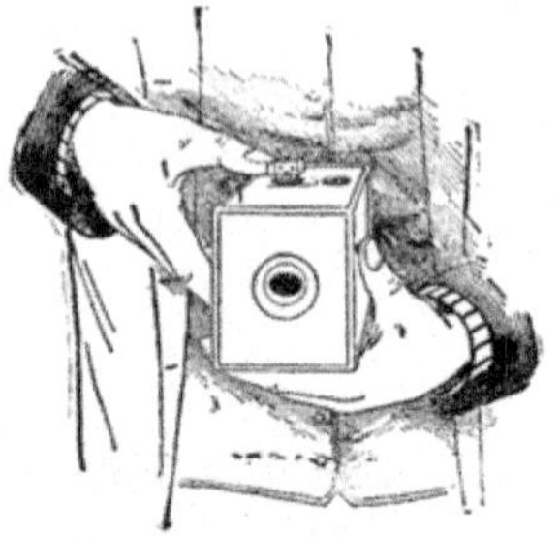

READY FOR EXPOSURE.

being placed to steady the camera, whilst the left thumb presses the button and releases the shutter. This done, the key is turned to the left until the indicator marks the correct length to wind off. This is here shown.

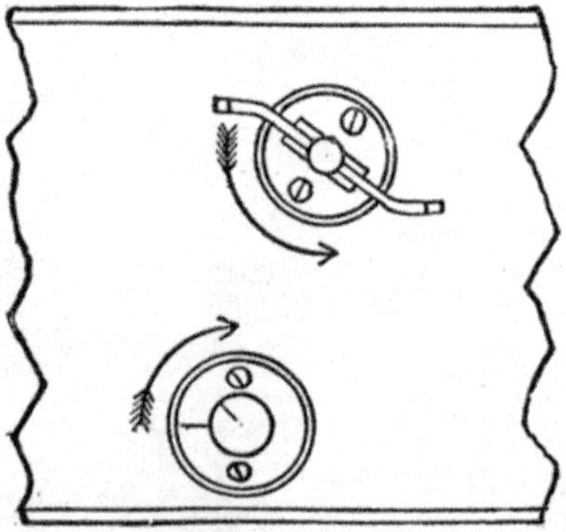

KEY AND INDICATOR DIAGRAM

As the key is turned to the left the indicator revolves to the right. When the two marks again correspond, the proper length has been wound before the lens.

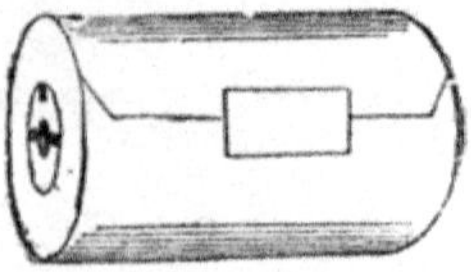

ROLL OF FILM.

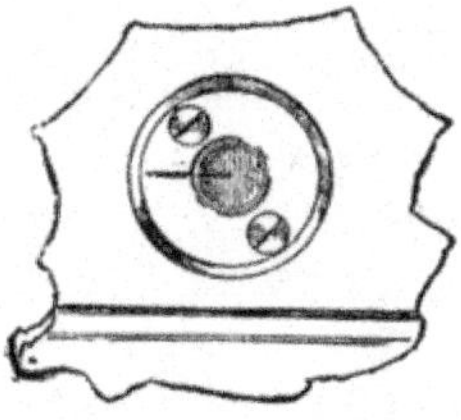

WHERE THE INDICATOR SHOULD BE.

The cord is pulled to wind the shutter spring, and the operator is ready again.

When it is required to remove the film already exposed, either to send away to be developed and finished, or at the end of the 100 pictures to insert a fresh roll, a photographic dark room is required. By this is meant one that is wholly dark—not a ray of light in it. Such a room can easily be secured at night almost anywhere. The reason a dark room is required is that the film is extremely sensitive to white light, either daylight or lamplight, and would be spoiled if exposed to it, even for a fraction of a second, while being removed from the Kodak. Having provided such a room or closet where, when the door is closed, no ray of light can be seen, set up on the table or shelf an orange candle lamp, which can readily be obtained from the makers or any photo dealers. The lamp gives a subdued yellow or orange light, which will not injure the film unless it is held close to it. The key is removed by turning to the right, and the back of the case pulled out. This part is roll-holder, and the key should be screwed in again.

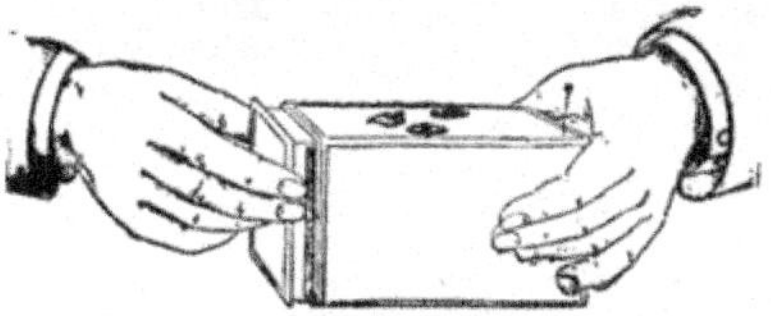

REMOVING BACK OF CASE.

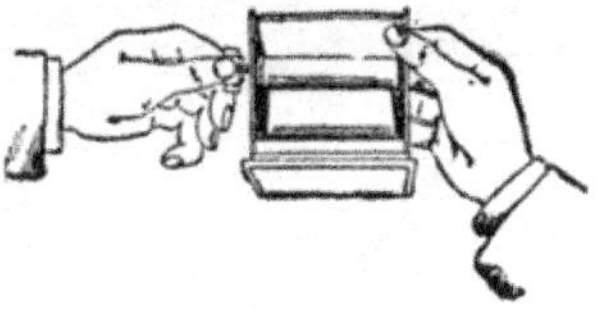

SCREWING IN THE KEY.

A sharp knife or pair of scissors is then used to cut the film on the side of the last exposure as indicated by the perforations.

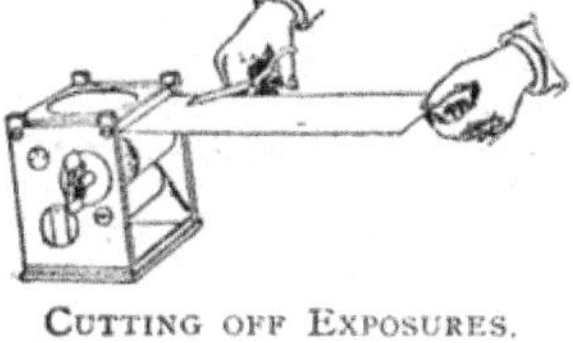

CUTTING OFF EXPOSURES.

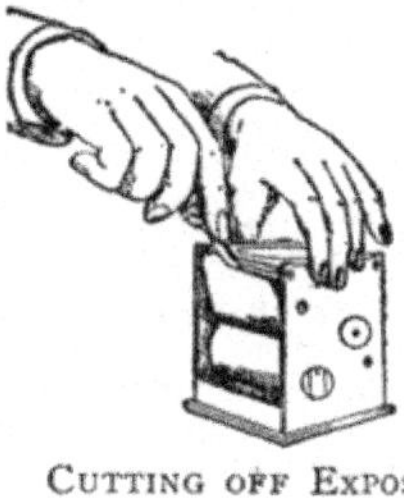

CUTTING OFF EXPOSURES.

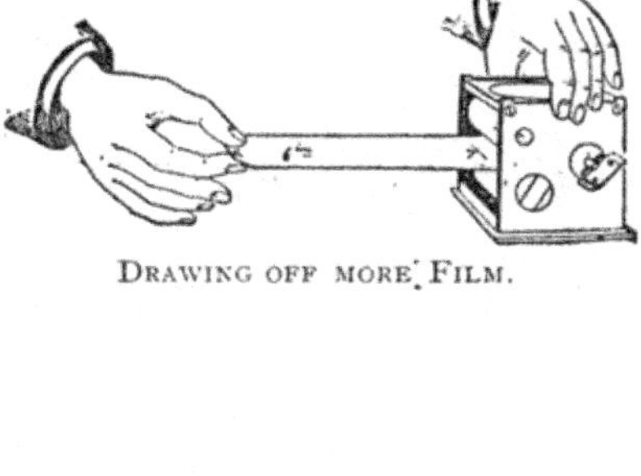

DRAWING OFF MORE FILM.

After removing the exposed film from the receiving roller, fresh film is drawn through from the stock roller over the exposure table, and is clamped down on

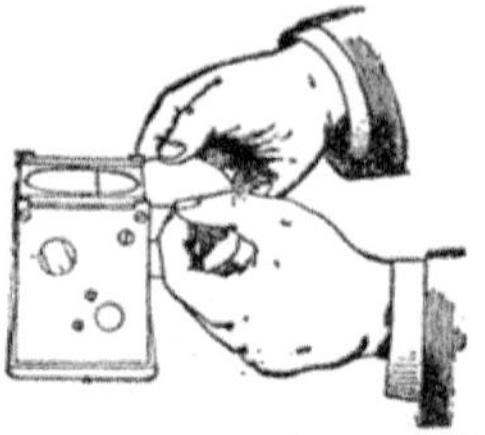
PASSING FILM OVER EXPOSURE TABLE.

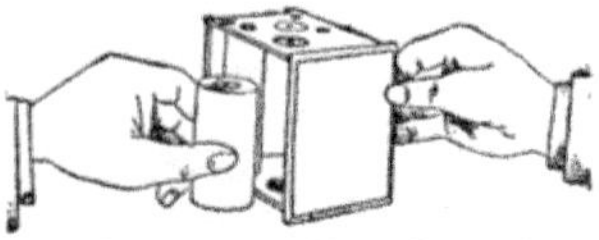
PUTTING IN NEW ROLL.

the receiving roller or reel. The back replaced, the camera is again ready for work. In putting in a new roll entirely the operations are much the same.

In the foregoing we have given very brief descriptions of the operations, as more details are not required in this article. When the camera is purchased, most elaborate and explicit directions are to be found in the instruction book included with the apparatus. We have endeavoured to merely describe the principal features of the Kodak and its method of use.

As regards using it for interiors, portraits, or other subjects, where an instantaneous exposure is not possible, or perhaps not requisite, the shutter is, so to speak, run down, the lens being left open. A small cap provided is then used for exposure, the camera being on some table or support, because it could not be held in the hand without movement

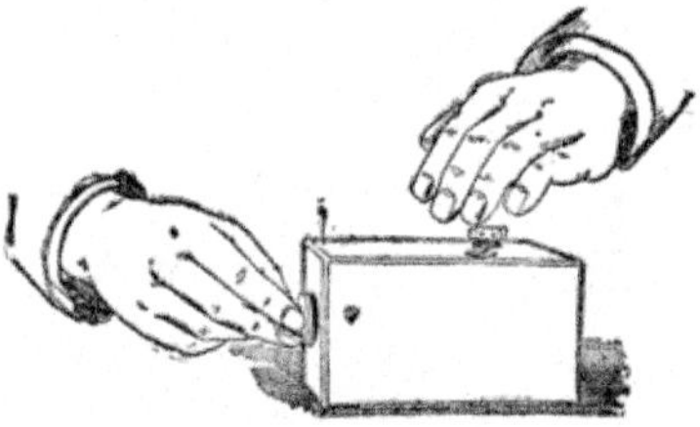
USING THE CAP.

The pictures at the commencement of the article, and on the opposite page, represent the exact size taken. The measurements given show the distance the operator was from the "operated upon." They are a trifle smaller than the standard size slide, but are none the worse for that.

Of course, it is needless to say that, in addition to the power the Kodak gives one of procuring in a cheap and easy manner *souvenirs* of pleasant journeys and happy hours, much amusement can be caused by its use as a detective camera. Space forbids our mentioning more than one as a sample A temperance friend of ours met an acquaintance in the street, and, as it happened, just at the door of a public-house We were there, so was the Kodak A night or two after, when showing some slides on the screen to a few friends, the man of temperance included, much laughter was caused by the sudden appearance of the picture, and the only explanation I could offer was—"Well, I can't help it, he must have been there, or else my camera would not have taken him in."

The makers of the Kodak are the Eastman Co, 115, Oxford Street, London, W.C, to whom we have pleasure in referring those interested. They will meet with every attention at their hands.

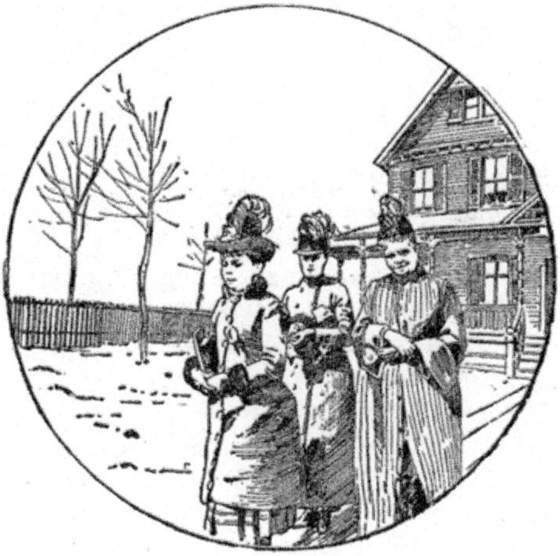

9 Feet.

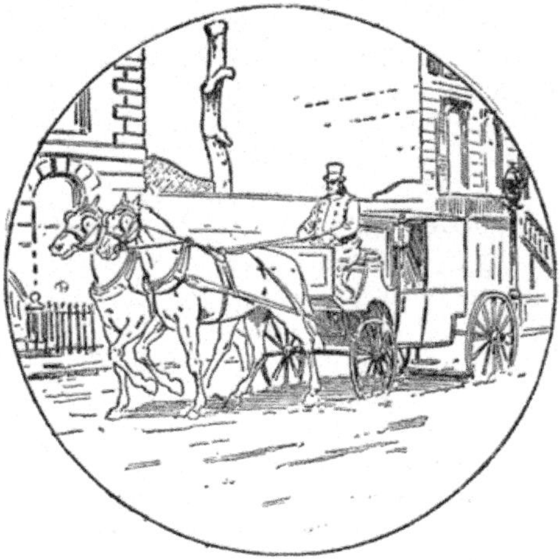

25 Feet

Addresses of Firms.

APPLETON & Co., 58, Manningham Lane, Bradford
ARCHER & SONS, W. F., Lord Street, Liverpool.
ARMSTRONG & BRO., 88, Deansgate, Manchester.
BANKS, W., Corporation Street, Bolton.
BRIN'S OXYGEN Co., Connaught Mansions, Victoria Street, Westminster.
BROOKS, W., Laurel Villa, Wray Park, Reigate, Surrey.
CATLIN, W., 17, Claremont Square, London.
CLARKSON, A., Bartlett's Buildings, Holborn Circus, London
CUBLEY & PRESTON, 4, High Street, Sheffield.
DALE, H. & E. J., 26, Ludgate Hill, London, E.C.
HOWARD, LANE & Co., 115, Palmerston Buildings, Old Broad St., London, E.C
HUGHES, W. C., Mortimer Road, Kingsland, London, N.
MASON, GEO., & Co., 186, Sauchiehall Street, Glasgow.
MEDLAND, J. B., 12, Borough, London Bridge, London, S E.
NEWTON & Co., 3, Fleet Street, London, E C.
NOAKES & SON, Billingsgate Street, Greenwich, London, S E.
OTTWAY & SON, 178, St. John's Street Road, London, E.C.
PEXTON CHATHAM, 22, Gray's Inn Road, London, W.C.
PERKEN, SON, & RAYMENT, 101, Hatton Garden, London, E C.
PLACE, J., 13, Bull Street, Birmingham.
PUMPHREY, A., Stanhope Street, Birmingham.
RILEY BROS., 5, Cheapside, Bradford.
SCIOPTICON Co., 26, Colebrook Row, London, N.
SCOTT, A. W., Marazion Villa, The Boulevard, Weston-super-Mare.
SOLGRAPH Co, Church Square, Guisborough, Yorkshire.
STEWARD, J. H, 406, Strand, London, W C.
TAYLOR, T. S. & W., Slate Street Works, Leicester.
THEOBALD & Co., J., 7, Bath Place, Kensington, London, W.
TYLER, W., 48, Waterloo Road, London.
UNDERHILL, ALFRED, 23A, Clarendon Road, West Croydon, Surrey.
VALENTINE & SONS, 154, Perth Road, Dundee.
WATSON & SONS, 313, High Holborn, London, W.C.
WEST, G., & SON, Palmerston Road, Southsea.
WILKINSON & Co., 15, Holmside, Sunderland
WILSON & Co, G. W., St. Swithin's Street, Aberdeen
WOOD, E. G, 74, Cheapside, London, E C.
WHOLESALE FIRM—For their goods apply to dealers.
YORK & SONS, 87, Lancaster Road, Notting Hill, London, W.

Appendix.

The following illustrations have been accidentally omitted, or have arrived too late for insertion in their proper place.

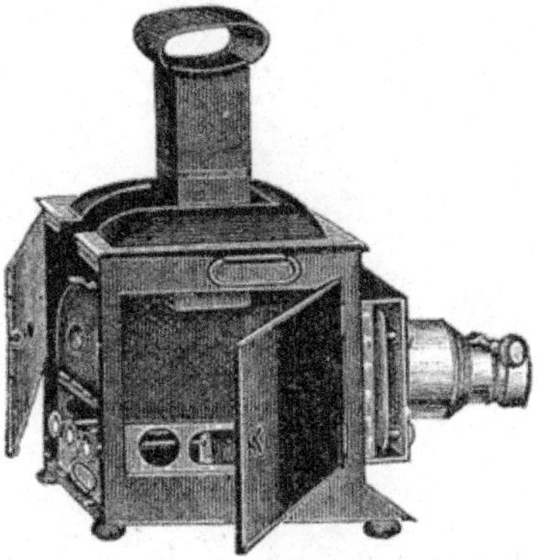

No. 32. PLACE'S No. 2 (page 22.)

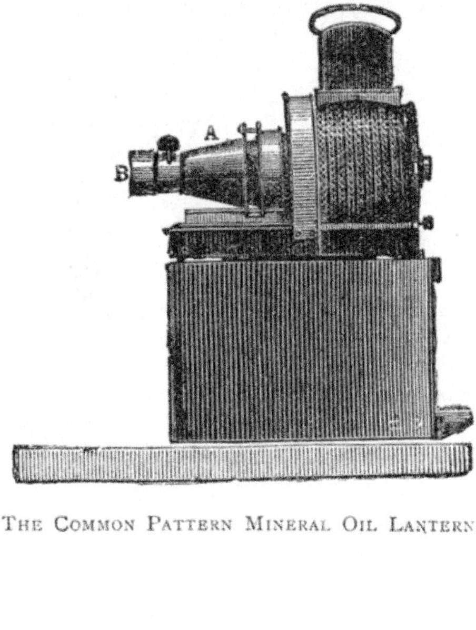

THE COMMON PATTERN MINERAL OIL LANTERN.

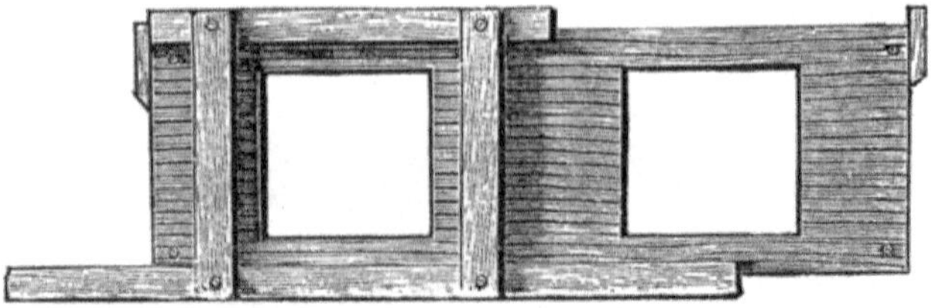

No. 410. Archer's Quick-action Carrier (page 226).

The above may also be taken as a sample of the usual double carrier for single lanterns.

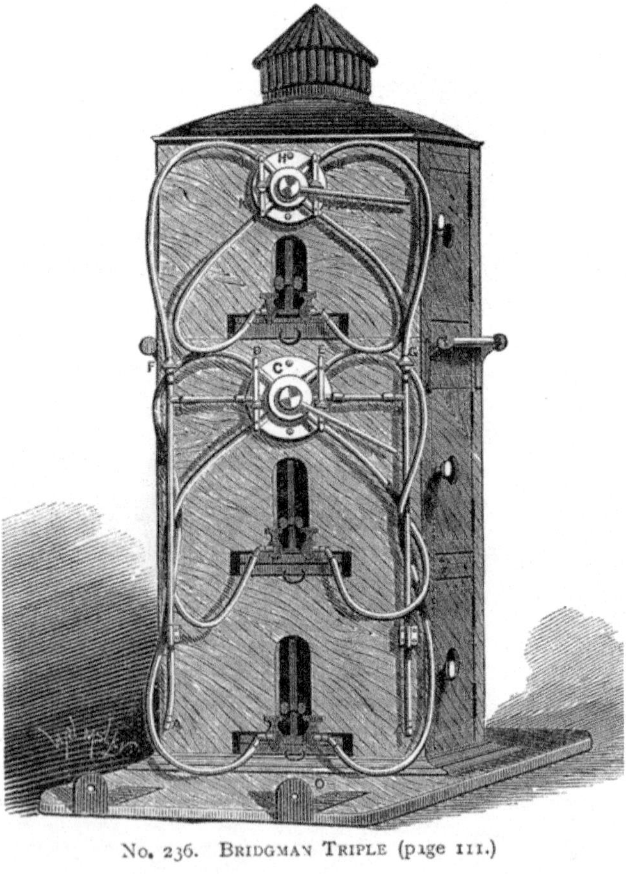

No. 236. Bridgman Triple (page 111.)

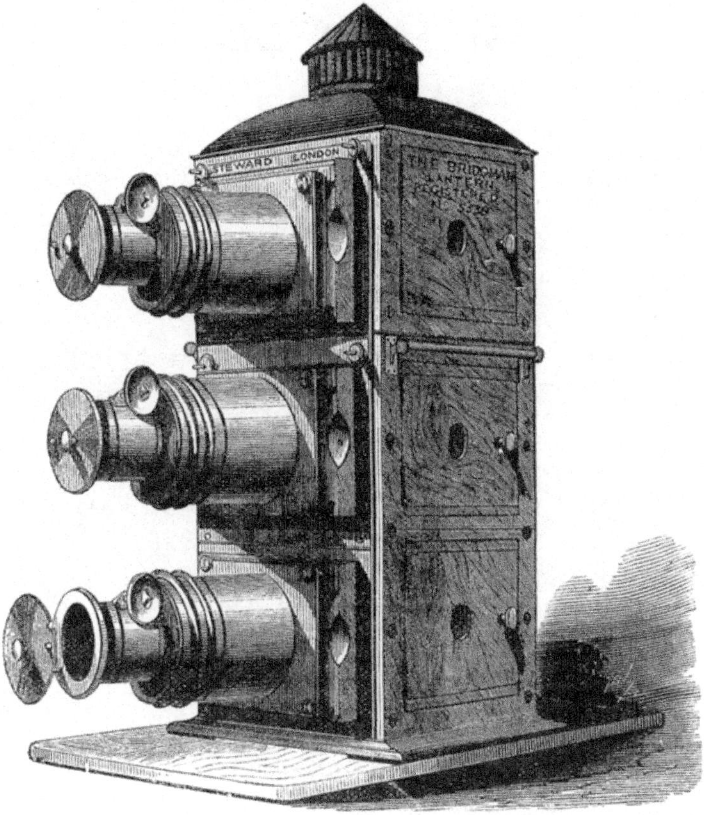

No. 236 BRIDGMAN TRIPLE (page 111.)

Sizes of Discs.

The following table gives the distances from the screen at which the Lantern should be placed to obtain the certain size disc required, the foci of the lens being known. It is calculated for foci from 3 ft. to 12 ft. and for each size of disc from 3 ft. to 12 ft. Larger sizes are as nearly as possible proportional, thus 16 feet is very nearly double 8 feet :—

En-largement.	12 times		16 times.		20 times		24 times		28 times.		32 times		36 times		40 times.		44 times.		48 times.	
Disc.	3 ft		4 ft.		5 ft		6 ft		7 ft		8 ft		9 ft.		10 ft.		11 ft.		12 ft.	
Focus In.	Ft.	In.	Ft	In.	Ft	In.	Ft.	In.	Ft	In.	Ft.	In.	Ft.	In	Ft	In.	Ft.	In.	Ft.	In.
3	3	3	4	3	5	3	6	3	7	3	8	3	9	3	10	3	11	3	12	3
4	4	4	5	8	7	0	8	4	9	8	11	0	12	4	13	8	15	0	16	4
5	5	5	7	1	8	9	10	5	12	1	13	9	15	5	17	1	18	9	20	5
6	6	6	8	6	10	6	12	6	14	6	16	6	18	6	20	6	22	6	24	6
7	7	7	9	11	12	3	14	7	16	11	19	3	21	7	23	11	26	3	28	7
8	8	8	11	4	14	0	16	8	19	4	22	0	24	8	27	4	30	0	32	8
9	9	9	12	9	15	9	18	9	21	9	24	9	27	9	30	9	33	9	36	9
10	10	10	14	2	17	6	20	10	24	2	27	6	30	10	34	2	37	6	40	10
11	11	11	15	7	19	3	22	11	26	7	30	3	33	11	37	7	41	3	44	11
12	13	0	17	0	21	0	25	0	29	0	33	0	37	0	41	0	45	0	49	0

Index to Lanterns, Slides, and Accessories.

Oil Lanterns, Better Class—

Oil Lanterns, Pairs of—

Pressure Boards—

J. THEOBALD & COMPANY,

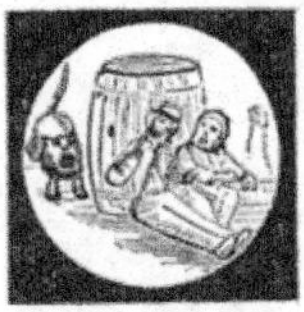

MANUFACTURERS OF THE

NEW LITHOGRAPHIC COLOURED LANTERN SLIDES.

Produced in BEAUTIFUL COLOURS, with Detail and Finish far superior to the best hand-painted slides, at less than one quarter the cost. All the following SLIDES done in 3¼ square glass; each Set comprises 12 SLIDES. Price 7/6 the Set free. Subjects:

Heroes of the Victoria Cross
Heroes of the Lifeboat
Palestine Views
The Life of Christ (2nd series)
Do do. (1st series)
Robinson Crusoe (2nd series)
Do do. (1st series)
Old London and our Ancestors
Jack the Giant Killer
Pilgrim's Progress
Cinderella
The Jackdaw of Rheims
Sinbad the Sailor
Natural History
Continental Views
Overland Route to India
Life of Joseph
Arctic Expedition
Gulliver's Travels
Dick Whittington
Bluebeard
Tiger and the Tub
Elephant's Revenge
Emigrant's Voyage
London Views
Aladdin
John Gilpin
Fun's Edition of the Poets.

Or Sheets ready for Transferring on Glass, each Sheet comprising one subject as above, are supplied at 4/- per Sheet.

COLOURED LITHOGRAPHIC

RACKWORK SLIDES

PREPARED BY THE SAME PROCESS.
MOST BEAUTIFUL WORK.

MANY ENTIRELY NEW REGISTERED DESIGNS.

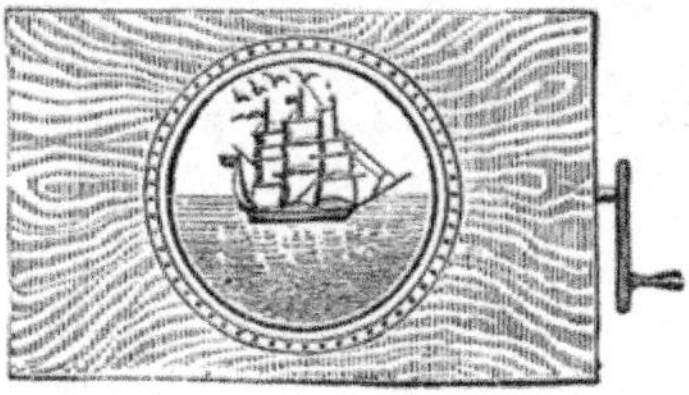

PRICE 6/- EACH FREE. SUBJECTS:

The Flirt
Man with Changing Heads
Lighthouse
Windmill
Water-mill
Water Cascade
Magician's Cave
Hair-brushing
Camp Fire
Vesuvius
House on Fire
Fountain
Rat-eater
Knife-grinder
Beehive
Aquarius
War Correspondent
Curtain
Leotard
Mail Boat
Children in Wood
Dinner Party

Best Hand-painted ditto 7/- each. Double Motion Rackwork Chromotropes 7/6 and 8/6 each

---

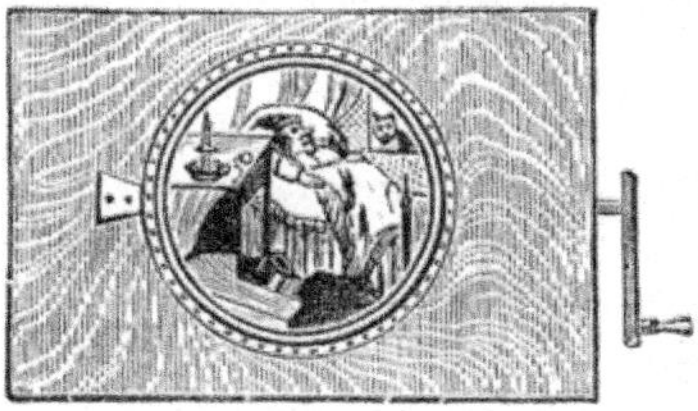

LITHOGRAPHIC COLOURED

COMIC SLIPPING SLIDES

Equal to those cost 2/6 each.

PRICE 12/- PER DOZ. ASSORTED.

Lever Slides, Panoramas, Photo Slides, Mechanical Slides of every description, and far cheaper than any other house in the trade See Illustrated Catalogues. Actual Manufacturers and not dealers. Magic Lanterns all prices from 1/- to £80 Magnificent 3-wick Patent Lantern, with 4in Compound Condensers, Special Achromatic Rackwork Front with Double Combination Lenses, £2 10s.

Biunial Lanterns having Mahogany Body, Special Double Combination Fronts, Rack Adjustment and 4in. Compound Condensers. Price with Japanned Fronts £3. Solid Brass Fronts £4. These are wonderful Instruments. Blow-through Jets 9/6 each Six-way Dissolving Taps 10/6. Ditto the Star Tap, with two Bye-passes 15/-. 4in. Compound Condensers 9/-. Double Combination Front Lenses with Rackwork Adjustment 12/6.

We do not ask you to buy of us right off. We only ask that you send for our list and compare our prices with any other house. We keep every necessary for Lantern work, Screens, Stands, Gas Bag, Gas Bottles, Gas.

Special Magic Lantern and Slide Catalogue, with description of over 20,000 Slides and Lanterns, up to £10 each. Post free 3d. Extra Special Lantern and Slide Catalogue, over 200 pages, price 6d.

J. THEOBALD & COMPANY,

MANUFACTURERS AND EXPORTERS,

7a, BATH PLACE, AND 20, CHURCH STREET, KENSINGTON, LONDON, W.

Printed in Dunstable, United Kingdom

91219941R00228